Rail Vehicle Mechatronics

Ground Vehicle Engineering
Dr. Vladimir V. Vantsevich
Professor and Director
Program of Master of Science in Mechatronic Systems Engineering
Lawrence Technological University, Michigan

Road Vehicle Dynamics
Fundamentals and Modeling Theory and Design, Second Edition
Georg Rill and Abel Arrieta Castro

Driveline Systems of Ground Vehicles
Theory and Design
Alexandr F. Andreev, Viachaslau Kabanau, Vladimir Vantsevich

Road Vehicle Dynamics
Fundamentals and Modeling
Georg Rill

Dynamics of Wheel-Soil Systems
A Soil Stress and Deformation-Based Approach
Jaroslaw A. Pytka

Design and Simulation of Heavy Haul Locomotives and Trains
Maksym Spiryagin, Peter Wolfs, Colin Cole, Valentyn Spiryagin, Yan Quan Sun, Tim McSweeney

Automotive Accident Reconstruction: Practices and Principles
Donald E. Struble

Design and Simulation of Rail Vehicles
Maksym Spiryagin, Colin Cole, Yan Quan Sun, Mitchell McClanachan, Valentyn Spiryagin, Tim McSweeney

Control Applications of Vehicle Dynamics
Jingsheng Yu and Vladimir Vantsevich

Rail Vehicle Mechatronics
Maksym Spiryagin, Stefano Bruni, Christopher Bosomworth, Peter Wolfs, Colin Cole

For more information about this series, please visit: https://www.crcpress.com/Ground-Vehicle-Engineering/book-series/CRCGROVEHENG

Rail Vehicle Mechatronics

Maksym Spiryagin, Stefano Bruni,
Christopher Bosomworth,
Peter Wolfs, and Colin Cole

CRC Press
Taylor & Francis Group
Boca Raton London New York

CRC Press is an imprint of the
Taylor & Francis Group, an **informa** business

First edition published 2022
by CRC Press
6000 Broken Sound Parkway NW, Suite 300, Boca Raton, FL 33487-2742

and by CRC Press

2 Park Square, Milton Park, Abingdon, Oxon, OX14 4RN

© 2022 Maksym Spiryagin, Stefano Bruni, Christopher Bosomworth, Peter Wolfs, and Colin Cole

CRC Press is an imprint of Taylor & Francis Group, LLC

Library of Congress Cataloging-in-Publication Data

Names: Spiryagin, Maksym, author. | Bruni, Stefano, author. | Bosomworth, Christopher, author. | Wolfs, Peter, author. | Cole, Colin, author.
Title: Rail vehicle mechatronics / Maksym Spiryagin, Stefano Bruni, Christopher Bosomworth, Peter Wolfs and Colin Cole.
Description: First edition. | Boca Raton, FL : CRC Press, 2022. | Series: Ground vehicle engineering | Includes bibliographical references and index.
Identifiers: LCCN 2021029755 (print) | LCCN 2021029756 (ebook) | ISBN 9780367464738 (hbk) | ISBN 9781032148601 (pbk) | ISBN 9781003028994 (ebk)
Subjects: LCSH: Railroad cars–Equipment and supplies. | Mechatronics. | Railroads–Electronic equipment.
Classification: LCC TF375 .S6955 2022 (print) | LCC TF375 (ebook) | DDC 625.2–dc23
LC record available at https://lccn.loc.gov/2021029755
LC ebook record available at https://lccn.loc.gov/2021029756

ISBN: 978-0-367-46473-8 (hbk)
ISBN: 978-1-032-14860-1 (pbk)
ISBN: 978-1-003-02899-4 (ebk)

DOI: 10.1201/9781003028994

Typeset in Times
by KnowledgeWorks Global Ltd.

Contents

Preface

Over the past ten years, a great number of books specializing on rail vehicle engineering design and dynamics have been published. A small number of those books cover some mechatronic aspects applicable to rail vehicles and technologies. It is also necessary to mention that there are many scientific and research papers on rail vehicle mechatronics published in journals and conference proceedings. However, for reasons that are not clear, there has not yet been a book that is fully dedicated to the questions related to rail vehicle mechatronics as well as providing the background required for students and engineers to fully understand the mechatronic concepts applicable in rail vehicle designs.

Considering that rail vehicle mechatronics is a high-tech approach that combines multidisciplinary engineering theories, numerical simulations, digital communications, data acquisition, instrumentation, control, and software development, it is not easy to collect and provide a coherent presentation of all the relevant information in just one book. The authors hope that the task was solved successfully with the introduction of the first mechatronic book related to rail vehicles. In this book, all relevant topics are covered, and the literature references presented provide detailed information and extensive answers to the background and current situation with mechatronics in the field of rail vehicle design and technologies.

We hope that readers find this rail vehicle mechatronic book useful. At the same time, we understand that there are always challenges for the further improvements of this book contents because the railway mechatronic science progresses very quickly, and it also seeks to implement all new design processes and technologies available on the market. The authors recommend that this book be used by students, practical engineers, and researchers interested in rail vehicle mechatronics designs.

Maksym Spiryagin
Stefano Bruni
Christopher Bosomworth
Peter Wolfs
Colin Cole
June 2021

Acknowledgments

The authors would like to thank their colleagues and supporters for their assistance with the preparation of this book. Particular mention must be made of:

- Tim McSweeney for the careful proofreading and valuable comments on this book.
- Qing Wu from the Centre for Railway Engineering for his help in the development of co-simulation interfaces between longitudinal train dynamics and multibody software packages.
- Igor Spiryagin and Valentyn Spiryagin for their long-term support and significant contribution in the development of content on locomotive traction and design issues.
- Bin Fu from Politecnico di Milano, for his help in the development of co-simulation examples to demonstrate active suspensions.
- Emanuele Zappa from Politecnico di Milano, for valuable comments on Chapters 6 and 10.

We also thank the following people involved in the development and support of specialized rail vehicle dynamics software products:

- Ingemar Persson from ABDesolver for his great support of our research developments and innovations, and for the implementation of our co-simulation ideas in the GENSYS rail vehicle dynamics software.
- Mark Hayman from Insyte Solutions for his persistence in the development of wheel-rail contact routines for traction studies and their benchmarking.

We would like to thank the CRC Press publisher's team who worked on this book for their acceptance of our proposal followed by ongoing support during the writing process.

Finally, we would like to thank our families for their understanding and support during the writing of this book.

About the Authors

Maksym Spiryagin is the Deputy Director of the Centre for Railway Engineering and a Professor of Engineering at Central Queensland University. He received his PhD in the field of Railway Transport in 2004. Professor Spiryagin's involvement in academia and railway industry projects includes many years of research experience in locomotive design and traction, rail vehicle dynamics, contact mechanics, wear, mechatronics, and the development of complex systems using various approaches. He has published four books, including *Design and Simulation of Rail Vehicles* in 2014 and *Design and Simulation of Heavy Haul Locomotives and Trains* in 2017 by Taylor & Francis, and he has more than two hundred other scientific publications and twenty patents as one of the inventors. Professor Spiryagin is a Chartered Professional Engineer and RPEQ in Australia and a Chartered Engineer in the UK.

Stefano Bruni, PhD, is full professor at Politecnico di Milano, Department of Mechanical Engineering, where he teaches applied mechanics and dynamics. He is the leader of the "Railway Dynamics" research group, carrying out research on rail vehicles and their interaction with the infrastructure. Prof. Bruni authored more than 270 scientific papers, mostly related to rail vehicle dynamics, train-track interaction, wheel/rail contact forces, damage and wear of wheels and rails, active control and condition monitoring of rail vehicles, and pantograph-catenary interaction. He is/has been the lead scientist for several research projects funded by the railway industry and by the European Commission. He is Vice-President of the IAVSD, the International Association for Vehicle System Dynamics, and was chairman of the IAVSD'05 International conference held in Milano in 2005. He is Editorial Board member for some international journals in the field of Railway Engineering.

Christopher Bosomworth, BMS (Computing), has worked for the Centre for Railway Engineering at Central Queensland University for over 15 years, firstly on software engineering for railway applications in direct employment and then as a subcontractor as a part of Insyte Solutions Pty Ltd on various simulation, instrumentation and mechatronic projects related to train, locomotive, and wagon dynamics. He has a deep expertise in high quality code writing, data acquisition, field testing, instrumentation, and microprocessor-based system design and development services.

Peter Wolfs is Adjunct Professor of Electrical Engineering at CQU. He is a Fellow of Engineers Australia, a senior member of IEEE and an associate member of the Centre for Railway Engineering. His special fields of expertise include electrical power distribution, power quality and harmonics, railway traction power supply, renewable energy supply, solar and hybrid electric vehicles, and intelligent systems applications in power systems and railways. He received his PhD in the area of High Frequency Link Power Conversion in 1992 from the University of Queensland. He has more than two hundred scientific publications, four book chapters and five patents as one of the inventors.

Colin Cole is the Director of the Centre for Railway Engineering at CQU. He has worked in the Australian rail industry since 1984, starting with six years in mechanized track maintenance for Queensland Railways. Since then, he has focused on a research and consulting career involving work on track maintenance, train and wagon dynamics, train control technologies and the development of on-board devices. He has been extensively engaged with industry via the past nationally funded Rail CRC programs and the Australasian Centre for Rail Innovation. His PhD was in Longitudinal Train Dynamics Modelling. He has authored and/or co-authored over two hundred technical papers, two books, numerous commercial research and consulting reports, and has developed two patents relating to in-cabin locomotive technologies.

1 Introduction to Rail Vehicle Mechatronics

1.1 HISTORICAL REVIEW

The establishment of general mechatronics as a discipline has taken a long period of time. If we consider the history timeline, it has a strong connection with the development of automation and it can be easily divided into the following phases:

Phase 1: Automation of direct processes. This phase is dated to the end of the 18th century. In some cases, it can also be the time when students of some colleges started to study the discipline referred to as "electromechanics."

Phase 2: Analogue automation. This phase started by the late 1920s and continued to be developed into the late 1940s.

Phase 3: Digital automation. Immediately after the first transistor was invented in 1947, the engineering world was focused on electronics and its application in a variety of industry fields.

Phase 4: Digital automation control. By the late 1960s, the Japanese engineer Tetsuro Mori from Yaskawa Electric Corporation was working on electronic controls for electric motors and at that time he introduced the term *mechatronics* which only covered a combination of *mechanics* and *electronics*. The major difference between earlier electromechanics and mechatronics as disciplines was that mechatronics provided much more flexibility in terms of system design and its operation.

Phase 5: Digital mechatronic control. This phase was introduced in the 1970s and it was still considered the digital automation control phase in most publications. However, the first 4-bit and 8-bit microprocessor chips introduced in 1971 and 1972, respectively, allowed moving away from the usage of mechanical mechanisms and devices and provided easy ways to program different tasks for mechatronic systems. The outcome of this was that mechatronic systems became more precise and faster in exercising control than their predecessors and, in addition, this made possible to introduce automatic data collection and reporting features in the design of mechatronic systems. At this stage, the transition of automatic and control engineering [1] to mechatronics can be observed.

Phase 6: The microprocessor mechatronic control. In the 1990s, mechatronics started to be more flexible by means of the usage of computerized systems that also included communication and networked technologies. This allowed mechatronics as a discipline to cover some additional knowledge areas such as information technologies, sensors, and actuators.

DOI: 10.1201/9781003028994-1

Mechatronics application for rail vehicles is a relatively new development in comparison with the general mechatronics discipline. Professor Roger Goodall published the history of railway vehicle design in 2009 [2], where the introduction of mechatronics as a separate design phase for railway vehicle design is referred to as the "Mechatronic Design Period" and is said to have started in the 1990s. He wrote [2]:

> During the 1990s, in other industries such as the aircraft and automotive industries, the power that became available from designing the mechanical system in conjunction with the electronics, computing and control, i.e., the use of mechatronics, was realised in a variety of research and development programmes, Although the railway industry is, perhaps naturally, somewhat behind these other two industries, nevertheless a variety of developments are being considered currently which imply that a 'mechatronic period' is close to happening for railway vehicles.

Since then, the development of mechatronics for rail vehicle design has been rapidly progressing with the implementation of new methods and modeling techniques in addition to hardware developments. The major historical rail vehicle mechatronic system design topics are [3–40]:

- Traction power systems
- Wheel adhesion systems
- Tilting systems
- Active suspension systems
- Active steering systems
- Braking systems
- Safety protection systems that provide protection against derailment
- Automatic train control systems
- Condition monitoring and fault detection
- Rail vehicle testing and roadworthiness acceptance

Considering the current dynamics in the application of modern communication and measurement technologies and advanced diagnostic and control algorithms in rail vehicle operations, including artificial intelligence and big data processing, it seems inevitable that rail vehicles will become "increasingly mechatronic" which confirms a prediction made in [13].

1.2   THEORETICAL ASPECTS FOR THE APPLICATION OF MECHATRONIC SYSTEM

Understanding design and modeling processes and the associated dynamic performance is fundamental for tasks associated with the rail vehicle mechatronic disciplines. The theoretical aspects for the practical application of mechatronic systems on rail vehicles can be defined based on the railway operational, safety, and performance requirements as presented in the next few subsections.

1.2.1   STABILITY AND CURVING

Stability and curving are two important aspects of a railway vehicle's running dynamics. Unfortunately, they often lead to conflicting design requirements for the running gear. The term running gear refers to the ensemble of components in the vehicle responsible for the vehicle's running behavior. The running gear includes components such as wheelsets, bearings, suspensions, bogie frames, brakes, and traction bars. The design of the running gear involves finding a trade-off between good stability at high speed and satisfactory curving behavior. Active vehicle control, particularly active suspensions, is a way to remove this design conflict, so this is one of the main areas for the use of mechatronics in railway vehicles.

1.2.1.1   Running Stability of a Railway Vehicle

Running stability is a term used by railway engineers in relation to a self-excited motion of the wheelsets and bogies consisting of the combination of lateral displacement and yaw rotation called *hunting* [41–43]. This behavior is typical of rail vehicles equipped with solid wheelsets, i.e., pairs of conical wheels rigidly connected to a common axle, while other types of lateral oscillation may arise for vehicles with independently rotating wheels.

The hunting motion is strongly affected by the speed at which the vehicle runs over the track: at low speed, an oscillation originated by an initial disturbance (e.g., from track imperfections) will be sufficiently damped so that the vehicle will soon return to the unperturbed condition. However, above a threshold speed called *critical speed*, the motion arising from the initial disturbance will have a growing amplitude until it will be limited by the wheel flanges making contact with the rails. In this condition, the hunting motion can be so violent as to produce permanent deformation of the track and even lead to the derailment of the vehicle. It is therefore extremely important that the vehicle is designed to have a critical speed sufficiently higher than the maximum operational service speed.

It should be noted that the critical speed of a railway vehicle is affected by a number of parameters, of which the most important to mention here is the conicity of the wheel/rail couple [41]. Conicity can be described in simple terms as the rate of variation of the wheel's rolling radius with the lateral displacement of the wheelset relative to the track centerline. For the same mechanical design of the running gear, a lower critical speed can be expected for higher wheel conicity. Conicity is in turn affected by some geometrical parameters of the wheel/rail couple, namely wheel and rail profiles, track gauge, and the distance between the back of the flanges of the wheels. Although some of these parameters are well defined, others are not known precisely or will be subject to significant variation during the vehicle's service life. In particular, wheel wear results in a modification of the wheel profile and increased conicity. Therefore, the design of the running gear for stability has to be performed considering different conditions of wheel wear, including the condition of maximal wear before wheel reprofiling.

The parameters involved in the design of the running gear having the largest impact on running stability are the stiffness of the primary suspensions in longitudinal and lateral directions, the bogie wheelbase and, as already mentioned, conicity.

In order to have a high critical speed, the vehicle should be designed to have stiff primary suspensions, a long bogie wheelbase and relatively low conicity. Unfortunately, all of these measures will have a negative effect on curving as discussed below. There are other design parameters of the running gear having influence on stability (one example being the inertia of the bogie), but they are not described here as they are less relevant to the design conflict between stability and curving.

1.2.1.2 Curving Behavior of a Railway Vehicle

A single, unconstrained solid wheelset with conical profiles running along a curve will naturally align its axis of revolution along a radial or nearly radial direction, minimizing the forces exchanged with the track [41]. However, some kind of connection between the wheelset and the car body shall be provided, either directly or through a bogie, to transfer efforts due to traction, braking and guidance and also, as discussed above, to meet requirements on vehicle stability. This connection is realized by suspensions establishing a flexible link between the wheelset and the car body directly (single stage of suspension) or between the wheelset and an intermediate body, the bogie frame, which is then connected to the car body by another stage of suspensions.

When two or more wheelsets are elastically connected with each other through the bogie or car body, the direction of their axes deviates significantly from the radial one and this is measured by the *angle of attack*, i.e., the angle formed by the axis of revolution of each wheelset and the local radial direction, as shown in Figure 1.1. The longer is the longitudinal distance L between two elastically connected wheelsets, the larger will be the angle of attack that can be expected for a given curve radius R.

Due to the non-zero angle of attack, lateral creepages arise at wheel/rail contact interfaces, affecting wheel/rail forces in the lateral direction, which will be larger

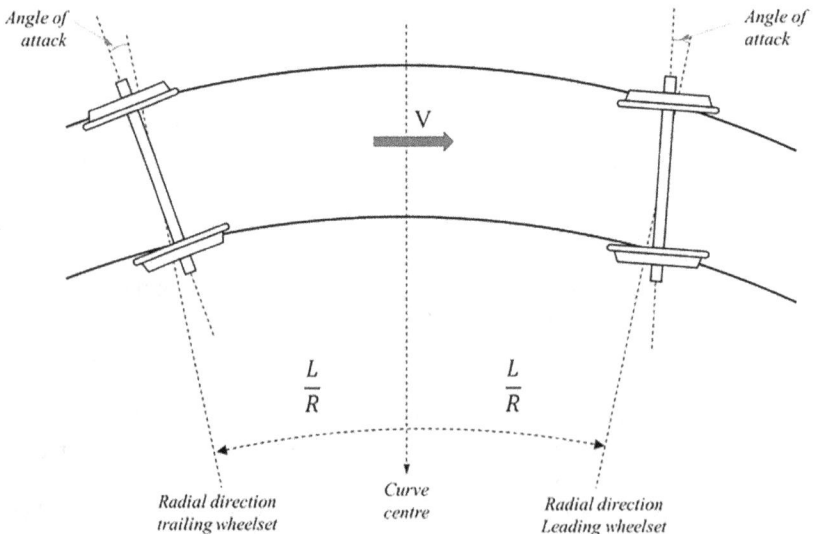

FIGURE 1.1 Angle of attack of a wheelset in a curve.

than what would be needed to balance the effect of centrifugal forces and track cant. Furthermore, the lateral force on the leading wheelset will push the wheelset out of the curve, causing contact of the flange of the outer wheel with the high rail. This in turn will produce a large variation of the rolling radius on the outer wheel, resulting in additional longitudinal creepage and hence longitudinal contact forces on both the inner and outer wheels of the leading wheelset and sometimes also in the trailing wheelset(s).

To sum up, having two or more wheelsets connected via elastic suspensions results in large wheel/rail contact forces that are not needed to balance other forces arising on the vehicle in the curve. These unnecessary forces may be harmful to the running safety of the vehicle, can produce additional damage of the rolling surfaces in terms of wear and rolling contact fatigue (RCF), may be the cause of permanent deformation of the track alignment, and may cause metal fatigue in the rails or damage to the rail fasteners. Since these forces are ultimately arising on account of the formation of an angle of attack, it follows that reducing the longitudinal distance between elastically connected wheelsets is a major measure to improve curving behavior. This is the primary reason why, since the early times of railways, bogies are used in railway vehicles as the distance between elastically connected wheelsets for a vehicle with bogies depends on the bogie wheelbase, which can be kept relatively short (typically 1.8–3 m), and not on the length of the car body.

From the above discussion, it also follows that using relatively soft primary suspensions is another effective measure to improve the curving behavior of the vehicle and, in general, it can be concluded that using more conical wheels also improves curving. We recall however that all these measures have a negative impact on the stability of the vehicle, revealing the design conflict between curving and stability.

For a more detailed discussion of running stability and curving in railway vehicles, the reader is referred to [41], while the use of mechatronic suspensions to remove the design conflict between curving and stability is discussed in Chapter 11 of this book.

1.2.2 Damage and Wear of Wheels and Rails

Contact forces arising at the wheel/rail interface are responsible for damage and degradation of the contacting bodies. Damage happens on the surface of the bodies in terms of wear and RCF and also in the subsurface layers of the contacting bodies, or even in portions of the vehicle and of the track not coming into direct contact with each other, examples being metal fatigue in wheels, axles, bogie frames, and rails, or track settlement and cracked sleepers. Mechatronic suspensions and control of traction and braking can reduce substantially many of these sources of damage.

In this subsection, the different types of damage mentioned above are concisely reviewed and the case for introducing mitigation measures based on mechatronic systems is outlined.

1.2.2.1 Wear of Wheels and Rails

According to tribological studies, wear of wheels and rails is related to the *wear number Tγ*. This is the scalar product of the creepage and creep force vectors at one wheel/rail contact point and can be seen as the energy per unit distance run dissipated by frictional forces in the considered contact point. Tests performed on twin disc machines have revealed the presence of different wear regimes depending on the value of the ratio $T\gamma/A$, with A being the area of the contact patch established between the two bodies [44, 45]. In each wear regime, a different relationship can be established between the frictional power density $T\gamma/A$ and wear rate, but generally a larger frictional power will lead to an increase of the wear rate.

Large values of the frictional power can be expected when a railway vehicle runs around a short radius curve, but also when large creep forces and creepages are caused by traction and braking efforts, especially if a full slip condition is reached. It follows that poor curving performance and poor traction and braking control are two typical causes of accelerated wear of wheels and rails. Improving the curving behavior through a proper design of passive suspensions is not always possible, as beyond some point this can only be done at the expense of running stability. However, a number of strategies making use of active control, active steering, and yaw relaxation to name just a few can be used to reduce wear dramatically without impairing stability and will be introduced in Chapter 11.

1.2.2.2   Rolling Contact Fatigue

RCF occurs in the surface and subsurface layers of the wheel and rail due to repeated nearly-cyclic local stresses caused by the rolling contact established between the two bodies. When the amplitude of stresses exceeds the shakedown limit [46], a phenomenon called "ratcheting" takes place, which means plastic deformation cumulates from one loading cycle to another, ultimately resulting in material failure. RCF effects appear on the rolling surfaces in the form of small cracks. This surface damage is normally not especially dangerous, as the damaged material is removed by wear. However, RCF may also give rise to subsurface cracks either in the wheel or in the rail head. Under some circumstances, these cracks will propagate until reaching a critical size that will produce the sudden detachment of a large portion of the wheel rim or of the rail head, potentially with very dangerous consequences.

Theoretical and experimental investigations have shown that ratchetting and hence RCF damage can be expected when the ratio of the maximum contact pressure over the material's yield strength is sufficiently large and, at the same time, the ratio of the tangential contact force to the normal one (sometimes called the "used friction coefficient") is large enough [46].

Another approach to assessing the severity of RCF damage was developed in the UK and, as for wheel and rail wear, is based on the value of the $T\gamma$ index [47]. According to this approach, when the $T\gamma$ is too low RCF damage will not develop while, for quite large values of $T\gamma$, wear will remove the surface damage caused by RCF, so that eventually RCF will only take place for intermediate values of the $T\gamma$ index.

Active control, by improving the curving performance of the vehicle, may lead to a reduction of the creepages and contact forces in curves which can be beneficial to reduce RCF. However, it is also possible that the reduction of $T\gamma$, while reducing wear, will promote RCF damage which would not develop in a vehicle equipped with passive suspension due to material removal caused by wear.

1.2.2.3   Metal Fatigue in Wheels, Axles, Rails, and Other Types of Damage

Metal fatigue is highly relevant to the durability of different components in the rolling stock and in the track. A fatigue failure occurring in a wheel or axle or bogie frame may have catastrophic consequences and these components are therefore designed for infinite life. However, some of these components, particularly railway axles, are regularly inspected using non-destructive techniques (NDTs). In a similar way, fatigue cracks developing in rails are highly dangerous and NDTs are being developed to check the integrity of rails and of fishplate joints. Active control can be used to reduce wheel/rail contact forces so that the stresses caused in these components can be kept below the threshold that would trigger the propagation of a fatigue crack. For wheels, axles, and rails, it is not only important to reduce the static stresses due to the vehicle's own weight and quasi-static stresses due to curving, but it is also extremely important to reduce dynamic stresses arising from defects on the rolling surfaces. Active suspensions in most cases will not be effective with mitigating the effect of such defects due to the high passband that would be required from the actuators to control fast disturbance caused by these defects, but still will be effective by reducing, e.g., RCF damage leading to spalling or shelling [48] which is one of the major causes of geometric defects arising on the surface of wheels and rails. Furthermore, braking control is effective to reduce the formation of wheel flats which are a main cause of large dynamic overloads at wheel/rail contact interfaces.

As far as fatigue effects in the bogie frame and car body are concerned, the most sensitive issue is fatigue in welded joints. Here again, mechatronics can be beneficial by reducing service loads thanks to the use of smart suspensions, or simply by reducing the vehicle mass thanks to a lighter mechanical design enabled by active suspensions [49, 50]. However, the use of active suspensions may also lead to increased fatigue loads for the bogie frame and car body. For instance, an active secondary suspension will introduce additional forces on the bogie frame and car body and the design of these components will need to consider the effect of these additional loads.

1.2.3   RIDE COMFORT

The term *ride comfort* refers to the effectiveness of the vehicle with ensuring that the passengers can travel the entire journey without feeling fatigued by the exposure to vibration caused by the vehicle's response to track imperfections. Generally, this term is used to refer to the exposure of passengers to low frequency vibration (in the range 0–30 Hz approximately), while the term *interior noise* is used to denote the effect of car body vibration at higher frequencies.

Despite the relationship between vehicle dynamics and passengers' well-being remaining difficult to establish due to its subjective nature and to the effect of environmental and even psychological factors, there is a general consensus that ride comfort is determined by the acceleration to which passengers are exposed. The quantitative evaluation of ride comfort can be performed according to different standards, among which ISO 2631 [51] and EN12299 [52] are particularly worthy of mention. The procedure for ride comfort evaluation consists in measuring car body acceleration at different locations in the car body, applying a weighting filter to consider the different human perception of harmonic components of vibration having different frequency, and finally computing one or more quantitative indicators (normally root mean square values of the weighted acceleration) that can be used to assess the comfort of passengers.

Mechatronic suspensions can be used to improve ride comfort in a railway vehicle, enabling a better ride comfort for the same track quality and train speed, or enabling the train to ride on a track with poorer quality for the same level of comfort. Indeed, a number of applications of mechatronic suspensions to improve ride comfort have been proposed and some are in use on vehicles in operational service.

There are two main ways in which an active or semi-active suspension can improve ride comfort compared to a passive one. Firstly, it is known from the concept of *transmissibility* which is often used to describe the performance of a vehicle's suspension that, in order to isolate the vehicle body from track irregularities, the viscous damping of the suspension should be as low as possible, at least for excitation falling in a frequency range well above the body's natural frequency [41, 53]. However, a suspension having too low a damping capability may behave poorly with certain low frequency features of track geometry, such as gradients, hills, and bumps. A mechatronic suspension can be used to adapt the amount of damping to the particular features of track geometry experienced by the vehicle in different running conditions, or even to introduce a frequency-dependent damping function for the purpose of ultimately reducing car body acceleration.

Secondly, in modern railway vehicles, a side effect of the use of lighter but more flexible car bodies is that the resonance frequency of some flexible modes of vibration of the car body will fall in a frequency range which is relevant to ride comfort. For instance, the natural frequency associated with the first bending mode of the car body may fall in a frequency range between 8 and 12 Hz. When the vehicle runs at particular speeds, the excitation produced by track irregularities on the flexible modes of the car body is maximized, leading to poor ride quality. Active or semi-active suspensions are more effective than standard passive suspensions with mitigating the effects of car body flexibility, or smart actuators can be used to attenuate the flexible vibration of the car body.

1.3   STRUCTURE OF THIS BOOK

The following text summarizes the content covered in other chapters of this book:

Chapter 2: Modeling of mechanical systems for mechatronic systems of rail vehicles requires consideration of the mechanical systems as a separate entity prior to dealing with a complete simulation modeling of mechatronic systems. The chapter covers all aspects and steps that should be considered in terms of advanced modeling of mechanical systems of rail vehicles starting from classification for theoretical- and experimental-based modeling approaches and finishing not only with the modeling of subsystems belonging to rail vehicles, but also including the description of approaches for modeling of wheel/rail contact processes, pantograph-catenary interaction, and track systems. In addition, the chapter provides a background for the modeling of traction and braking systems as a part of mechanical system investigation studies as well as the modeling of inter-train connections when it is necessary to consider the working of rail vehicle mechanical systems in various train configurations.

Chapter 3: The designs of electric power transmission systems that are in use on rail vehicles are covered in this chapter. Detailed descriptions of the main electric power components, energy storage systems, and machines are provided in detail. The basic principles of the modeling techniques for various types of main components are introduced, worked examples are presented as case studies which are analyzed, and numerical results are delivered for a bogie level slip control system.

Chapter 4: This chapter is focused on the application of theoretical aspects of control applicable in mechatronic system designs used in rail vehicles. The application of two common control systems of open-loop and closed-loop control systems for mechatronic rail vehicle designs is discussed. The background and fundamentals are provided as a high-level guidance for classical and modern control approaches applicable in rail vehicle mechatronic systems. Details of the major area of control method applications in the railway field are also addressed in this chapter through the provision of an extensive list of relevant references.

Chapter 5: This chapter is focused on actuators, which play a pivotal role in mechatronic systems, allowing the implementation of active control. This chapter aims to present different types of actuators used in mechatronic railway vehicles, discussing their relative advantages and drawbacks. The other important aim of the chapter is to introduce mathematical models, which can be used to consider the effect of actuator dynamics on the design and performance of a mechatronic vehicle.

Chapter 6: This chapter is focused on sensor designs and applications. Sensors are used in mechatronic systems to provide the control unit with measured signals describing the state of the system that can be used to define the desired control action. Signals measured by sensors can also be used to monitor the system and to identify the occurrence of faults. Particularly relevant to mechatronic systems is the measurement of quantities related to the motion of the system: displacement, velocity, acceleration, and the measurement of forces and torques. This chapter provides an overview of sensors frequently used in mechatronic railway vehicles, presenting their working principles and discussing the relative advantages of different principles of transduction.

Chapter 7: This chapter is focused on the complex system modeling techniques and their implementation in a computer environment and software packages designated for rail vehicle multibody and control system studies. The co-simulation technique and methodology required for the modeling of complex systems is described in detail. Two worked examples for utilizing rail vehicle multibody models with a wheel slip control technique and an advanced longitudinal train dynamics modeling are provided, and the delivered results are analyzed and explained.

Chapter 8: This chapter is focused on the microprocessor computer design and application aspects for rail vehicle mechatronic systems. Possible rail vehicle application areas for microprocessor-based systems are reviewed. The design concepts of different microprocessor computer architectures are discussed. The chapter also presents a case study for the integration of a microprocessor-based system for a wagon dynamics condition monitoring device with the formulation of the power problem and the energy balance solution being provided.

Chapter 9: This chapter is focused on the description of communication principles, architectures, and technologies used in individual rail vehicles and train consists. The chapter describes major communication architectures that are currently in use in rail vehicles and contains a review of related railway standards used for enabling communication technologies in rail vehicle design. The case study presented in this chapter looks at the advancement in freight train braking possible through the introduction of the end of train device and the development of a robust train communication network.

Chapter 10: This chapter provides an overview of data acquisition and digital data treatment in the framework of mechatronic systems. The general configuration of a digital controller is introduced, and the main issues related with interfacing the digital controller to analog sensors and actuators are reviewed. Then, the layout of a data acquisition and data processing system is presented, describing the role of its main components. The chapter also provides a non-specialistic review of the problems related to analog-to-digital and digital-to-analog conversion and to some techniques used to manipulate digital signals, particularly digital filters and frequency analysis for digital signals.

Chapter 11: Mechatronic suspensions are one of the most relevant areas where mechatronics has impacted the design of railway vehicles. Active or semi-active control can be used to improve ride quality for passengers, to remove the design conflict between stability and curving which is typical of passive suspension railway vehicles, and to mitigate the impact of the vehicle on the track, e.g., reducing rolling contact fatigue and wear in the rails. This chapter reviews the rationale underlying the use of active or semi-active suspensions, considering the fundamental distinction between primary and secondary suspensions, and provides a general overview of the present state-of-art. Different functions that can be assigned to active and semi-active suspensions are presented, including steering, vehicle stabilization and guidance, improvement of ride quality, and active tilt. Control strategies to realize these functions are described and main configurations for mechatronic suspensions are reviewed.

Chapter 12: Real-time simulations have found wide application for the investigation of the behavior of complex mechatronic systems. The explanation on real-time systems and the development of programming code for a real-time application are discussed in this chapter. The design of mechatronic systems of rail vehicles requires performing verification and validation in the real-time mode running on a real-time system The validation tools are commonly based on the application of software-in-the-loop and hardware-in-the-loop simulation approaches. Both approaches require the development of a real-time model of the physical system as a part of the design process. The real-time model of the rail vehicle created in multibody software is provided as the case study in this chapter.

Chapter 13: The development of a rail vehicle mechatronic product involves multiple disciplines, and it results in multiple stakeholders being involved in the design of the appropriate mechatronic systems. In terms of engineering capabilities, different types of engineers might be involved in the process related to the development of a new design or the modification of an existing design. This chapter discusses the integration concept for multidisciplinary engineering design tasks and the challenges related to the specific views on rail vehicle mechatronic system design, and the applicable standards for design and validation tasks.

Chapter 14: To demonstrate the process of a mechatronic system design and investigation, four rail application case studies are described in this chapter, one consisting of a simplified analytical model and the other three being multi-body simulation cases. The first case study presents a classic 2 degrees of freedom (DOF) model for a single wheelset with primary suspension and a 6-DOF linear model of a 2-axle bogie with primary and secondary suspensions. Both models are based on linearized wheel/rail contact forces and are designated for defining active control strategies in railway vehicle mechatronic systems. The second case study describes the development of an active primary suspension steering system of a rail vehicle for better dynamic performance on curves. The third case study presented addresses the development of a heavy haul diesel-electric locomotive model with a bogie traction control strategy. The fourth case study describes the development of a model for an energy storage system of a hybrid locomotive. Simple mathematical modeling is used in the first case, while the other cases are modeled with an advanced simulation technique based on the co-simulation modeling approach between multibody software products and the Matlab®/Simulink software package.

REFERENCES

1. F. T. Barwell, Automation and Control in Transport, Pergamon Press, Oxford, UK, 1973.
2. R. Goodall, Control of Rail Vehicles: Mechatronic Technologies, In: W. Schiehlen (Ed), Dynamics Analysis of Vehicle Systems, Theoretical Foundations and Advanced Applications, International Centre for Mechanical Sciences, Springer, Udine, Italy, 231–235, 2009.
3. M. Spiryagin, C. Cole, Hardware-in-the-loop simulations for railway research, Vehicle System Dynamics, 51(4), 497–498, 2013.
4. S. Iwnicki, M. Spiryagin, C. Cole, T. McSweeney (Eds), Handbook of Railway Vehicle Dynamics (2nd Edition), CRC Press, Boca Raton, FL, 2020.
5. R. Mathew, F. Flinders, W. Oghanna, Locomotive" total systems" simulation using SIMULINK. Proceedings of International Conference on Electric Railways in a United Europe, IET, 202–206, 1995.

6. S. Müller, R. Kögel, R. Schreiber, Simulation of a Locomotive as a Mechatronical System, 4th ADAMS/Rail Users' Conference, Utrecht, the Netherlands, 1999.

7. H. P. Kotz, Simulation of Effects Based on the Interaction of Mechanics and Electronics in Railway Vehicles, SIMPACK User Meeting, Freiburg, Germany, 2003.

8. H. P. Kotz. A Toolkit for Simulating Mechatronics in Railway Vehicles, SIMPACK User Meeting, Freiburg, Germany, 2003.

9. M. Spiryagin, K. S. Lee, H. H. Yoo, Control system for maximum use of adhesive forces of a railway vehicle in a tractive mode, Mechanical Systems and Signal Processing, 22(3), 709–720, 2008.

10. N. Bosso, M. Spiryagin, A. Gugliotta, A. Soma, Mechatronic Modeling of Real-Time Wheel-Rail Contact, Springer, Heidelberg, Germany, 2013.

11. M. Spiryagin, S. Simson, C. Cole, I. Persson, Co-simulation of a mechatronic system using Gensys and Simulink, Vehicle System Dynamics, 50(3), 495–507, 2012.

12. M. Spiryagin, P. Wolfs, C. Cole, V. Spiryagin, Y. Q. Sun, T. McSweeney, Design and Simulation of Heavy Haul Locomotives and Trains, Ground Vehicle Engineering Series, CRC Press, Boca Raton, FL, 2016.

13. R. M. Goodall, W. Kortüm, Mechatronic developments for railway vehicles of the future, Control Engineering Practice, 10(8), 887–898, 2002.

14. R. M. Goodall, Tilting trains and beyond – the future for active railway suspensions part 1: Improving passenger comfort, Journal of Computing and Control Engineering, 10(4), 153–159, 1999.

15. R. M. Goodall, Tilting trains and beyond – the future for active railway suspensions part 2: Improving stability and guidance, Journal of Computing and Control Engineering, 10(5), 221–230, 1999.

16. K. Tanifuji, S. Koizumi, R. Shimamune, Mechatronics in Japanese rail vehicles: Active and semi-active suspensions, Control Engineering Practice, 10(9), 999–1004, 2002.

17. R. Goodall, Mechatronics in motion – Some railway applications, IFAC Proceedings, 37(14), 543–548, 2004.

18. H. Zamuri, A. C. Zolotas, R. M. Goodall, Intelligent control approaches for tilting railway vehicles, Vehicle System Dynamics, 44(Supp 1), 834–842, 2006.

19. A. C. Zolotas, J. Wang, R. M. Goodall, Reduced-order robust tilt control design for high-speed railway vehicles, Vehicle System Dynamics, 46(Supp), 995–1011, 2008.

20. H. Zamuri, A. C. Zolotas, R. M. Goodall, Tilt control design for high-speed trains: A study on multi-objective tuning approaches, Vehicle System Dynamics, 46(Supp), 535–547, 2008.

21. H. Li, R. M. Goodall, Linear and non-linear skyhook damping control laws for active railway suspensions, Control Engineering Practice, 7(7), 843–850, 1999.

22. T. X. Mei, R. M. Goodall, Recent development in active steering of railway vehicles, Vehicle System Dynamics, 39, 415–436, 2003.

23. T. X. Mei, R. M. Goodall, Practical strategies for controlling railway wheelsets independently rotating wheels, Journal of Dynamic Systems, Measurement, and Control, 125(3), 354–360, 2008.

24. T. X. Mei, H. Li, Control design for the active stabilization of rail wheelsets, Journal of Dynamic Systems, Measurement, and Control, 130(1), 011002, 2008.

25. T. X. Mei, Z. Nagy, R. M. Goodall, A. H. Wickens, Mechatronic solution for high-speed railway vehicles, Control Engineering Practice, 10,1023–1028, 2002.

26. M. Spiryagin, V. Spiryagin, Modelling of Mechatronic Systems of Running Gears for a Rail Vehicle, East Ukrainian National University, Lugansk, Ukraine, 2010. ISBN 978-966-590-871-5 (in Ukrainian).

27. S. A. Simson, C. Cole, Control Alternatives for Yaw Actuated Force Steered Bogies, Proceedings of 17th World Congress of the International Federation of Automatic Control (IFAC), Seoul, South Korea, 8281–8286, 2008.

28. S. A. Simson, C. Cole, Simulation of traction curving for active yaw – Force steered bogies in locomotives, Journal of Rail and Rapid Transit, 223(1), 75–84, 2009.
29. S. Bruni, R. Goodall, T. X. Mei, H. Tsunashima, Control and monitoring for railway vehicle dynamics, Vehicle System Dynamics, 45, 733–779, 2007.
30. R. M. Goodall, S. Bruni, T. X. Mei, Concepts and prospects for actively-controlled railway running gear, Vehicle System Dynamics, 44(Supp 1), 60–70, 2006.
31. R. Johansson, Dependability characteristics and safety criteria for an embedded distributed brake control system in railway freight trains, Report No. 8, Chalmers Lindholmen University College, Göteborg, Sweden, 2001.
32. R. Johansson, A Fault Tolerant Architecture for Brake-by-Wire in Railway Cars, Technical Report 15, Chalmers Lindholmen University College, Göteborg, Sweden, 2003.
33. L. Pugi, M. Malvezzi, A. Tarasconi, A. Palazzolo, G. Cocci, M. Violani, HIL simulation of WSP systems on MI-6 test rig, Vehicle System Dynamics, 44(Supp 1), 843–852, 2006.
34. C. G. Kang, H. Y. Kim, M. S. Kim, B. C. Goo, Real-Time Simulations of a Railroad Brake System Using a dSPACE Board, Proceedings of the ICROS-SICE International Joint Conference 2009, Fukuoka, Japan, 4073–4078, 2009.
35. B. Fu, R. L. Giossi, R. Persson, S. Stichel, S. Bruni, R. Goodall, Active suspension in railway vehicles: A literature survey, Railway Engineering Science, 28(1), 3–35, 2020.
36. C. P. Ward, P. F. Weston, E. J. C. Stewart, H. Li, R. M. Goodall, C. Roberts, T. X. Mei, G. Charles, R. Dixon, Condition monitoring opportunities using vehicle-based sensors, Journal of Rail and Rapid Transit, 225(2), 202–218, 2011.
37. C. P. Ward, R. M. Goodall, R. Dixon, Contact Force Estimation in the Railway Vehicle Wheel/Rail Interface, Proceedings of 18th International Federation of Automatic Control Congress, Milano, Italy, 4398–4403, 2011.
38. C. P. Ward, R. M. Goodall, R. Dixon, Wheel-Rail Profile Condition Monitoring, Proceedings of 2010 UKACC International Conference on Control, Coventry, UK, 1184–1189, 2010.
39. C. P. Ward, R. M. Goodall, R. Dixon, G. Charles, Condition Monitoring of Rail Vehicle Bogies, Proceedings of 2010 UKACC International Conference on Control, Coventry, UK, 1178–1183, 2010.
40. N. Watanabe, Y. Maki, T. Shimomura, K. Sasaki, T. Tohtake, H. Morishita, Hardware-in-the-loop simulation system for duplication of actual running conditions of a multiple-car train consist, Quarterly Report of RTRI, 52(1), 1–6, 2011.
41. K. Knothe, S. Stichel, Rail Vehicle Dynamics, Springer International, Cham, Switzerland, 2017.
42. A. H. Wickens, Fundamentals of Rail Vehicle Dynamics: Guidance and Stability, Swets & Zeitlinger, Lisse, the Netherlands, 2003.
43. A. A. Shabana, K. E. Zaazaa, H. Sugiyama, Railroad Vehicle Dynamics: A Computational Approach, CRC Press, Boca Raton, FL, 2007.
44. R. Lewis, U. Olofsson, Mapping rail wear regimes and transitions, Wear, 257, 721–729, 2004.
45. F. Braghin, S. Bruni, R. Lewis, Railway Wheel Wear, In: R. Lewis, U. Olofsson (Eds), Wheel/Rail Interface Handbook, Woodhead Publishing, Cambridge, UK, 172–210, 2009.
46. A. Ekberg, Fatigue of Railway Wheels, In: R. Lewis, U. Olofsson (Eds), Wheel/Rail Interface Handbook, Woodhead Publishing, Cambridge, UK, 211–244, 2009.
47. J. R. Evans, M. C. Burstow, Vehicle/track interaction and rolling contact fatigue in rails in the UK, Vehicle System Dynamics, 44(Supp 1), 708–717, 2006.
48. R. Lewis, U. Olofsson, Basic Tribology of the Wheel-Rail Contact, In: R. Lewis, U. Olofsson (Eds), Wheel/Rail Interface Handbook, Woodhead Publishing, Cambridge, UK, 34–57, 2009.

49. A. Pacchioni, R. M. Goodall, S. Bruni, Active suspension for a two-axle railway vehicle, Vehicle System Dynamics, 48(Supp 1), 105–120, 2010.
50. R. L. Giossi, R. Persson, S. Stichel, Gain Scaling for Active Wheelset Steering on Innovative Two-Axle Vehicle, In: M. Klomp, F. Bruzelius, J. Nielsen, A. Hillemyr (Eds), Advanced Dynamics of Vehicles on Roads and Tracks, Proceedings of the 26th Symposium of the International Association of Vehicle System Dynamics, IAVSD 2019, Gothenburg, Sweden, 57–66, 2019.
51. International Organization for Standardization, ISO 2631 – Mechanical vibration and shock – evaluation of human exposure to whole-body vibration: Parts 1 to 5, 1997–2018.
52. European Committee for Standardization, European Standard CEN. EN-12299, Railway applications – ride comfort for passengers – measurement and evaluation., Brussels, Belgium, 2009.
53. L. Meirovitch, Fundamentals of Vibrations, McGraw-Hill, New York, NY, 2001.

2 Modeling of Mechanical Systems for Rail Vehicles

2.1 INTRODUCTION

Railway vehicles are complex dynamic systems which include components such as wheels, axles, bogies, carbodies, and suspensions following the laws of mechanics, but also other components governed by physical laws pertaining to other branches of physics, particularly electromagnetism (for electric traction units, electromechanical and electrohydraulic actuators), fluid dynamics (for hydraulic and pneumatic actuators including pneumatic actuation of braking), and thermodynamics (for thermal traction units). This chapter focusses on the modeling of mechanical components in a railway vehicle and, together with Chapters 3–6, lays the foundations required for the modeling of a railway vehicle or a train as a mechatronic system.

Mechanical models of railway vehicles can be used to forecast the vehicle's behavior in different running conditions, supporting the design of a new vehicle and exploring ways to improve the running behavior under different scenarios. The theoretical foundation for mathematical models of railway vehicles started to be laid in the late 19th century, with notable contributions from J. Klingel and H. Hertz. The first modern treatment of wheel/rail contact forces introducing the notion of creepage is due to F. W. Carter and dates to the third decade of the 20th century. From the 1950s to the late 1970s, substantial progress was made defining advanced models of wheel/rail contact that are still in use nowadays. Eminent scholars involved in these developments were A. D. de Pater, K. L. Johnson, and J. J. Kalker. Meanwhile, the first numerical models allowing a quantitative study of curving and stability started to appear. In the late 20th century, multi-body system (MBS) approaches were developed and applied to railway engineering, leading to the use of more detailed and complex numerical models. In the past 20 years, MBS models have been further refined, especially in view of multi-physics simulation and co-simulation, allowing to incorporate the effect of actuation and control systems, pneumatic suspensions, pneumatic braking circuits, etc., into MBS models.

Two approaches currently can be used to define a mechanical model of a rail vehicle. On one hand, relatively simple semi-analytical models can be defined; these are generally characterized by a reduced number of degrees of freedom (DOF) and often resort to linearization of the equations of motion for the system. Models of this kind can be used to obtain a qualitative understanding of vehicle dynamics and of the effect of design variables on the vehicle's running behavior. Simple linear models can also be used for mechatronic systems, such as active suspensions, in the context of model-based control. On the other hand, more complex numerical models can be defined using an MBS approach. In this latter case, software packages exist which enable the user to define and use the model without the need to give regard to writing

DOI: 10.1201/9781003028994-2

and solving the equations of motion; the main task of the user in this case is to choose a proper level of detail of the MBS model and to validate the model prior to its use.

The aim of this chapter is to provide a concise introduction to the modeling of a railway vehicle as a mechanical system and clarify the interfaces and relationships to mechatronic systems being part of the vehicle (e.g., active or semi-active suspensions, traction/braking control). Due to space limitations, some details about the mathematical formulation of models will be omitted; for this, the reader is referred to some of the excellent textbooks already existing such as [1–3]. A significant part of this chapter is allocated to describing models for wheel/rail contact forces, which are a distinctive feature of rail vehicle models, but again some details are left to other textbooks focusing on this specific topic.

This chapter starts with the classification of approaches for theoretical and experimental-based modeling of railway vehicles (Section 2.2). A summary of models for wheel/rail contact is then provided in Section 2.3; this is relevant to the scope of this chapter because mechatronic systems can be used to enhance the vehicle's performance/ running safety in regard to issues involving the forces arising at the wheel/rail interface, particularly in regard to vehicle stability, curving, traction, and braking.

In Section 2.4, models of the railway track and of track irregularities relevant to the study of rail vehicle dynamics are introduced

In Section 2.5 the behavior of passive suspension components in a railway vehicle is described and needs for complementing or partially replacing passive suspensions by semi-active or active ones are identified. Section 2.6 provides a short overview of the pantograph-catenary system and related models. In Section 2.7, the traction and braking systems of a railway vehicle are described considering different possible design variants.

In Sections 2.8, 2.9 and 2.10, the effect of train dynamics, pneumatic brakes and inter-car forces on vehicle dynamics is discussed and models for forces at the buffers and draw gear are presented.

A practical example for the development and simulation of simple 2 DOF and 6 DOF models of a railway based on the techniques introduced in this Chapter is presented in Chapter 14 as Case A, while more complex multi-body and multi-physics models and their use for mechatronic system studies are presented in Chapter 14, Cases B–D.

2.2 CLASSIFICATION FOR THEORETICAL AND EXPERIMENTAL-BASED MODELING APPROACHES

Different approaches can be followed to define the mathematical model of a system like a railway vehicle. In this Section, an outline of modeling approaches is proposed as an introduction to further development of mathematical models throughout this book, particularly in Chapters 3–6.

A mathematical model is a quantitative description of the behavior of a physical system defined using a suitable mathematical language, usually in the form of a set of equations. The equations describing the mathematical model can be either ordinary equations or differential equations. For instance, ordinary equations can be used to describe the steady-state curving behavior of a railway vehicle, while a set of differential equations can be used to describe the motion of a vehicle running on an irregular track.

Mathematical models can be physical-based models, i.e., formulated based on the laws of physics governing the behavior of the system considered, or can be defined as black-box models, i.e., mathematical relationships representing the system's behavior in terms of input-output relationships. A simple, non-exhaustive categorization of mathematical models based on the basic distinction between physical-based and black-box models is presented in the sub-sections below.

2.2.1 PHYSICAL-BASED MODELS

Physical-based models use the governing laws of physics to describe the system's behavior. This can be done using different mathematical instruments such as partial-derivative differential equations (PDEs), ordinary differential equations (ODEs) and mixed differential algebraic equations (DAEs). The form of the resulting equations depends on the features of the system modeled, e.g., the model of a flexible mechanical system may be based on the use of PDEs, while a model consisting of one or more rigid bodies can be described using either ODEs or DAEs.

Here, we focus on a mechanical system whose motion can be described in terms of the variation with time of a finite set of independent kinematic coordinates. This includes all systems formed by rigid bodies, but is also applicable to systems including flexible bodies, provided a discretization of the flexible bodies is performed using modal synthesis and/or the finite element method (FEM). We also assume the system is subject to holonomic constraints [4]. In this case, the equations describing the motion of the system, known as the system's *equations of motion,* take the general form:

$$M(\overline{q})\ddot{\overline{q}} - Q_v + C_Q^T\lambda = F\left(\overline{q},\dot{\overline{q}},u(t)\right)$$
$$C(\overline{q}) = 0 \tag{2.1}$$

where $\overline{q}$ is a vector collecting m scalar kinematic coordinates describing the motion of the system, $M(\overline{q})$ is the square, m-th order configuration-dependent mass matrix of the system, Q_v is the quadratic velocity vector, describing quadratic inertia terms in the equations such as centrifugal and Coriolis effects, F is the vector of generalised forces acting on the system (including internal forces from e.g. suspension components), C is a vector representing the constraint conditions that the system's coordinates $\overline{q}$ shall satisfy to comply with constraints applied to the system, C_Q is a Jacobian matrix representing the derivative of the constraint vector C with respect to the coordinate vector $\overline{q}$, λ is the vector of Lagrange multipliers representing the effect of constraint forces on the motion of the system, and u is a vector of time-dependent external inputs exciting the dynamics of the system, consisting of time-dependent forces/displacements, e.g., due to track irregularities.

Equation (2.1) takes the form of a system of DAEs. However, the algebraic conditions C describing the constraints can be used to identify a minimum set of coordinates q with size $n < m$ so that the elements of vector q, called the *independent coordinates*, are not subject to any kinematic condition. The size n of vector q is called the number of DOFs of the system. It shall be emphasized that, for a given mechanical system, the choice of the set of independent coordinates is not unique, but the number of DOFs is uniquely determined. Based on the choice of a minimum set of coordinates q, and assuming the constraints acting on the system are non-dissipative (i.e., the virtual work of constraint forces is zero), the system's equations of motion can be rewritten in the form:

$$m(q)\ddot{q} + \frac{dm}{dt}\dot{q} - \frac{1}{2}\frac{\partial}{\partial q}\left(\dot{q}^T m\dot{q}\right) = f\left(q,\dot{q},u(t)\right) \tag{2.2}$$

where $m(q)$ is the square, n-th order mass matrix of the system, and f is the vector of generalized forces acting on the system.

The advantage of writing the system's equations of motion in the form of Equation (2.1) is that the identification of a minimum set of coordinates is not required in this case. As a consequence, a general method known as *multi-body system dynamics* can be established to write a system of equations of motion in the form of Equation (2.1). This method is implemented in a number of software packages (including commercial packages, free software, and research-oriented academic software), allowing the user to define a mathematical model of a complex mechanical system, solve the equations for given inputs and initial conditions, and visualize the results with limited effort. On the other hand, significant effort is generally required to derive the system's equations of motion in the form of Equation (2.2) from the fundamental laws of mechanics, so this approach is generally used for relatively simple mechanical systems and is often performed manually.

In the case where a system of equations in the form of Equation (2.2) is obtained, a further simplification can be obtained if the motion of the system can be seen as a small perturbation in the neighborhood of a steady-state configuration, corresponding to a constant value q_0 of the independent coordinates. In this case, the equations of motion can be linearized, i.e., the terms in the equation can be approximated to Taylor's series expansions truncated to the first order, resulting in a system of n second order linear ODEs with constant coefficients:

$$m\ddot{q} + c\dot{q} + kq = bu(t) \tag{2.3}$$

where m, c, and k are the (constant, square, n-th order) mass, damping, and stiffness matrices of the linearized system and b is a constant matrix transforming the inputs u into a vector of generalized time-dependent forces acting on the system.

In control theory, it is customary to express the linearized system of equations of motion (Equation 2.3) in the form of a system of $2n$ first order equations, introducing a *state vector x* defined as:

$$x = \left[\begin{array}{cc} \dot{q}^T & q^T \end{array}\right]^T \tag{2.4}$$

so that Equation (2.3) is rewritten as:

$$\dot{x} = Ax + Bu \tag{2.5}$$

where A is a square matrix of order $2n$ called the *state matrix* of the system. A second equation is often used in combination with Equation (2.5), providing a linear relationship between the system's state x, the input vector u, and a vector of observed variables y representing the outputs of the system:

$$y = Cx + Du \tag{2.6}$$

Altogether, Equations (2.5) and (2.6) describe the state equations for a linear dynamic system.

So far, mathematical models were defined in terms of differential equations or mixed algebraic-differential equations with time t as the independent variable. Models of this type are said to be *time domain* models.

Laplace transforms or Fourier transforms can be used to transform a set of ODEs defined in the time domain to a set of ordinary equations defined in the domain of the complex-valued Laplace s variable or in the frequency domain, respectively. For linear systems, this leads to a simple and effective formulation of the mathematical model in terms of an input-output relationship expressed by a *transfer function* (TF) defined in the s variable domain or by a frequency response function (*FRF*) defined in the frequency domain. An input-output model can be derived directly from the state equations of the system using the Laplace or Fourier transform, but can also be obtained from experimental data, see the next sub-section.

Finally, it may be useful in some cases to formulate the mathematical model in terms of relationships between the values of variables describing the state of the system at discrete time intervals rather than as a continuous function of time, thereby resulting in a discrete-time model. This is particularly advantageous for use in digital controllers which operate on signals sampled at discrete time intervals. The equivalent of the Laplace transform for discrete-time models is the Z-transform which enables to formulate the mathematical model in terms of an input-output relationship taking the form of a rational function of the complex variable z.

2.2.2 BLACK-BOX MODELS

Black-box models are meant to describe the system in terms of an input-output relationship which can be defined in the frequency domain or in the discrete time domain. To set up a model of this kind, the knowledge of the physical laws governing the system is not required. The model is instead defined based on the knowledge of the outputs produced by a given sequence of inputs fed to the system. Inputs and outputs can be generated by a complex physical-based mathematical model or measured from experiments performed on the physical system for which the mathematical model is defined. In the first case, the black-box model is a simplified model which is *trained* and then used to replace the complex physical-based model in cases where a large number of simulations is required (e.g., Monte-Carlo optimization problems). In the latter case, the black-box model is an alternative to a physical-based model and is particularly useful in cases when the system to be modeled is too complex or partly unknown.

One way to define a black-box model from an experiment is to measure the inputs and outputs of the system and to derive from these measurements an estimate of the system's FRF. Considering, for the sake of simplicity, a single-input single-output system (SISO) with input $u(t)$ and output $y(t)$, the estimate of the system's FRF is obtained according to one of the following two alternative forms:

$$H_1 = \frac{G_{UY}}{G_{UU}}$$

$$(2.7)$$

$$H_2 = \frac{G_{YY}}{G_{UY}}$$

where G_{UU} and G_{YY} are the auto-spectra of the system's input and output signals, respectively, and G_{UY} is the cross-spectrum of the two signals [5]. The H_1 estimate of the FRF is insensitive to uncorrelated noise on the output, while the H_2 estimate is insensitive to uncorrelated noise on the input.

Black-box models can also be defined as discrete-time models, i.e., models establishing a relationship between the values of the input and output variables at discrete times $t - k\Delta t$ with $k = 0, 1, ..., N$. These models are called auto-regressive moving-average (ARMA) models [6] and take the form:

$$y(t) + a_1 y(t - \Delta t) + \cdots + a_n y(t - n\Delta t) = b_0 u(t) + b_1 u(t - \Delta t) + \cdots + a_m u(t - m\Delta t) \quad (2.8)$$

More complex formulations, including, e.g., non-linearities in the input-output relationship, can also be considered but are not addressed here as they are out of the scope of this book.

2.3 MODEL OF WHEEL/RAIL CONTACT

The model of wheel/rail contact is a distinctive feature of any mechanical model of a railway vehicle. Depending on the scope of the overall model, the mathematical representation of wheel/rail contact can be introduced at different levels of complexity but, in general, some common steps have to be dealt with.

Firstly, for a given position of the wheel relative to the rail, the number and position of wheel/rail contact points shall be determined. This first step will be referred to hereafter as *geometric analysis* and, depending on the complexity of the approach followed, may result in determining more than one contact point taking place at the same time between a single wheel and rail couple. Details about the geometric analysis are provided in the following sub-section.

At each contact point, a common normal direction to the wheel and rail surfaces and a local tangential plane can be defined. The contact force at each contact point is then defined as a vector having a component along the normal direction, called the normal contact force N, and a component in the tangent plane called the *tangential contact force*. The tangential contact force is further decomposed into a *longitudinal force* F_x parallel to the wheel's mid-plane and a *transversal force* F_y perpendicular to F_x. Furthermore, a moment of the frictional forces acting around the normal direction, called the *spin moment M_z* is present but is often neglected due to its small effect on the vehicle's dynamics. The two scalar components F_x and F_y of the tangential force and the spin moment M_z are called the *generalized creep forces*. Figure 2.1 shows the different components of wheel/rail contact forces.

The different components of wheel/rail contact forces defined above can be expressed as functions of the motion of the wheel relative to the rail. This is usually dealt with in two stages, hereafter called the *normal contact analysis* and the *tangential contact analysis* [7, 8]. The normal analysis consists of relating the normal contact force to the relative motion of wheel and rail surfaces in the normal direction and determining the size and shape of the contact patch (and optionally the

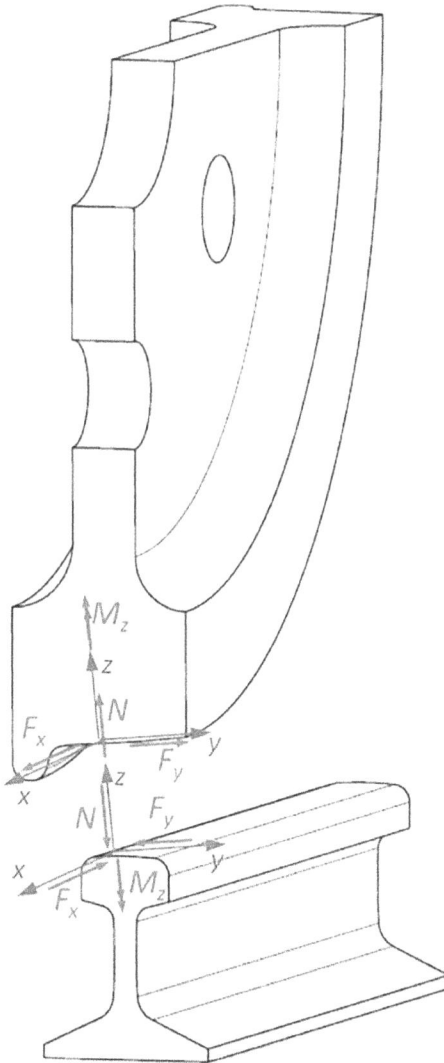

FIGURE 2.1 Components of wheel/rail contact forces and creepages at wheel/rail contact.

distribution of the normal pressure in the contact patch). The tangential analysis consists of evaluating the components of the frictional force. If the wheel and rail bodies can be assimilated to elastic semi-half spaces (and are made of the same material), the contact problem is said to be *quasi-identical* [7] which means the normal and tangential problems are decoupled. In this case, the normal problem can be solved first and the solution will not depend on the distribution of tangential stresses in the contact patch. Then, the tangential problem can be solved using as inputs the quantities obtained from the solution of the normal problem.

2.3.1 Geometric Analysis of Wheel/Rail Contact, Equivalent Conicity

The contact between a railway wheel and a rail takes place in a small region having a size of a few square millimeters, called the *contact patch*. Due to the shape of the wheel and rail profiles, the position of the contact patch is changing with the motion of the wheel relative to the rail and, in some cases, multiple contact patches may occur (multiple contact condition). The problem of determining the number and position of the contact patches is normally performed considering the wheel and the rail as rigid bodies, resulting in the determination of single points of contact between the two bodies and is therefore called the *contact point search*. Once the position of the contact point(s) is/are determined, a number of geometrical parameters involved with wheel/rail contact can be obtained which are relevant to the normal and tangential contact analysis; these parameters include the inclination of the plane tangent of contact with respect to the top-of-rail plane (contact angle), the difference of the rolling radius between the two wheels (rolling radius difference RRD), the local curvatures of the wheel, and rail surfaces at the contact point.

Different algorithms can be used to perform the contact point search, and a summary of available methods is provided in [9]. In the context of numerical simulation of the vehicle's running behavior, a basic distinction can be established between methods performed *offline* as a pre-processor of the simulation and methods in which the contact point search is performed *online* as part of the calculation of wheel/rail contact forces at each time step. If the offline approach is followed, the position of the contact points is computed for different values of the relative lateral position of the wheelset with respect to the track centerline (and, sometimes, of other quantities such as the relative yaw angle of the wheelset with respect to the track, or the track gauge) and the geometric parameters relevant to the calculation of wheel/rail forces are stored in contact tables which are then interpolated at each time step of the numerical integration of the model. This approach is more efficient from a computational point of view than the one resorting to an online contact search but, to keep the dimensionality of the contact table within reasonable limits, some simplifying assumptions need to be made, e.g., neglecting the effect on the contact point search of railhead rotation due to the flexibility of rails, the relative rotation of the wheel profiles due to axle bending, the longitudinal variation of rail profiles, and sometimes even neglecting track gauge variations along the track.

Figure 2.2 shows an example of the results of an offline contact search for ORE S1002 wheel profiles and UIC 60 rails with track gauge 1435 mm and 1:40 rail inclination. The upper sub-plots show the position of the contact points on the wheel

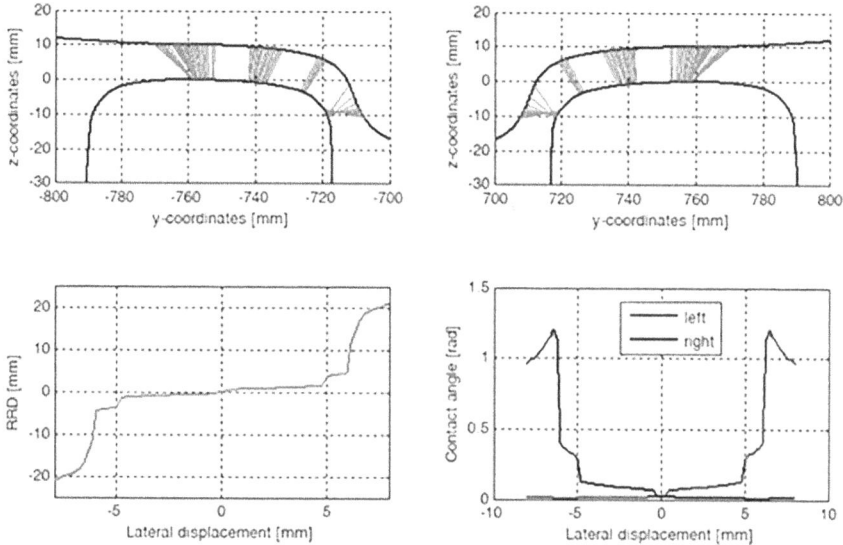

FIGURE 2.2 Top: Position of the wheel/rail contact points for different lateral shift of the wheelset relative to the rails; bottom: RRD and contact angle diagrams. The results shown are for ORE S1002 wheel profiles and UIC 60 rails with track gauge 1435 mm and 1:40 rail inclination.

and rail transversal profiles for different lateral shift of the wheelset relative to the track, the lower sub-plots show the RRD and contact angles as function of the lateral shift. The two lower diagrams show sudden variations of the contact parameters at some values of the wheelset shift which occur due to *jumps* in the position of the contact points which are observed in the upper plots. This effect is consistent with the assumption of rigid profiles in the contact point search and may lead to numerical disturbances in the numerical integration of the equations of motion for an MBS model of the vehicle. To avoid this issue, local deformability of the two profiles in the region close to the contact point can be introduced in a simplified way, resulting in a smoother transition of the contact point position and in the formation of multiple contacts between one wheel and the contacting rail.

2.3.2 The Normal Contact Analysis: Normal Force, Contact Patch, and Normal Stresses

Once the contact points are found, the relationship between the normal contact force and the contact patch can be determined. Two approaches are possible, known respectively as *rigid contact* and *flexible contact* models. In the rigid contact case, the contact between the wheel and the rail is treated as a constraint and the normal force exchanged between the wheel and the rail is obtained from dynamic equilibrium equations or from Lagrange multipliers introduced in the MBS formulation [4]. In this case, the normal force is a known input used to determine the shape and size of the contact patch which is then required to perform the tangential contact analysis.

The distribution of normal stresses in the contact patch is normally not required if the aim of the analysis is just to determine the motion of the vehicle over the track, but the knowledge of the normal stresses may be of interest if the analysis is also intended to consider surface damage effects, rolling contact fatigue in particular. If a flexible contact approach is followed, the normal force is computed as a function of the kinematic approach of the wheel and rail surfaces in the normal direction, defined as the inter-penetration of the rigid profiles which is compensated by local elastic deformability leading to the formation of a contact patch. In this case, wheel/rail contact is not modeled as a constraint, but rather the normal force becomes a function of the relative position of the wheel and the rail at the considered time step.

The methods used to study the normal contact do not differ much whether a rigid or flexible contact model is used. The simplest but still widely used method is Hertz theory. The assumptions on which Hertz theory is based are:

 i. In terms of relationship between normal pressures applied and local defor-
    mation, the two bodies in contact behave as elastic half-spaces.
 ii. The curvature of the bodies is large compared to the contact size.
 iii. The curvature of the surface of the bodies is constant over a region large
      enough to enclose the entire contact patch.

Under these assumptions, the contact patch is found to be an ellipse defined by the value of the two semi-axes a (in the longitudinal direction) and b (in the transversal direction). For this elliptic patch, an *ellipticity* parameter g is defined as the ratio of the length of the two semi-axes:

$$g = \frac{b}{a} \tag{2.9}$$

The formulae defining the parameters a and b as functions of the local curvature of the contacting surfaces can be found in [8] and in many other books, so they are not repeated here.

Furthermore, according to Hertz's theory, the relationship between the inter-penetration δ and the normal force N is expressed as:

$$N = C_H \delta^{\frac{3}{2}} \tag{2.10}$$

where C_H is a coefficient depending again on the local curvature of the surfaces and on the Young's modulus and Poisson ratio of the material.

Finally, the distribution of normal pressure in the elliptic contact patch is defined by the semi-ellipsoid having the equation:

$$p(x,y) = \frac{3}{2} \frac{N}{\pi ab} \sqrt{1 - \left(\frac{x}{a}\right)^2 - \left(\frac{y}{b}\right)^2} \tag{2.11}$$

where x and y are the coordinates spanning the elliptic contact region.

Hertz theory is simple and allows a very fast calculation of the parameters involved in the normal contact analysis, therefore it is very widely used in the context of MBS models for rail vehicle dynamics. However, the assumption of constant curvature of the wheel and rail across the entire contact patch often turns out to be not fully adequate to describe the real contact condition [1, 8]. For this reason, other approaches have been proposed to solve the normal problem.

In order of increasing complexity, simplified non-Hertzian methods aim at providing more accurate solutions by just removing assumption iii) stated above, i.e., constant curvature of the surfaces. In this way, a more general solution of the normal contact problem can be found characterized by a contact patch having a non-elliptic shape at the price of a small increase of the computational effort required. Among simplified non-Hertzian methods worthy of mention are the method due to Piotrowski and Kik [10] and methods based on partitioning the contact patch into strips such as the STRIPES method [8] and the one by Sichani et al. [11].

The algorithm CONTACT originally proposed by Kalker [7] and then improved by Vollebregt [12] can account for any shape of the wheel and rail surface in the contact region but still retains assumption i) above. It is more accurate than the simplified non-Hertzian method but requires a considerably higher computational effort and, at the time when this book is written, it is believed to be much too computationally intensive to be used for vehicle dynamics simulation. It is used instead for accurate analysis of contact stresses in relation to damage problems and as a validity reference for faster but less accurate methods. Finally, if all the simplifying assumptions inherent to Hertz theory are removed, the normal problem can only be solved using the FEM [13], but this results in a very heavy computational effort which usually cannot be afforded in the context of vehicle dynamics simulation in the time domain.

2.3.3 THE TANGENTIAL CONTACT ANALYSIS: CREEPAGE VERSUS CREEP FORCE RELATIONSHIP

Tangential contact analysis consists in evaluating the components of the wheel/rail contact force in the tangent plane (frictional forces, also known as creep forces) as a function of the motion of the wheel and rail. It is assumed that Coulomb's dry friction law applies locally to each point in the contact patch, giving:

$$
\begin{cases}
\vec{s} = 0 \text{ if } \|\vec{\tau}\| < \mu p \text{ (adhesion)} \\
\vec{\tau} = \mu p \dfrac{\vec{s}}{\|\vec{s}\|} \text{ if } \vec{s} \neq 0 \text{ (slip)}
\end{cases}
\tag{2.12}
$$

where $\vec{s}$ is the slip between the two surfaces, p is the local value of normal stress, μ is the adhesion coefficient, and $\vec{\tau}$ is the vector of local tangential stress. If the tangential stresses (and hence the resulting frictional forces) are low, adhesion prevails in the entire contact patch and the stresses and total forces can be shown to be linearly dependent on the creepage components defined as follows:

$$v_x = \frac{V_{1,x} - V_{2,x}}{V_m} \quad \text{(longitudinal creepage)}$$

$$v_y = \frac{V_{1,y} - V_{2,y}}{V_m} \quad \text{(lateral creepage)} \tag{2.13}$$

$$\varphi = \frac{\omega_{1,z} - \omega_{2,z}}{V_m} \quad \text{(spin creepage)}$$

where $V_{i,x}$, $V_{i,y}$ ($i = 1$ for the wheel, $i = 2$ for the rail) are the scalar components in longitudinal and lateral directions of the absolute speed of the two bodies at the contact point and V_m is a reference speed which, for a railway wheel, is chosen as:

$$V_m = \frac{1}{2}(V + \omega r_0) \tag{2.15}$$

where V is the longitudinal speed of the wheelset, ω is the angular speed of the wheelset around its axis of symmetry, and r_0 is the nominal rolling radius of the wheels. If no traction or braking forces are applied on the wheels, $V_m = V$.

For larger values of the creepages, a region of slip is formed in the contact patch. The slip region becomes progressively larger with increasing creepages, until slip occurs over the entire contact region and a full slip condition is established at the contact (a condition also called *saturation*). In this condition, the magnitude of the tangential force equals the product of the normal force N times the friction coefficient and the direction of the tangential force is opposite to the direction of the relative sliding velocity of the bodies. The transition from the full adhesion to the full slip condition is qualitatively shown in Figure 2.3 considering a simplified case where only longitudinal creepage is present.

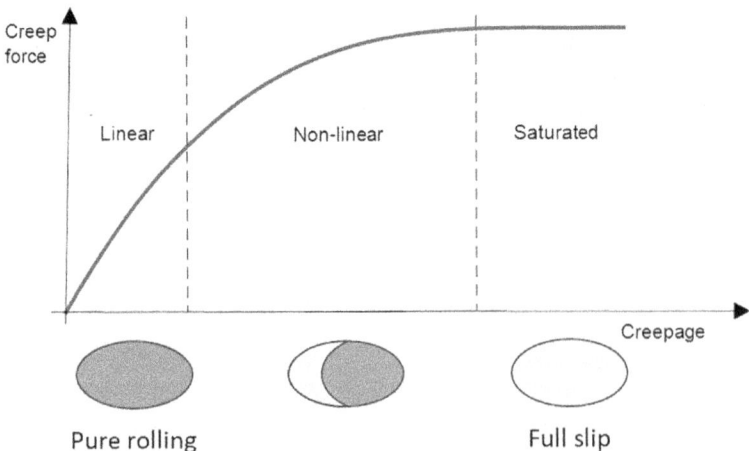

FIGURE 2.3 Transition from pure rolling to full slip (simplified one-dimensional case).

Models having different levels of complexity can be used to describe the relationship between creepages and creep forces and some of the most relevant ones are described below.

2.3.3.1 Kalker's Linear Theory

This model is valid for small creepages, under the assumption of an elliptic contact patch and assuming the contacting bodies behave as elastic half-spaces in terms of the relationship between surface stresses applied and deformation in the tangent plane. According to the linear theory, the linear creep forces F_x^{lin} and F_y^{lin} are expressed as:

$$F_x^{\text{lin}} = -f_{11}v_x$$
$$F_y^{\text{lin}} = -f_{22}v_y - f_{23}\varphi$$

(2.16)

where f_{11}, f_{22}, and f_{23} are the coefficients defining the linear law. These coefficients are in turn expressed as:

$$f_{11} = Gabc_{11}; \; f_{22} = Gabc_{22}; \; f_{23} = G(ab)^{\frac{3}{2}} c_{23}$$

(2.17)

where G is the shear modulus of the material, a and b are the semi-axes of the elliptic contact patch, c_{11}, c_{22}, and c_{23} are three non-dimensional coefficients given by Kalker which can be obtained from the interpolation of tabular values as a function of the ellipse's shape factor g defined by Equation (2.9), see [8] for example. It is worthy of mention that, since the size of the ellipse semi-axes depends on the normal load, the creep forces defined by Equation (2.16) are functions not only of the creepages but also of the normal load.

2.3.3.2 Heuristic Saturation Laws

Kalker's linear theory is used in all cases when a linear model of creep forces is needed, e.g., for a stability analysis based on the calculation of the system's eigenvalues or when a linear model is used in the context of a model-based control scheme (e.g., control with observers). However, if a time domain simulation of the vehicle running behavior or a steady-state curving analysis shall be performed, the linear theory may turn out to be inadequate as it is not suited to represent a condition close to saturation. The simplest way to consider saturation is to apply a scaling factor to the linear forces defined by Equation (2.16) so that the creep force components are increased in a less than proportional way with increasing creepages and the resultant force does not exceed the product of the friction coefficient μ times the normal force N. Several saturation laws have been proposed in the past, some of which can be justified based on simplified analytical models. For the sake of brevity, only the formulae proposed by Shen, Hedrick, and Elkins in [14] are reported here, while the reader is referred to [8] for other saturation laws.

According to the formulae by Shen, Hedrick, and Elkins, the saturated forces in longitudinal and transversal directions, F_x and F_y, respectively, are expressed as:

$$F_x = \varepsilon \frac{\mu N}{F^{\text{lin}}} F_x^{\text{lin}}$$
$$F_y = \varepsilon \frac{\mu N}{F^{\text{lin}}} F_y^{\text{lin}}$$

(2.18)

where F_x^{lin} and F_y^{lin} are the linear creep forces defined by Kalker's linear theory (Equation 2.16) and ε is a scaling coefficient. Defining the total magnitude of the linear creep force F^{lin} as:

$$F^{\text{lin}} = \sqrt{\left(F_x^{\text{lin}}\right)^2 + \left(F_y^{\text{lin}}\right)^2}$$

(2.19)

the scaling factor ε used in Equation (2.18) is computed as:

$$\varepsilon = \begin{cases} \dfrac{F^{\text{lin}}}{\mu N} - \dfrac{1}{3}\left(\dfrac{F^{\text{lin}}}{\mu N}\right)^2 + \dfrac{1}{27}\left(\dfrac{F^{\text{lin}}}{\mu N}\right)^3 & \text{for } F^{\text{lin}} \leq 3\mu N \\ 1 & \text{for } F^{\text{lin}} > 3\mu N \end{cases}$$

(2.20)

In this way, when $F^{\text{lin}} \ll \mu N$, the creep forces from Kalker's linear theory are preserved but, when the magnitude of the linear force becomes comparable or even higher than the maximum frictional force μN, a saturation effect is progressively introduced so that the total magnitude of the creep force cannot exceed the limit value μN. A drawback of this saturation law is that the saturated creep force vector points along the direction of the linear creep force even in case of large saturation, while the actual direction of the frictional force in full slip is opposite to the direction of the relative sliding velocity. This drawback can be easily overcome by a modified formulation of the Shen, Hedrick, and Elkins theory as follows:

$$F_x = \varepsilon \mu N \cos(\theta)$$
$$F_y = \varepsilon \mu N \sin(\theta)$$

(2.21)

where θ is an angle defining the direction of the creep force vector with respect to the longitudinal direction x, defined as:

$$\theta = \tan^{-1}\left(\frac{F_y^{\text{lin}}}{F_x^{\text{lin}}}\right)(1-\varepsilon) + \tan^{-1}\left(-\frac{v_y}{v_x}\right)\varepsilon$$

(2.22)

In this way, for $F^{\text{lin}} \ll \mu N$, both the magnitude and direction of the saturated creep forces will be close to the ones from Kalker's linear theory but, in a condition close to full slip, the direction of the saturated force will be opposite to the relative sliding velocity of the two bodies.

2.3.3.3 The Fastsim Method

Another simplified method suitable for performing the tangential contact analysis considering saturation effects is also originally due to Kalker [15] and goes under the name of *Fastsim*. The method was originally developed for elliptic contact patches but was later adapted to non-Hertzian contact patches [10]. In this method, the contact patch is firstly divided into a finite number of strips in the longitudinal direction and then each strip is divided into an equal number of cells. The unsaturated tangential stress components in longitudinal and transversal directions, τ_x^{lin} and τ_y^{lin}, respectively, are obtained as functions of the creepage components under the simplifying assumption that the two bodies behave elastically as two perfectly rigid bodies connected by a Winkler foundation, assuming zero tangential stress at the leading edge of each strip. In this way, the elastic deformation of the bodies at one point in the contact patch only depends on the stresses exchanged by the bodies at the same point through some flexibility coefficients.

At each cell of the discretized contact patch, a traction bound is defined assuming a parabolic or semi-elliptic distribution of the normal stress along each strip and, in case the total magnitude of the shear stress vector exceeds the traction bound, the scalar components of the vector are all reduced by the same scaling factor so that the magnitude of the saturated shear vector equals the traction bound at the cell considered and the direction of the saturated shear vector is the same as for the unsaturated shear vector. At cells where saturation of the shear stress is found, the slip and the frictional work can also be computed. The process is repeated for all cells in a strip and for all strips in the contact patch and the total creep forces are obtained from the sum over the discretized contact patch of the shear stresses multiplied by the area of the cells. The total frictional work can also be obtained as the sum of the frictional work values obtained at all cells where saturation of the contact stresses takes place. Full details regarding the formulation of this algorithm are given in [8].

The Fastsim method, as the name suggests, is sufficiently efficient from a computational point of view to be used in time domain simulations of the running behavior of a railway vehicle. However, it may lead to significant deviations from other, more accurate methods such as CONTACT (see below). One advantage of this method is that a distribution of the shear stresses and frictional power can be easily obtained, which can be useful to study damage and wear phenomena at the wheel/rail interface.

2.3.3.4 Kalker's CONTACT Algorithm

In his PhD thesis, Kalker developed an algorithm to solve the tangential problem under the sole simplifying assumption that the relationship between the shear stresses in the contact patch and the deformation of the bodies in the contact region are described by the Boussinesq–Cerruti formulae, valid for elastic half-spaces. This method goes under the name CONTACT which is also used for a method devised by Kalker for the solution of the normal problem, see above. The reason is that both methods were coded in a software originally developed by Kalker and then improved by Vollebregt. The formulation of the CONTACT algorithm is very complex and is out of the scope of this chapter, so the interested reader is referred to the classic book by Kalker [7]. It is just worth mentioning here that the CONTACT algorithm is today too computationally intensive to be used for time domain simulation of MBS models of rail vehicles and its use is mostly as a validity reference for simplified methods or to generate lookup tables, see below.

2.3.3.5 Use of Lookup Tables

A further alternative method to compute creep forces is the use of lookup tables. This approach was firstly developed by British Rail Research for elliptic contacts. The method consists of computing the creep forces for different creepage conditions and for different shapes of the contact patch, storing the results in a multi-dimensional table and obtaining the creep forces for the case of interest through the interpolation of the table. Since the calculations required to generate the contact table are performed only once, an accurate method (typically the one involving the software CONTACT) can be used, while the interpolation of the table is very fast and fully suitable for the calculation of the contact forces in the context of time domain numerical simulations. The use of lookup tables was recently extended to a special case of non-elliptic contact patches called simple double elliptic region (SDEC) [16]. In this case, one additional dimension has to be added in the table to consider a shape factor parameter that, together with the ellipticity parameter g from Equation (2.16), defines the geometry of the SDEC region.

2.3.4 WHEEL/RAIL CREEP FORCE MODELS FOR TRACTION AND BRAKE STUDIES

The models mentioned above cannot represent the creep force characteristics at large creep in agreement with the typical measurements on locomotives and other traction vehicles (Figure 2.4). The paper [17] provides the following description of this phenomena: "A justification of model parameters in regard to the slope at a very small creepage requires rather moderate values of the factor used to reduce the coefficients of Kalker's linear theory, resulting in the maximum adhesion occurring at creep values smaller than in measurements (point A in Figure 2.4). A justification of model parameters in regard to the maximum adhesion (point B) requires a very small value of the reduction factor, which is too low compared with values reported from measurements. In addition, it is possible to see that the results obtained from experiments are not covered by the adaptation of the friction coefficient. Furthermore, as shown

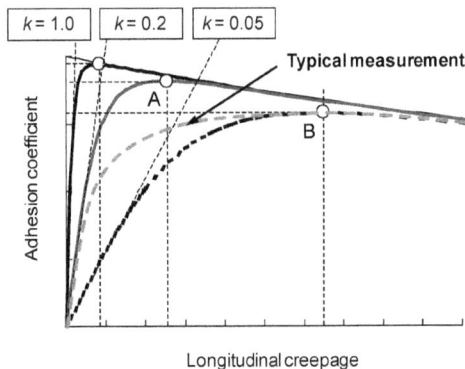

FIGURE 2.4 Modeling of creep force characteristic using falling friction coefficient and different reduction factors k. There is disagreement between the modeled creep force characteristic shape and its typical shape from measurements on wet rail. (From [17], with permission.)

in Figure 2.4, the reduced initial slope and the reduced friction coefficient at high creepages are not sufficient to achieve a good agreement with typical measurements."

Recently performed traction research studies with the application of vehicle dynamics show the use of contact approaches as classified in [18]:

- Fast-computing and linear contact mechanics theories;
- Modified Simplified Contact Theory (based on Fastsim as described in Subsection 2.3.3.3);
- Exact Theory (Kalker's CONTACT algorithm as described in Subsection 2.3.3.4) integrated in a vehicle dynamics model as the shared library called Extended Contact.

In this book, only two creep force calculation approaches for traction and braking are presented because they are most widely in use in rail vehicle mechatronic studies:

- Polach model published in [19]
- Modified Fastsim published in [17]

2.3.4.1 Polach Model

The modeling of a variable friction coefficient [19] is defined as:

$$\mu = \mu_s\left((1-A)e^{-Bw} + A\right) \tag{2.23}$$

where μ_s is the maximum coefficient of friction, A is the ratio of the limit friction coefficient at infinity slip velocity to the maximum friction coefficient μ_s, w is the magnitude of the slip (creep) velocity vector, m/s, and B represents the coefficient of exponential friction decrease, s/m.

The slip velocity used for the calculation of the slip velocity-dependent friction coefficient can be expressed as:

$$w = V - V_w \tag{2.24}$$

where V is the ground speed of a rail vehicle, m/s, V_w is the linear wheel speed, m/s.

The Polach creep force model [19] is also needed to find the adhesion force between a wheel and rail. This model has low computational needs and is perfectly suited for a simplified approach. For longitudinal adhesion forces on one wheel, we have:

$$F_x = \frac{2Q\mu}{\pi}\left(\frac{k_A e}{1+(k_A e)^2} + \arctan(k_S e)\right) \quad k_S \le k_A \le 1 \tag{2.25}$$

$$e = \frac{G\pi ab C_{11}}{4Q\mu}s \tag{2.26}$$

where a and b are the length of the semi-axes of the elliptic contact patch, r and l are indexes for right and left wheels of the wheelset, respectively, Q is the wheel load, C_{11} is the Kalker's linear theory coefficient, k_A and k_S are reduction factors in the area of adhesion and the area of slip, respectively, and G is the shear modulus.

A value of the longitudinal slip (creep) is defined as:

$$S = \frac{V_w - V}{V} \tag{2.27}$$

2.3.4.2 Modified Fastsim

The variable friction coefficient is calculated as [19]:

$$\mu = \mu_s((1 - A)e^{-Bw} + A) \tag{2.28}$$

where μ_s is the maximum coefficient of friction, A represents the ratio of the limit friction coefficient at infinity slip velocity to the maximum friction coefficient μ_s, w is the magnitude of the slip (creep) velocity vector, and B is the coefficient of exponential friction decrease. The slip velocity used for the calculation can be expressed as:

$$w = V\nu \tag{2.29}$$

where V is the locomotive speed, m/s, and ν is the total creep. The total creepage for Equation (2.29) then is:

$$\nu = \sqrt{\nu_x^2 + \nu_y^2} \tag{2.30}$$

where ν_x is the longitudinal creepage and ν_y is the lateral creepage.

The contact flexibility coefficient L^* in this modification of the Fastsim algorithm is calculated as [17]:

$$L^* = \frac{L}{k} \tag{2.31}$$

where L is the contact flexibility coefficient defined by Kalker [15]. The variable stiffness reduction factor k is defined as [17]:

$$k = k_0 \left(\alpha_{inf} + \frac{1 - \alpha_{inf}}{1 + \beta\varepsilon} \right) \tag{2.32}$$

where k_0 is the initial value of Kalker's reduction factor at creep values close to zero, $0 < k_0 \leq 1$, α_{inf} is the fraction of the initial value of the Kalker's reduction factor at creep values approaching infinity, $0 \leq \alpha_{inf} \leq 1$, β is a non-dimensional parameter related to the decrease of the contact stiffness with the increase of the slip area size, $0 \leq \beta$, and ε is a parameter describing the gradient of the tangential stress in the stress distribution transformed to a hemisphere [17]:

$$\varepsilon = \frac{1}{4} \frac{G\pi abk_0 c_{11}}{Q\mu} s_\varphi \tag{2.33}$$

where G is the shear modulus, a is the semi-axis length of the contact ellipse in the longitudinal direction, b is the semi-axis length of the contact ellipse in the lateral direction, C_{11} is the Kalker's coefficient, Q is the wheel load, μ is the variable friction coefficient, and s_φ is the total creep which can be found as:

$$s_\varphi = \sqrt{v_x^2 + v_\varphi^2} \qquad (2.34)$$

In this case, the lateral creepage also considers the contribution of spin:

$$v_\varphi = v_y + \varphi a \qquad (2.35)$$

where φ is the relative spin, rad/m. If the values of lateral creep and spin have opposite signs and the total lateral creepage is lower than the pure lateral creep, the higher absolute value of v_φ and v_y is selected.

2.3.4.3   Example of Identification of Creep Force Model Parameters from Measured Data

The identification of model parameters has been done by means of comparison with measurements as described in [19, 20].

Figure 2.5 presents an example with the comparison of the modified Fastsim with measurements, and with the calculated results using the Polach model with parameters identified in [19]. The input parameters of the variable friction coefficient are identical in both models and the parameters used in the models are

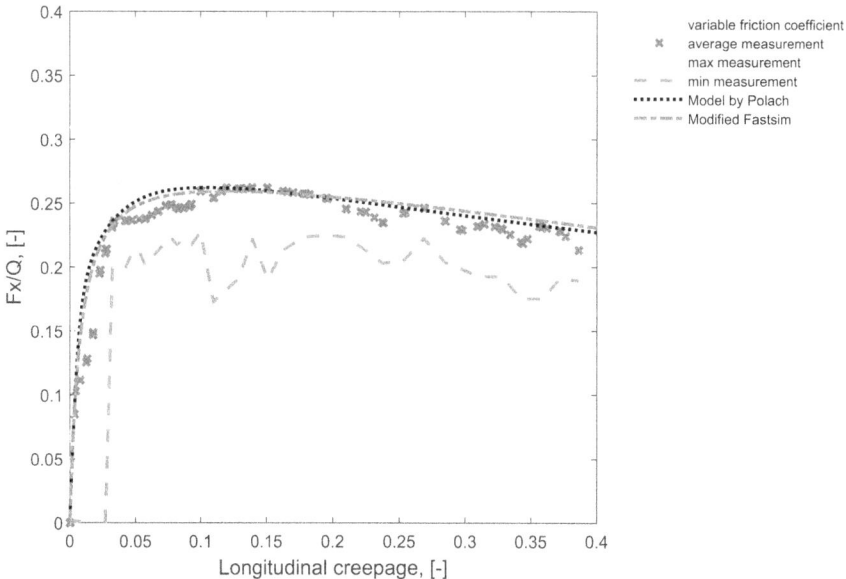

FIGURE 2.5 Comparison of results using modified Fastsim and using Polach model [19] with measurements for locomotive SBB 460 [20]. (From [17], with permission.)

TABLE 2.1

Parameters of the Creep Force Models for SBB460 Running on Wet Track at the Speed of 40 km/h

μ_S	A	B, s/m	k_A	k_s	k_0	α_{inf}	β
			Polach model				
0.31	0.5	0.16	0.16	0.07	-	-	-
			Modified Fastsim				
0.31	0.5	0.16	-	-	0.14	0.025	0.8

presented in Table 2.1. The friction coefficient curve is also displayed in Figure 2.5. The comparison shows good agreement between calculated results using both models and experimental results. The curves displaying the results of the modified Fastsim method and the Polach model are nearly identical.

In summary, it is possible to say that the Polach model computes creep force results significantly faster than the modified Fastsim. However, the modified Fastsim can be more precise in comparison with the Polach model because it is based on the original Fastsim [21].

2.4 MODELING OF TRACK AND TRACK IRREGULARITIES

Although the focus of this chapter is on modeling of rail vehicles, it is important to introduce the main features of the track system, as the track provides support and guidance to the vehicles and its geometric and mechanical features significantly affect the vehicles' running behavior.

2.4.1 THE TRACK SYSTEM

The term track system, or track in short, is used to denote the rails and the components that are used to support the rails in a way that is functional to provide suitable performance in terms of capability to withstand the loads caused by the rail vehicles while maintaining proper geometry and stiffness properties that are required to ensure running safety and ride quality. The design of the track is also concerned with reducing air borne and ground borne noise, with routing trains along different branches of the network at switches and crossings (S&C) and the track can also be part of the signaling system used for train command and control. The track system is laid over a supporting structure which can be different depending on the specific features of the railway line and on local topography. Typical cases for the supporting structure are an embankment made of compacted soil, the concrete invert of a tunnel, a bridge, or a viaduct. The design of the track system reflects not only the type of supporting structure, but also the design speed of the line, the design loads, and other features such as maintenance requirements, noise and vibration, etc.

The simplest track system is *direct fixation*. The rails are directly connected to a rigid construction like a tunnel invert by means of a fastening system. More frequently,

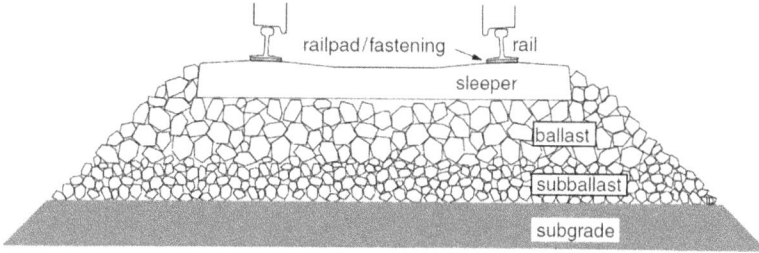

FIGURE 2.6 The ballasted track with its different components: Rails, rail pads, fastenings (not shown), sleepers, ballast, sub-ballast, and sub-grade. (From [22].)

the ballasted track system is used (see Figure 2.6), where the rails are supported by sleepers laid at regular intervals over a layer of stones called the ballast bed [22]. Sometimes a lower layer of stones with different granularity and mechanical properties called *sub-ballast* is included. The function of the sleepers is to control the track gauge (see below) and to provide filtering of vibrations transmitted from the rails to the ground. The function of the ballast bed is to support the sleepers while providing the required amount of supporting stiffness, to facilitate the drainage of water from the track, and to mitigate airborne noise as a result of the non-reflective surface of the stones. Below the ballast bed, one or more layers of compacted earth or other material form the sub-grade.

A third track system is called slab track. In this case, the rails are fastened to a concrete slab which can be either rigidly fixed to the ground or elastically suspended from the ground by means of a layer of elastomers. In the first case, the construction is similar to direct fixation, the difference being that the slab track system can be installed over an embankment, while direct fixation can be used only in tunnels or bridges. In the second case, the slab floating over a resilient elastomer layer forms a mass-spring foundation which is particularly effective with attenuating the transmission of noise and vibration. Compared to a ballasted track, the slab track has a higher construction cost but has superior performance with maintaining track geometry and generally has reduced need for maintenance.

All the abovementioned track systems use a fastening system to hold the rails in place. The fastening system can be realized according to different designs but typically consists of a steel tie plate, fixed to the invert, sleeper or slab by screws, a rail pad placed under the rail foot and usually made of some elastomeric material, and a rail foot retention system which may consist of spikes, anchors, or clips. A simplified drawing of a fastening system is shown in Figure 2.7.

2.4.2 Nominal Track Geometry

The track is laid according to a nominal geometry which includes longitudinal gradients and curves. The geometry of curves and curve transitions is particularly relevant to the running dynamics of rail vehicles and therefore an outline is provided here. A railway track consists of a sequence of tangent track sections, full curve sections and curve transitions. This can be represented by two companion diagrams showing

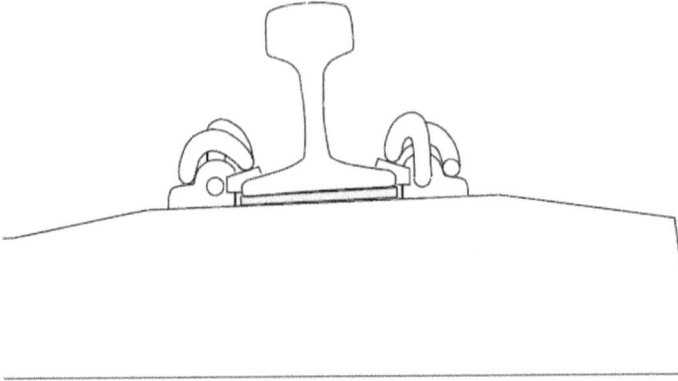

FIGURE 2.7 Schematic drawing of a fastening system connecting a rail to a sleeper. (From [22].)

respectively track curvature and cant (super-elevation h_t of the outer rail) as a function of the distance from a reference point. Tangent track segments have zero curvature, full curve segments have constant non-zero curvature and curve transitions are designed to provide a linear variation of curvature with the position. Therefore, the diagram representing track curvature as a function of distance takes the shape of a sequence of trapezoidal patterns, see the upper part of Figure 2.8. In curves, a cant angle is introduced in the track, i.e., the outer rail is super-elevated with respect to the inner one to provide some amount of compensation of centrifugal effects acting on the vehicles and on the passengers. Track cant in the full curve is set to a constant

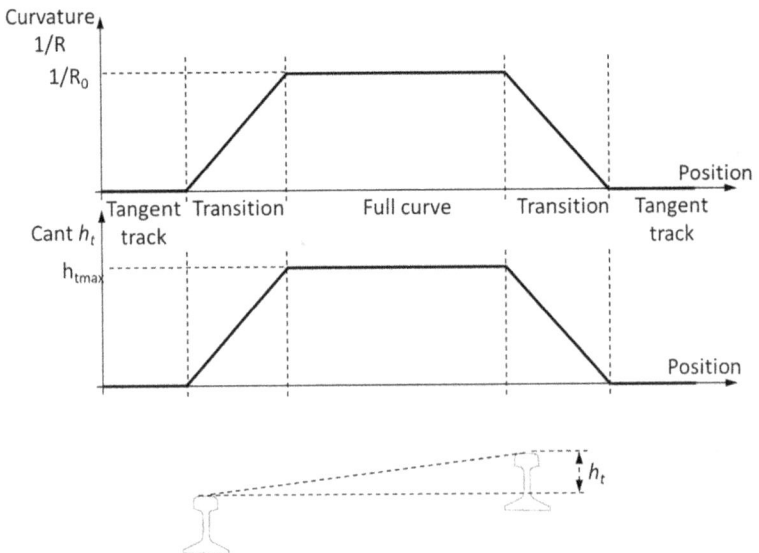

FIGURE 2.8 Curvature and cant as function of distance (top) and definition of track cant (bottom).

value and develops from zero to full cant linearly with the position in curve transitions, therefore leading to a diagram of cant versus distance of the type shown in the central portion of Figure 2.8.

It is worth noting a peculiar geometric feature of curve transitions known as *track twist*: due to the variation of cant angle along the transition, the super-elevation of the leading and trailing wheelsets in a bogie (or in a two-axle vehicle) is different and therefore the contact points of the four wheels belonging to the same bogie or vehicle do not lie in a common plane. This affects the forces in the primary suspensions (see Section 2.5) and hence the vertical component of wheel/rail contact forces. In particular, the wheel running on the high rail and undergoing the larger super-elevation will see an increase of the vertical contact force while the other wheel running on the high rail will see a decrease of the contact force with increased risk of derailment. The effect of track twist is obviously larger for a larger gradient of track cant, therefore curve transitions are designed to keep the cant gradient within acceptable limits (often set to 3 mm per m) while the vehicle/bogie shall be designed to avoid excessive sensitivity to track twist, which is usually achieved by ensuring a suitable degree of flexibility of the suspension in the vertical direction. Track twist also produces a relative roll rotation between the front and rear bogie of a vehicle, hence producing a re-distribution of vertical forces transmitted by the secondary suspensions, thereby requiring that the secondary suspensions of the vehicle are also designed to keep the sensitivity to track twist within acceptable limits.

2.4.3 Track Irregularity

Besides deterministic track features, the effect of random track irregularities on the running dynamics of a railway vehicle needs to be considered. Track irregularities are deviations in the level and alignment of rails. In a newly built track, irregularities are the consequence of errors and tolerances in the construction phase, but with the progression of loads produced by passing trains, further deviations from the ideal geometry accumulate due to permanent deformation taking place in some track components, particularly the ballast and the sub-grade. Track irregularities are a very significant source of excitation for the dynamics of a train running on the track and affect safety, ride quality, and wheel/rail interaction. Therefore, maintenance measures are regularly taken by infrastructure managers to keep the irregularity level within acceptable limits, so that the monitoring and maintenance of track geometric quality takes a large share of maintenance costs for railway lines.

Four components of track irregularities are relevant to vehicle dynamics and are shown in Figure 2.9:

- **Longitudinal level:** a deviation from the ideal geometry of the track centerline in the vertical direction
- **Lateral alignment:** a deviation from the ideal geometry of the track centerline in the lateral direction
- **Cross level:** a difference in the vertical position of the two rails; cross level in curves is not including the deterministic (and desired) effect of cant
- **Gauge variation:** a deviation of the local gauge from the design value

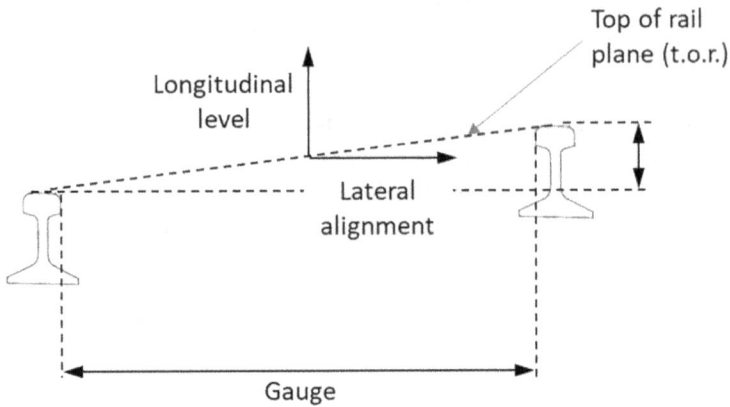

FIGURE 2.9 Components of track irregularity.

Track irregularity is a broad band random phenomenon: a typical irregularity profile includes components with long wavelengths (in the order of tens or hundreds of meters) and components associated with short wavelengths in the order of meters or below 1 m. Long wavelength irregularities have larger amplitude and excite vehicle dynamics in the low frequency range (0÷30 Hz approximately) and mainly affect ride quality. Short wavelength irregularities excite vehicle dynamics at higher frequency and may be the cause of large dynamic variation of wheel/rail contact forces leading to damage of the rolling surface, noise and vibration, and potentially affecting running safety.

Besides track irregularity, rail roughness, and rail corrugation also need to be mentioned. Rather than being caused by incorrect positioning of the rails, both roughness and corrugation are deviations of the railhead profile from its ideal geometry. Rail roughness arises as a consequence of rail manufacturing processes and is increased during the life of the rail by surface damage phenomena such as shelling and spalling [23, 24], while rail corrugation is the effect of irregular wear taking place on the rail head and producing a wavy pattern of crests and valleys with wavelengths ranging from tens to hundreds of millimeters.

Rail roughness and rail corrugation excite vehicle dynamics in the high frequency range, causing rolling noise and vibration but also inducing dynamic loads on the vehicle which can lead to accelerated damage of components like bearings, springs, and brake caliper mounts.

2.4.4 TRACK MODELS FOR VEHICLE DYNAMICS SIMULATION

Train/track interaction effects are highly relevant to the running dynamics of a railway vehicle. Therefore, the mathematical model of a railway vehicle needs to include a model of the track as well. Depending on the scope of the analysis, the model can be a simple or more refined one. Below, typical track models used in combination with railway vehicle models are described.

2.4.4.1 Rigid Track Model

This is the simplest track model and assumes the rails to be rigidly fixed to the ground. No DOFs are introduced to represent the motion of the track, but the excitation effects caused by track deterministic features and irregularities are considered introducing a fixed profile for the position of the rails. It shall be mentioned that there is frequent evidence in the scientific literature, see [25, 26] for example, that the use of a rigid track model may lead to inaccurate results in the simulation of rail vehicle dynamics in terms of running stability, shape of the contact patch, local distribution of contact pressure, etc. Besides this, the rigid track model is obviously not suited to represent noise and vibration effects arising from train/track interaction.

2.4.4.2 Co-Following Sectional Models

Sectional models provide a simplified representation of track flexibility effects. These models consist of a lumped parameter system like the one shown in Figure 2.10 representing the masses of the rails and sleepers and the flexibility of the track in vertical and lateral directions. Further to the example shown in the figure, additional DOFs can be introduced to model the motion of the rails relative to the sleepers and/ or to include a lumped mass below the sleeper representing the ballast. One sectional model of the type shown in the figure is introduced for each wheelset in the vehicle and the sectional model is assumed to travel with the same speed as the wheelset. Track deterministic features and irregularities are represented by a space profile similar to the rigid track model described above, but in this case the model can also describe the dynamic displacements of the rails caused by track flexibility.

Sectional models involve a limited number of DOFs and hence represent a simple and efficient way to avoid the shortcomings of the rigid track model but are only adequate in the low frequency range.

2.4.4.3 Finite Element Models

In cases when train/track interaction effects in the high frequency range (i.e., above say 30 Hz) are of interest, a more detailed model of the track is required and in this regard the use of finite element models is becoming increasingly popular. These models reproduce to a high level of detail a portion of the track and specific methods are used to consider the interaction between the train as a moving dynamic system

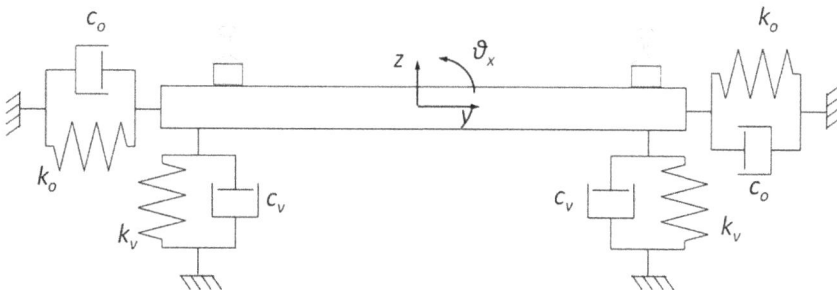

FIGURE 2.10 Co-following sectional model of the track.

and the track as a non-moving dynamic system contacting the train at wheel/rail interfaces [27]. Depending on the frequency range addressed by the analysis, the rails and sleepers can be modeled using beam-type or solid finite elements. To avoid spurious effects related with the finite size of the portion of track represented in the model, special techniques can be adopted, e.g., applying circular boundary conditions to the track model, or defining the track model in a moving reference.

Finite element models of the track are substantially more complex and computationally intensive than co-following models, but provide a detailed representation of train/track interaction effects that can be required for specific purposes, including the design of the track system and of supporting structures like bridges and viaducts, transitions between different track systems, etc.

2.4.4.4 Model of Switches and Crossings

Switches and crossings (S&C) are singularities in the track at which large wheel/rail contact forces may occur due to impact caused by the wheel being transferred from one rail to another. Wear and plastic deformation of the rails caused by accumulated service leads to even larger impacts, and S&Cs are therefore particularly relevant to the evaluation of extreme service loads for the wheelsets and to noise and vibration issues. Mathematical models of train/track interaction at S&Cs require that the variable profile of the rails in the crossing panel and the variation in space of the track stiffness are considered. The latter requirement can be satisfied using either a co-following sectional model of the track or a more detailed finite element model of the crossing, and nowadays most MBS software packages allow the simulation of a train passing over S&Cs as a standard feature. For more details, the reader is referred to [28].

2.5 MODEL OF SUSPENSION COMPONENTS

Another peculiar feature of mechanical models of a railway vehicle is the model of the vehicle's suspension system. Although in many cases the entire suspension can be modeled quite simply using a combination of linear springs and viscous dampers, specific and more complex models may be needed for some components. The aim of this section is to provide a concise introduction to modeling suspension components in railway vehicles. For a more comprehensive treatment of this subject, the reader is referred to [29].

2.5.1 PRIMARY AND SECONDARY SUSPENSIONS IN RAILWAY VEHICLES

In most passenger railway vehicles, the carbody sits over two intermediate bodies called *bogies* which in turn are supported by wheelsets directly making contact with the track. In this vehicle arrangement, shown in Figure 2.11, two stages of suspension are present: a *primary suspension* connecting the wheelsets to the bogie frame and a *secondary suspension* connecting the bogie to the carbody. Although in the simple scheme shown in Figure 2.11 springs are shown as acting only in the vertical direction, the actual design of primary and secondary suspensions provides stiffness in all directions and also includes devices to transfer traction and braking efforts from one body to another.

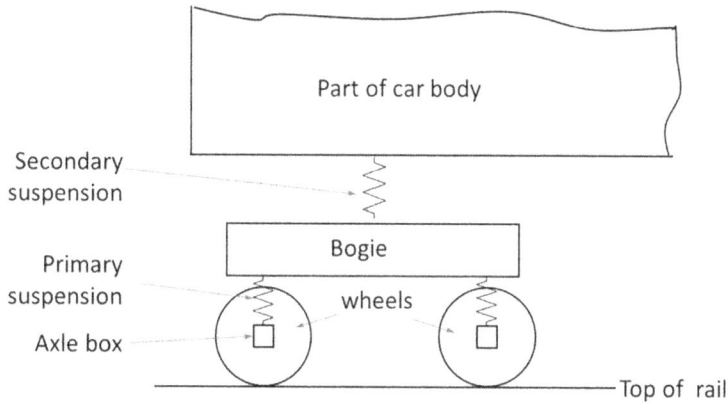

FIGURE 2.11 Schematic of the primary and secondary suspensions in a vehicle with bogies.

The functions performed by the primary suspensions are:

- Transfer the static loads from the bogie frame to the track and ensure an even distribution of static loads
- Keep the quasi-static variations of wheel loads due to track twist within acceptable limits
- Ensure satisfactory stability properties for the vehicle
- Ensure satisfactory curving behavior of the vehicle
- Reduce the dynamic forces applied to the bogie frame due to track irregularity and wheel out-of-roundness

To perform these functions correctly, the primary suspension needs to have sufficient flexibility in the vertical direction, which is obtained using either coil or rubber springs. A vertical damper is also used in many cases to ensure sufficient dissipation of vibration. The stiffness of the primary suspension in the horizontal plane, i.e., in longitudinal and lateral directions, is otherwise designed to meet requirements on stability and curving. This involves a trade-off between contrasting requirements, as a stiff primary suspension will be generally favoring stability (at least to some extent) but might lead to poor curving performance of the vehicle.

The functions performed by the secondary suspension are:

- Transfer and equalize the static loads from the carbody to the bogies
- Reduce carbody vibration to ensure ride comfort
- Keep carbody movements within the dynamic swept envelope gauge limits and avoid excessive movement of the carbody relative to the bogies

Therefore, the design of the secondary suspension has relatively little impact on the vehicle's running performance in terms of stability and curving but it is critical for passengers' comfort. This calls for a relatively soft suspension, but two factors need to be considered. Firstly, excessive variation of carbody height above the rails due to variations in the payload is undesirable as it will lead to a misalignment of the carbody floor with respect to the station platform when passengers are embarking or disembarking. Furthermore, too large a flexibility of the secondary suspension might have the consequence that the vehicle violates the infrastructure clearance gauge limits due to excessive movements in response to quasi-static and dynamic forces arising during vehicle motion in response to centrifugal forces in curves, track cant, track irregularities, or action of crosswind.

In relatively old vehicle designs (but still largely in use) the secondary suspension is realized by *flexicoil* springs, i.e., coil springs designed to ensure enough flexibility and to withstand relatively large deformation, complemented by lateral dampers. In many modern design vehicles, flexicoil springs are replaced by a pneumatic suspension consisting of two air springs, i.e., rubber bellows filled by pressurized air. The advantage of this solution is that it is easy to incorporate in the pneumatic suspension an automatic leveling system that will ensure a constant height of the carbody independently from the loading condition. The pneumatic suspension can be then designed to provide the desired amount of flexibility, leading in general to superior ride comfort compared to flexicoil springs. Since the stiffness of the pneumatic suspension is not sufficient to avoid excessive roll rotation of the carbody due to curving and to crosswind, an anti-roll bar is additionally included in the secondary suspension.

2.5.2 COIL SPRINGS, RUBBER SPRINGS, AND BUSHINGS

The simplest way to model a coil spring is by a one-dimensional linear stiffness corresponding to the stiffness of the spring in the axial direction. A model consisting of a spring element providing stiffness in three directions can be used to consider shear stiffness of the coil spring. In some cases, however, additional care has to be paid to consider the effect of compressive preload in the spring (second order stiffness) which plays a relevant role in determining quasi-static and dynamic variations of wheel loads due to lateral and roll movements of the carbody. In general, there is no need to consider damping or inertial effects taking place in coil springs. Energy dissipation due to material hysteresis is very low and can be neglected, while inertial effects are meaningful only in a frequency range involving internal resonances of the coil spring, which is often higher than the frequency range of interest of the overall vehicle model.

A simple one-dimensional or a three-dimensional model based on springs and dashpots in parallel is often also suitable to model rubber springs and bushings. In this case, damping effects are not negligible and can be modeled in an approximate way by means of viscous dashpots. However, this model cannot reproduce two notable features in the behavior of a rubber component. Firstly, the stiffness will be varying with the frequency of the deformation applied. Secondly, the dissipative behavior of rubber materials tends to be of a hysteretic rather than of a dissipative nature. This means the amount of energy dissipated for a cyclic deformation of the component will be weakly depending on the frequency of the deformation applied, while energy dissipation is directly proportional to frequency for a viscous dashpot. To reproduce with better accuracy the behavior of rubber components,

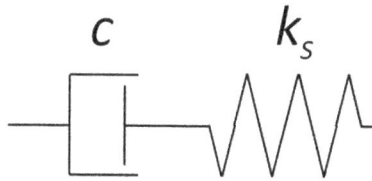

FIGURE 2.12 Simple Maxwell element.

more complex models can be used, consisting of combinations of springs and dash-pots with serial stiffness (e.g. Maxwell element; see Figure 2.12).

Another important point with rubber springs is that the range of linearity of these components is generally small, so it may be required to use a non-linear spring model. This can be easily defined through the interpolation of a non-linear force versus a deformation characteristic curve defined by discrete points. The characteristic curve, in turn, can be obtained from a measurement performed on the component or from a detailed model of the single component, e.g., a solid finite element model of the rubber element using a suitable non-linear constitutive law for the material. When rubber elements are used as bumpstops, the effect of the play shall also be considered introducing a dead zone in the non-linear characteristic of the component.

2.5.3 FRICTION-BASED SUSPENSION COMPONENTS

In many railway vehicles, especially freight wagons and passenger vehicles with old design, friction damping is used in place of other damping devices. Friction elements are quite challenging from the point of view of establishing a suitable model, as the Coulomb model for dry friction is non-smooth, i.e., it implies a discontinuous variation in the transition from sticking to sliding. The approach usually followed to deal with this issue is to consider a linear spring in series with a dry friction element [29]. The linear spring may be intended to represent the elastic flexibility of the end mountings of the dry friction component or can be simply a numerical expedient adopted to regularize the model.

Particularly challenging is the modeling of a dry friction component when sliding can occur in two directions, as the saturation of the friction force in this case will be determined by two scalar components of the force vector and the model must be able to reflect the fact that the force generated in the sliding condition should be opposite to the sliding velocity, while the direction of the friction force vector in the sticking condition is only determined by equilibrium equations. Examples of two-dimensional models of dry friction suitable for use in the modeling of rail vehicle suspensions can be found in [30, 31].

Specific models for different types of friction-based suspension components such as leaf springs, friction wedges, and the Lenoir damper are provided in [29] and are not repeated here, also because friction damping is generally only used in suspension components of old designs which are not likely candidates for the use of mechatronic technology.

2.5.4   HYDRAULIC DAMPERS

Hydraulic dampers are sophisticated damping devices used in modern passenger railway vehicles. Energy dissipation is produced by the passage of a fluid (generally oil) through a restricted orifice. In order to meet conflicting requirements on comfort, safety, and reduced train/track interaction forces, hydraulic dampers are often designed to have a non-linear behavior which is described by a characteristic curve defining the relationship between the rate of deformation and the force generated by the damper. However, one should be aware that, due to the internal compressibility of oil and the elastic deformation of the end mounts, the actual behavior of hydraulic dampers is frequency-dependent, i.e., the force generated by the damper is not uniquely defined by the rate of deformation, and actually depends on both the amplitude and frequency of deformation, with different combinations of these parameters producing the same rate of deformation but not the same force. The damper's characteristic curve is typically derived from measurements performed on the damper using cycles of deformation with relatively large amplitude and low frequency, so it is appropriate to describe the damper's low-frequency behavior.

Although it is possible to define a physical model of a hydraulic damper, the use of these models is mostly confined to special applications including damper design. In the context of defining a complete model of the railway vehicle, simplified lumped parameter models of hydraulic dampers are usually adopted. The simplest model is a linear or non-linear viscous dashpot. In the linear viscous dashpot, a linear relationship is assumed between the damper force F_d and the rate of deformation $\dot{d}$, through a damping constant c:

$$F_d = c\dot{d} \tag{2.36}$$

This model describes the damper's behavior over a limited range of deformation rates so that non-linearities play a minor role. In case non-linear effects need to be reproduced, a non-linear relationship between damper force and rate of deformation can be introduced instead:

$$F_d = F_d\left(\dot{d}\right) \tag{2.37}$$

and this non-linear relationship is often defined by the measured characteristic curve of the damper using linear interpolation.

In order to reproduce the frequency-dependent behavior of the damper, a serial stiffness k_s is added to the linear/non-linear viscous dashpot, so that the model is a Maxwell element, either linear or non-linear, like the one shown in Figure 2.12. The quantification of the serial stiffness can be based on the physical properties of the damper's end mounts or can be identified by matching the results of tests performed on the damper considering different combinations of amplitude and frequency of deformation. The use of a correctly tuned serial stiffness is important to represent the damper's behavior at a high frequency of deformation, which is particularly relevant to the transmission of high-frequency vibrations through the damper. The serial stiffness value should be carefully chosen also in the case of yaw dampers, as an incorrect estimation of this parameter may lead to significant inaccuracy in the estimation of the vehicle's critical speed. For more details about modeling of hydraulic dampers, see [29].

2.5.5 AIR SPRING SUSPENSION

The rationale for use of a pneumatic suspension (air spring suspension) as part of the secondary suspension of a railway vehicle has been provided in Section 2.5.1. In this section, options for modeling an air spring suspension are reviewed.

The arrangement of an air spring secondary suspension is shown in Figure 2.13. The total volume of air contained in the suspension is split into two main parts: the volume of the bellows (component 1) and the volume of the auxiliary tanks (component 2). This is due to the fact that the volume of air required to provide the suspension with the desired amount of flexibility is too large to be accommodated in the bellows and additional air volumes are therefore contained in tanks which can be placed in the carbody (either in the underframe or on the roof) while only the bellows will be fitted in the small space available between the bogie frame and the carbody. Each bellows is connected to a tank by a pipe (component 3) with an orifice (component 4). The function of the orifice is to provide energy dissipation, so that secondary vertical dampers are not needed.

One important effect with this arrangement is that the stiffness of the suspension will be variable with the frequency of the deformation applied. When the deflection of the suspension takes place at low frequency, the air will be able to move freely from the bellows to the tank or vice-versa and this will lead to a minimum value of stiffness. However, with increasing frequency, the friction loss in the pipe and orifice will increase, eventually impeding the exchange of air between the bellows and the tank. When this condition is reached, the air contained in the tank will become inactive and the stiffness of the suspension will be determined by the volume of air contained in the bellows only and hence it will be increased significantly. This

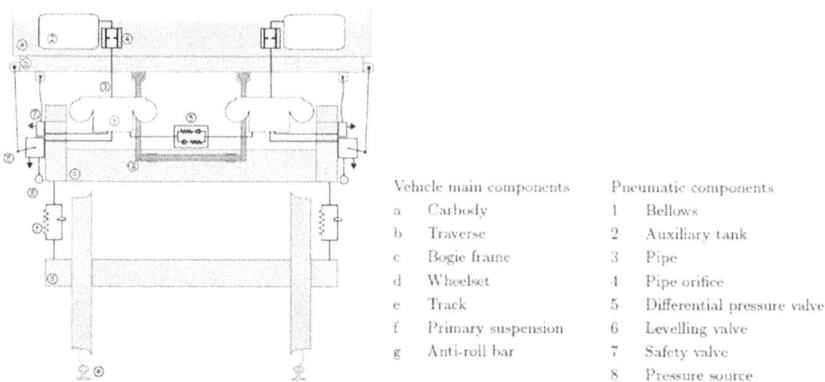

Vehicle main components		Pneumatic components	
a	Carbody	1	Bellows
b	Traverse	2	Auxiliary tank
c	Bogie frame	3	Pipe
d	Wheelset	4	Pipe orifice
e	Track	5	Differential pressure valve
f	Primary suspension	6	Levelling valve
g	Anti-roll bar	7	Safety valve
		8	Pressure source

FIGURE 2.13 Schematic and main components in an air spring secondary suspension. (From [32]).

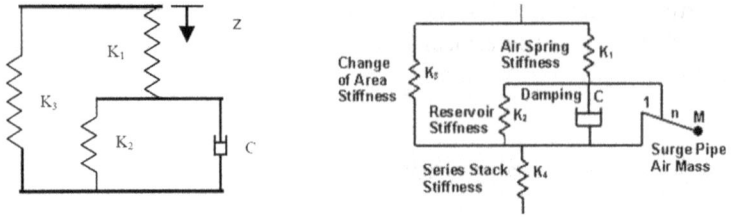

FIGURE 2.14 Left: Nishimura air spring model. Right: Vampire air spring model. (Both images adapted from [29].)

behavior can be reproduced using a simple equivalent model that goes under the name of the Nishimura model [29], shown in Figure 2.14 on the left. This model consists of a linear spring k_1 in series with the parallel arrangement of a spring k_2 and a viscous damper c. Finally, a spring k_3 is placed in parallel. The springs k_1 and k_2 represent respectively the stiffness of air contained in the bellows and in the tank, while the viscous damper represents the effect of pressure drops in the pipe and orifice. Finally, the spring k_3 represents an additional stiffening effect provided by the change of the effective area caused by a change of the height of the air spring; this last term is however relatively small compared to the others. For a sufficiently low frequency of deformation, the viscous damper will be nearly inactive, and the total stiffness will be mostly determined by the series of springs k_1 and k_2. However, at higher frequency, the force in the viscous damper will become increasingly large and will have the effect of reducing the deformation of spring k_2 compared to spring k_1 so that, at a sufficiently high frequency, the stiffness of the model will approach the sum of k_1 and k_3.

If the tanks are placed far from the bellows requiring relatively long pipes to be used, the inertial effects related to the flow of air in the pipes becomes important and requires that a model different from the one by Nishimura is used. In this case, two approaches are possible. On one hand, a different, more complex equivalent model can be used, such as the one used in the software package Vampire; see Figure 2.14 on the right, where additional springs and one mass are used to reproduce affects not included in the Nishimura model. On the other hand, a state model can be defined for the thermodynamic transformations occurring to the air contained in the bellows, pipes, and tanks. This is referred to as a *thermodynamic model* of the pneumatic suspension and can be used to relate in state-space form the deformation applied to the suspension to the variation of air pressure in the bellows and ultimately to the force generated by the air spring. More details on the different modeling options and comparison of dynamic stiffness versus frequency diagrams for the different models can be found in [29, 32].

The advantage of a thermodynamic model of the air spring suspension is that it can be easily extended to also describe the behavior of the leveling system, along with the differential pressure valves and safety valves that are part of the air spring suspension arrangement. With modern multi-body systems software, the co-simulation of a mechanical model of the vehicle and a pneumatic model of the air spring arrangement is also a viable option.

Before closing this section, it is worth mentioning that the pneumatic secondary suspension can easily be designed as a semi-active or full-active suspension. For instance, an orifice with variable size can be introduced to provide the suspension with variable damping capability [33], or additional pressurized tanks and compressors can be added to actively inflate and deflate the bellows, introducing a limited amount of active carbody tilt in the vehicle [34]. These options will be further discussed in Chapter 10.

2.6   PANTOGRAPH-CATENARY INTERACTION

When electric power is used to move trains, the flow of electric power from wayside substations to the train is achieved through a sliding electric contact between current collectors on the train and a fixed infrastructure that can be either a third rail or an overhead contact line (OCL). The use of an OCL is usually preferred because it is inherently safer than a third rail in terms of preventing electrocution of passengers and personnel operating the train. This means that higher voltage in AC or DC can be used while the third rail is restricted to relatively low voltage DC currents.

In systems making use of an OCL, the electrical power is transmitted to the train by means of one or more pantographs. This is an articulated system with pneumatic actuation and carrying a head with collector strips that can be placed in contact with the overhead line.

Although various OCL systems are in use, we refer here only to the most frequently used one (Figure 2.15), which is formed by hanging a messenger wire from trackside masts by suspension systems and a contact wire with which the pantograph collector strips are raised against under pressure to achieve the transfer of electric power to the train. The contact wire is attached to the messenger wire by droppers and both wires are tensioned to provide the required stiffness. Due to the system's own weight, the messenger wire takes an approximate catenary shape so that this OCL system is called a standard (or simple) catenary. More details about the OCL and the pantograph and issues related to their design and operation can be found in [35].

Due to its geometry, the OCL is characterized by a nearly periodic spatial variation of stiffness and deforms under the action of the load applied by the pantograph, having a maximum deflection at mid span and minimum deflection at the masts.

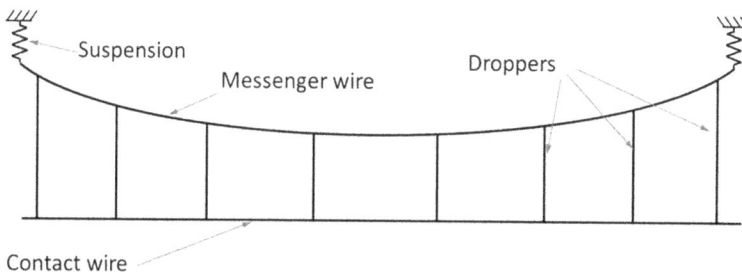

FIGURE 2.15   Standard catenary wire OCL system.

Therefore, the pantograph head is subject to a vertical vibration which causes a fluctuation in the contact force exchanged with the contact wire. When the speed of the pantograph is sufficiently high, the fluctuation of the force may lead to loss of contact between the collector strips and the contact wire, and frequent contact losses result in unsatisfactory current collection and also in accelerated damage of the contacting surfaces due to electric arcs. Therefore, the pantograph-catenary system needs to be accurately designed to ensure proper performances and, in some cases, may turn out to be a limiting factor in raising the service speed of an electrified railway line. Another important problem of pantograph-catenary systems, especially in Europe, is concerned with inter-operability because national railway networks in Europe historically developed using different standards, not only for voltage and current type (AC/DC), but also in the design of catenaries so that the unification of existing OCLs to a common standard is not feasible. This means a train travelling from one country to another, or even from one network to another in the same country (e.g., in countries like Italy or Spain from the conventional to the high-speed network) needs to be equipped with more pantographs, each one suited to enter in contact with a specific OCL.

Active and semi-active pantograph control has been proposed to improve pantograph-catenary interaction, allowing to raise service speed on existing lines, reduce construction and maintenance costs for OCLs, and improve inter-operability. For these reasons, pantograph-catenary interaction is fully relevant to the scope of this book.

Mathematical models of pantograph-catenary interaction consist of three main components:

- The model of the OCL
- The model of the pantograph
- The model of pantograph-catenary contact

For the OCL, simple lumped parameter models were used initially but nowadays the standard model is based on a finite element schematization using tensioned string or tensioned beam elements to represent the messenger and contact wires, non-linear spring elements to represent the droppers and lumped spring/dashpot elements to represent the suspensions attaching the catenary to the masts. Finite element models can be either 2D, i.e., approximate the catenary as a system laying in the vertical plane, or 3D. In the latter case, the stagger of the contact wire is also reproduced, leading to a coupling of vertical and lateral vibration of the system.

For the pantograph, lumped models with 2 or 3 DOFs are frequently used. These models only describe the vibration of the pantograph in the vertical direction and their parameters can be identified from the direct measure of the pantograph's FRF measured on a test stand. Non-linear spring or friction elements can be included in lumped mass models to represent amplitude-dependent effects in pantograph motion. Alternatively, a detailed multi-body model of the pantograph can be used (Figure 2.16).

Pantograph-catenary contact is a sliding unilateral constraint. This can be modeled using either a kinematic joint formulation or using a force element which enforces the sliding constraint through a contact force whose value becomes progressively larger for increasing violation of the constraint; this latter approach is

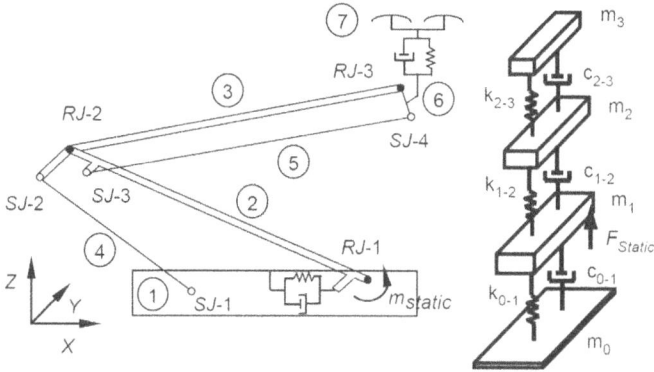

FIGURE 2.16 Multi-body model (left) and lumped mass model of a pantograph (right). (From [36].)

known as the *penalty method*. In the first case, the complete pantograph-catenary model is defined as a system of differential-algebraic equations (DAEs), while the penalty method allows a simple mathematical formulation of the model as a system of non-linear ODEs. For these reasons, the penalty method is more widely used compared to the kinematic joint formulation; see [37] for a survey and benchmarking of some existing pantograph-catenary models.

2.7 TRACTION AND BRAKING DYNAMICS, CONTROL AND MODELING

When investigating the movement of a powered rail vehicle, it is necessary to understand the principles and control techniques that are in use for the control of traction motors and brake systems based on the consideration of the interaction between the wheel and the rail. The systems involved in implementing such control represent a complex mechatronic interaction between a traction motor and its control mechanism, motion dynamics and the adhesion at the wheel/rail interface.

2.7.1 PRINCIPLES OF TRACTION BRAKING DYNAMICS

In the traction mode, the powered rail vehicle produces an external force to move a train along the track. This force is considered as the sum of traction forces acting at the wheel/rail interface for each wheel of a powered rail vehicle. Figure 2.17 presents a single wheel force and torque diagram with relevant loads acting under a traction scenario.

The torque T_{wheel} acting on the wheel from the traction motor can also be represented as an equivalent force at the rim of the wheel, which can be defined as:

$$F_{wheel} = \frac{T_{wheel}}{r} \tag{2.38}$$

where r is the wheel radius.

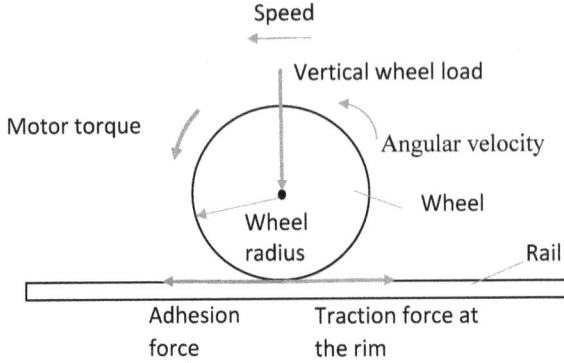

FIGURE 2.17 Traction case.

This force is opposed by the wheel/rail adhesion force F_a which also acts along the track and, in order to produce movement of a rail vehicle, the traction force F_{rim} should be greater than F_a. In this case, the torque T acting on a wheel can be defined as:

$$T = T_{wheel} - T_a \tag{2.39}$$

where T_a is the adhesion torque, also called the load torque, which can be calculated as:

$$T_a = F_a \cdot r \tag{2.40}$$

The vertical wheel load Q establishes a limit for the traction force that can be realized, and it determines the maximum adhesion available. This limit condition under traction can be defined as:

$$F_{wheel} \leq \mu \cdot Q \tag{2.41}$$

where μ is the friction coefficient available at the wheel/rail interface.

The adhesion force in this case can be defined as:

$$F_a = \mu \cdot Q \tag{2.42}$$

For the braking mode, the single wheel force and torque diagram is shown in Figure 2.18.

The wheel is braked by means of a braking torque acting from the traction motor (so-called dynamic braking), which operates as a generating device instead of a power device, or from friction braking generated by a pneumatic air system with brake blocks (brake shoes). The main operational condition in this case is that the braking torque applied to the wheel should not exceed the product of the maximum adhesion force and the wheel radius. This condition is necessary to avoid wheel/rail

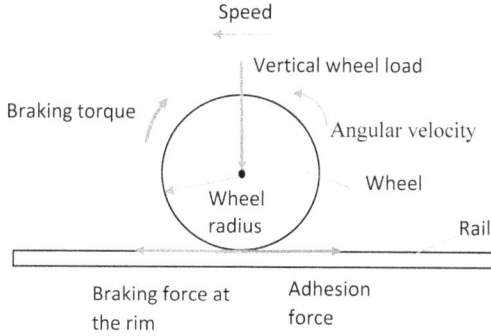

FIGURE 2.18   Braking case.

damage and provide an appropriate safe braking performance. The maximum adhesion that can be achieved is the fundamental limitation of the hauling capability of modern powered rail vehicles and the physical fundamentals of the processes in the wheel/rail contact interface still need to be further investigated.

2.7.2   DESIGN PRINCIPLES OF TRACTION AND BRAKING CONTROL

Considering all the points above, the best solution is to control the traction drive operating in traction or braking modes, although it is worth noting that the mechanical brake control system and auxiliary equipment should also be monitored in detail in order to provide satisfactory operational outcomes.

The traction/braking control systems used on rail vehicles should achieve an optimal adhesion between wheels and rails and avoid any potential damage caused by exceeding the maximum allowable traction torque applied to the wheels. These systems can be classified according to the control method used to achieve this [38]:

- Adhesion control strategies
- Adhesion/creep control algorithms
- Design configurations

Adhesion control strategies are fundamentals, but are commonly excluded for some predefined train operational modes [39], namely:

- Starting strategy, used when the locomotive commences movement
- Safety mode strategy, used when other strategies have failed

The starting strategy is designed for starting the hauling movement of the train and it usually does not require the detection of the ground speed because the accuracy of the data obtained from radar at slow speeds is very low. The algorithm for this strategy usually relies on the traction motor speeds and their computed accelerations. This strategy applies for a low-speed range only, i.e., when speed is less than 5 km/h.

The safety mode strategy is used when the failure of any of the equipment components occurs and the rail vehicle cannot function in the normal operational mode.

The adhesion/creep control algorithms are used at a speed higher than 5 km/h, and most of them require information about the ground speed of the locomotive and the traction motor/wheelset rotational speeds. The minimum implementation has adhesion strategies that rely on the monitoring of traction drive behavior (e.g., when the adhesion limit is reached at the wheel rail interface, the system experiences angular fluctuations that can cause severe vibrations with relatively high frequencies (4565 Hz), particularly if only one wheel loses contact with the rail) or on the comparison of traction motor currents. The latter is problematic to introduce in real practice due to the possible differences in wheel rail diameters that cause unbalance in the system. However, the development of some advanced algorithms is possible to compensate for such unbalanced conditions.

The adhesion control algorithms, which require the ground speed of the locomotive and the traction motor/wheelset rotational speeds as inputs, are based on two principles [38, 40, 41]:

- Slip-based approaches
- Combined technique

In slip-based approaches, the data obtained from angular velocity sensors are processed in order to find a minimal angular velocity for each wheel or bogie, or for the whole locomotive, and then to compare same with the locomotive ground speed. A value of the longitudinal slip (creep) is estimated based on the following relation:

$$s = \frac{w \cdot r - V}{V} \tag{2.43}$$

where w is the real angular velocity of a wheel, and V is the rail vehicle speed.

The values of rolling radii r for each of the wheels are updated on the computerized management system of a rail vehicle at each service interval, and rotational speeds of traction motors on some vehicles are recalibrated daily under non-slip conditions to ensure avoiding accumulative errors during rail vehicle operations. While the terms *slip* and *creep* are often used interchangeably, slip is the additional speed that a wheel may have due to its relative motion at its contact point with the rail, while creep is characterized as the slip speed divided by the vehicle speed.

Adhesion control is one of the critical systems for powered rail vehicles and its function is to control the traction drive operating in traction or braking modes. This system should therefore achieve an optimal adhesion between wheels and rails, as shown in Figure 2.19, and avoid any potential damage caused by exceeding the maximum allowable traction torque applied to the wheels, i.e., it should not allow the system to act in the unstable zone. However, in practice, there are two types of controllers in use. These controllers use conventional and extended slip control techniques. The conventional systems usually operate with a predefined wheel slip threshold independent of wheel/rail interface friction condition as shown in Figure 2.19 that provides a nominal (practical optimal) level of wheel slip that

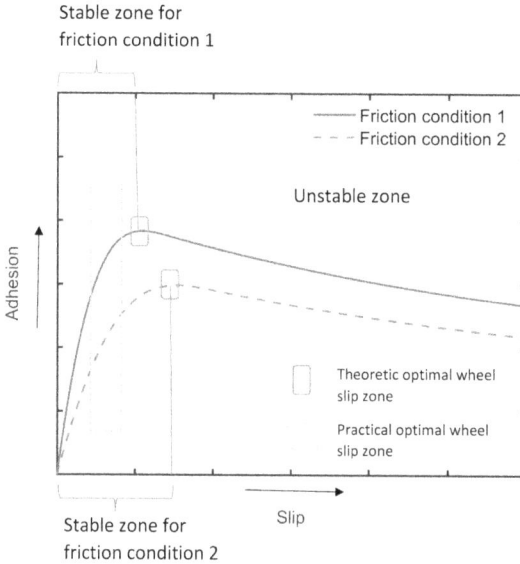

FIGURE 2.19 Example of difference between theoretical and practical optimal slip in the longitudinal direction for two friction conditions.

should be positioned in the stable zone. Meanwhile, the extended slip control technique seeks ways to operate at the peak of the adhesion-slip curve, i.e., at the point of the theoretic optimal wheel slip zones as shown in Figure 2.19 for two different friction conditions.

The combined technique assumes the application of a slip-based approach in parallel with another method. Two common methods are:

- Prevention of unexpected wheel acceleration or deceleration by setting relevant threshold values for these parameters and their control through traction drive control or brake control systems;
- The application of sand through a sanding system installed on a rail vehicle or some other friction modifier at the wheel/rail interface.

The abovementioned techniques can be changed, re-configured or further improved depending on the mechatronic system design decisions taken on individual powered rail vehicles considering train configuration parameters and systems in use on other vehicles included in the train.

The design configuration strategies to control adhesion processes are strongly dependent on rail vehicle design variants. For example, adhesion can be influenced through the selection of traction power system design variants (independent rotating wheel traction control, individual wheelset traction control, bogie traction control, or locomotive control) or the application of special actuators to control the wheel load distribution between wheelsets. A great many variants exist, but all of them should be considered from the point of view of costs, maintenance, and safety aspects.

2.7.3 MODELING OF THE TRACTION SYSTEMS

There are four common approaches on how to introduce the traction system in rail vehicle models:

- A constant torque value is applied to each wheelset or wheel;
- A time- or distance-dependent torque value applied to each wheelset or wheel;
- A simplified modeling approach; and
- An advanced modeling approach.

The first two approaches neglect any physical representation and dynamic response of the traction system, so they can be used for the investigation of other mechatronic sub-systems of a rail vehicle model rather than just the traction control. For example, the active suspension system's response under different traction or dynamic braking operational scenarios can be performed with both these approaches.

The simplified approach is based on the presentation of power equipment dynamic responses through a low-pass filter modeling approach. For example, the traction torque applied from the traction motor to the wheelset or wheel taking into account the inverter and traction motor dynamics can be presented as:

$$T_{\text{motor}} = \frac{1}{\tau s + 1} T_{\text{in}} \qquad (2.44)$$

where T_{in} is an input traction torque obtained from tractive or dynamic braking effort characteristics, and τ is a time constant. The value of this time constant can be chosen based on the analysis of existing rail vehicle log files or assumed from results of advanced modeling when the works are at the rail vehicle design stage. This approach works well for studying of dynamic rail vehicle response under traction and dynamic braking at the initial stages for the development of slip control algorithms and a further validation of workability of other mechatronic sub-systems integrated in rail vehicle models.

An advanced modeling approach for the electric traction system for a classical rail traction vehicle will include the following elements [39]:

- A physics-based model for the traction machine and, in the case of the induction machine, this is implemented using a coupled inductor model
- A model for the power electronics-based inverter and its controls
- A model for the inverter DC bus that will include the DC storage capacitors and dynamic brake resistors

If it is needed, the power plant can be also modeled in this approach as a part of the electrical traction system.

The advanced traction system modeling approach should be applied for the study of traction slip control systems and it requires the usage of co-simulation modeling approaches. The advanced modeling approach provides much more stable workability of the whole rail vehicle mechatronic model and, at the same time, it significantly benefits in terms of providing accuracy to studies focused on the processes at the wheel/rail interface as well as vibration dynamics of traction motor suspension.

It is also necessary to mention that the verification of simplified models for traction control system designs using advanced models is recommended because the values of time constants used in low-pass filter equations for power system components must be tuned in order to get results that are the most accurate for any given scenario [39].

2.8 TRAIN DYNAMICS

Adhesion at the wheel/rail interface is considered a link between train dynamics and wheel dynamics. However, the dynamics of the train are quite separate from the dynamics of the wheel and both systems are commonly studied as separate items. However, considering recent trends in the development of complex mechatronic systems such as automated train driving and monitoring, both sets of dynamics need to be combined for advanced studies. In this section, we need to understand what train dynamics is and why it is important for the development of rail vehicle mechatronic systems.

2.8.1 TRAIN DYNAMICS FOR A SINGLE VEHICLE

In the case of a single vehicle dynamics study, the train in motion is represented by means of all relevant forces acting on the single vehicle as shown in Figure 2.20. The traction effort is represented by the sum of the traction forces acting on all wheelsets. The resistance to the rail vehicle's motion is a result of other forces (train load, friction in bearings, wind resistance, etc.) acting on the vehicle during train operations in a simplified form. In addition, the gravity resistance force is also acting on the vehicle when the train is running uphill (ascending slope) or downhill (descending slope).

2.8.2 LONGITUDINAL TRAIN DYNAMICS

To control train movements in traction or braking, it is necessary to clearly understand the resistive forces coming from train configurations and the effects that the produced tractive or braking efforts have on the whole system and environment. In this case, the train dynamics should not only consider traction, dynamic braking,

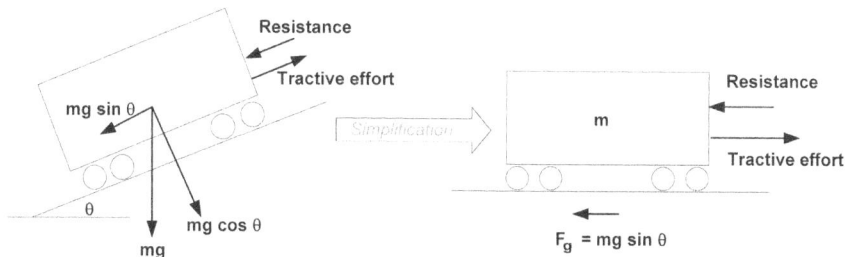

FIGURE 2.20 Simplified train dynamics for a single rail vehicle.

air braking, grades, rolling, and air resistance force inputs, but should also consider the motion of the train as a whole and any relative motions between vehicles allowed due to the looseness and travel allowed by the coupling and spring and damper connections between vehicles. For consideration of such issues, it is necessary to use longitudinal train dynamics [42]. The study and understanding of longitudinal train dynamics was probably initially motivated by the desire to reduce these relative vehicle motions in passenger trains and, in so doing, improve the general comfort of passengers. The practice of *power braking*, which is the seemingly strange technique of keeping locomotive power applied, while a minimum air brake application is made, is still used on older type passenger trains. Power braking is also used on partly loaded mixed freight trains to keep the train stretched during braking and when operating on undulating track.

Interest in train dynamics in freight trains increased as trains became longer, particularly for heavy haul trains as evidenced in some early published technical papers. In the late 1980s, considerable activity in the measurement and simulation of in-train forces on such trains was reported in Australia [43, 44] and in South Africa [45]. From these studies, an understanding of the force magnitudes was developed along with an awareness of the need to limit these forces with appropriate driving strategies [38, 42–48]. The core of longitudinal dynamics studies is the modeling and simulation of the train vehicles as connected masses. In particular, the non-linearity of the wagon connections involving draft gears or buffers and draw gear components needs to be modeled accurately to get representative results for force transients and accelerations. For this reason, this aspect is described in more detail in Section 2.10 of this chapter. Steady forces in longitudinal dynamics studies depend on steady force inputs from traction, braking, grades, and resistances (rolling, curving, and air) – the so-called co-dimensionality of the problem. In other words, the longitudinal behavior of trains is a function of train control inputs from the locomotive, train brake inputs, track topography, track curvature, rolling stock and bogie characteristics, and vehicle connection characteristics.

The longitudinal dynamic behavior of a train can be described by a system of differential equations. For the purposes of setting up the equations for modeling and simulation, it is usually assumed that there is no lateral or vertical movement of the vehicles. This simplification of the system is employed by all known commercial rail specific simulation packages and by texts such as Dukkipati and Garg [49]. The governing differential equations can be developed by considering the generalized three mass train shown in Figure 2.21. It will be noticed that the in-train vehicle, whether locomotive or wagon, can be classified as one of only three connection configurations, lead (shown as m_1), in-train, and tail. All vehicles are subject to retardation and grade forces. Traction and dynamic brake forces are added to powered vehicles. Wagon connection models are usually defined as a non-linear stiffness and damper with a gap element for coupler slack.

In Figure 2.21, a is vehicle acceleration in m/s², f_{wc} is the non-linear function describing the wagon connection model including stiffness and gap components, m is vehicle mass in kg, v is vehicle velocity in m/s, x is vehicle displacement in m, F_g are gravity force components due to track grade in N, F_r is sum of retardation forces in N, and $F_{t/db}$ are traction and dynamic brake forces from a locomotive unit in N.

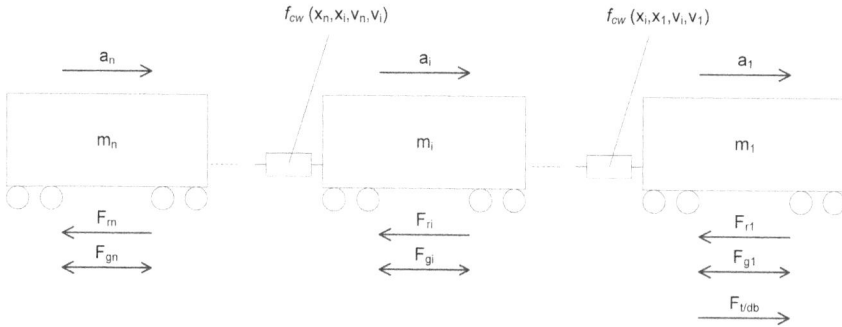

FIGURE 2.21   Generalized train model. (From [42], with permission.)

It will be noted on the model in Figure 2.21 that the grade force can be in either direction. The sum of the retardation forces F_r is made up of rolling resistance, curving resistance or curve drag, air resistance, and braking (excluding dynamic braking which is more conveniently grouped with locomotive traction in the $F_{t/db}$ term). Rolling and air resistances are usually grouped as a term known as propulsion resistance F_{pr} and the resistances to motion are grouped as:

$$F_r = F_{pr} + F_{cr} + F_b \tag{2.45}$$

where F_{pr} is the propulsion resistance, F_{cr} is the curving resistance, and F_b is the braking resistance due to pneumatic braking.

The three-mass train allows three different differential equations to be developed as:

- For the lead vehicle:

$$m_1 a_1 + f_{wc}(v_1, v_2, x_1, x_2) = F_{t/db1} - F_{r1} - F_{g1} \tag{2.46}$$

- For the ith vehicle:

$$m_i a_i + f_{wc}(v_i, v_{i-1}, x_i, x_{i-1}) + f_{wc}(v_i, v_{i+1}, x_i, x_{i+1}) = F_{t/dbi} - F_{ri} - F_{gi} \tag{2.47}$$

- For the nth or last vehicle:

$$m_n a_n + f_{wc}(v_n, v_{n-1}, x_n, x_{n-1}) = F_{t/dbn} - F_{rn} - F_{gn} \tag{2.48}$$

where f_{wc} is the non-linear function describing the full characteristics of the vehicle connection.

Note that a positive value of F_g is taken as an upward grade, i.e., a retarding force. The equations also allow the locomotives to be placed in any position in the train by adding the term $F_{t/dbi}$ which can be set to zero for unpowered vehicles.

Solution and simulation of the above equation set is complicated by the need to calculate the force inputs to the system, i.e., $F_{t/db}$, F_r, and F_g. The traction-dynamic brake force term $F_{t/db}$ must be continually updated for driver control adjustments and any changes to locomotive speed. The retardation forces F_r are dependent on braking settings, vehicle velocity, track curvature, and rolling stock design. Air brake systems have their own time response and accurate calculation of brake forces therefore requires a brake simulator running in parallel. Gravity force components F_g are dependent on track grade and therefore on the position of the vehicle on the track.

Based on the physical representation of processes described by train dynamics, it is possible to say that longitudinal dynamics in terms of aspects of a mechatronic system have significant implications for the control of longitudinal forces, accelerations, vehicle instability, and energy consumption. In terms of heavy haul train operations, the study of longitudinal dynamics is applicable both for improvement of designs and the assessment of train configurations on various haulage routes and autonomous or driverless control strategies. For passenger trains and short freight trains, the mechatronic approaches considering longitudinal train dynamics can be applicable to areas such as improving passenger comfort, minimizing damage to sensitive freight, or controlling energy use.

2.9 PNEUMATIC BRAKE MODELS

The pneumatic brake systems of rail vehicles consist of:

- **The pneumatic sub-system:** compressors, control valves, pipes, reservoirs, pneumatic cylinder (actuators)
- **The mechanical sub-system:** brake levers and braking friction pairs (braking shoes and wheel tread or friction pads and brake discs)
- Electronic control units

The modeling of such systems can be done in different ways:

- A simplified approach based on the braking force calculation attached to a rail vehicle's carbody.
- A simplified approach based on the braking torque calculation attached to a wheelset or a wheel.
- A detailed full model approach that models all components of the pneumatic brake system except brake levers and the braking friction pairs. In this case, the braking torque value is attached to a wheelset or a wheel.
- An advanced modeling approach that covers all components included in the pneumatic brake system. In some cases, when it is necessary to consider train behavior in brake studies, the advanced model might include inter-car forces and detailed modeling of pipe connections in a train configuration.

As in the case of the modeling of traction and dynamic braking control systems, the simplified approaches might be used for the study of dynamic behavior and performance of other mechatronic systems integrated in a rail vehicle model (e.g., active suspension systems), and for the investigation of instrumented onboard systems and braking control strategies. An example of such an approach can be found in [50].

The detailed full model approach, similar to models described in [51], is designated to adjust the obtained results to selected design stages with different levels of accuracy and to consider air-wave phenomena including air viscosity, the influence of the brake pipe branches, heat transfer in the brake pipes and reservoirs, air flows in the brake valves, and control principles.

The advanced modeling approaches in most cases replicate the detailed approach, but it also allows to consider the dynamics of moving mechanical parts and friction processes between friction braking pairs. If it is necessary to consider a whole train braking system (also called the brake pipe simulation), it can be modeled as described in [48] and then integrated in the rail vehicle model. The output from the brake pipe simulation is the brake cylinder force which is converted by means of rigging factors and shoe friction coefficients into a retardation force that is one term of the sum of retardation forces.

The very simplified model presented assumes air flow through an orifice (in practice, much more precise controls exist using the complicated brake triple valve with flows precisely controlled with an arrangement of orifices and chokes), giving:

$$v = K(\Delta P)^{0.5} \tag{2.49}$$

where v is flow velocity, K is the proportionality constant, and ΔP is the difference between reservoir pressure and cylinder pressure. Pressure on the cylinder can be estimated from Boyle's law, $PV = mRT$, where P is pressure, V is volume, m is mass, R is the gas constant for air, and T is absolute temperature. As V is cylinder volume which can be taken as constant after initial movement, and T is assumed constant, the equation can be written as:

$$P = mRT/V \text{ or in the differential form } dP = dmRT/V \tag{2.50}$$

By integrating the differential form of Equation (2.50) and using the gas flow rate from Equation (2.49), the pressure rise in the cylinder can be modeled. An example of the simplified brake pipe model is shown in Figure 2.22 and example simulation results are shown in Figure 2.23. The brake pressure can be easily transferred to the brake cylinder force when the design characteristics of the brake cylinder are known.

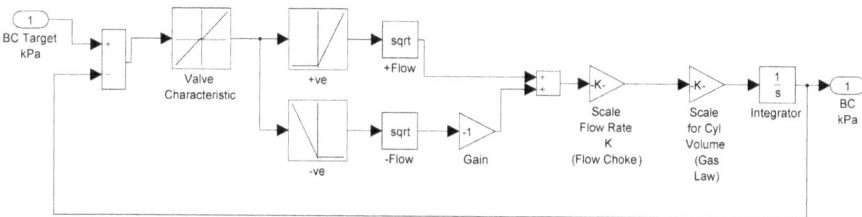

FIGURE 2.22 Example of simplified triple valve cylinder brake model – Implemented in Simulink as a sub-system. (From [48], with permission.)

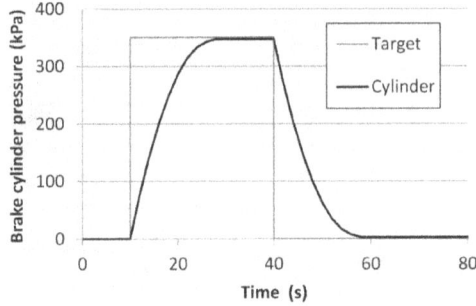

FIGURE 2.23 Example of results for simplified triple valve cylinder brake model. (From [42], with permission.)

An electronically controlled brake system allows making required adjustments to the brakes via the brake pipe considering that it takes time to propagate along the train by means of control valves designed with special valve characteristics.

The main issue with the consideration of longitudinal train dynamics in rail vehicle mechatronics simulation is the determination of the braking force and the timing of its application. More information on how the train pneumatic brake should be modeled considering such aspects can be found in [42, 46].

2.10 MODELING OF INTER-CAR FORCES

Perhaps the most important component in any studies that consider train dynamics is the connection element between vehicles. A common wagon connection model is something similar to the schematic in Figure 2.24 [48]. Modeling the coupler slack is straightforward by representing it as a simple dead zone. Modeling of the steel components including wagon body stiffness can be provided by a single linear stiffness. The train could be either in a tensile or compressed condition. The stiffness corresponding to the fundamental vibration mode observed was defined as the locked stiffness of the wagon connections. The locked stiffness for the freight trains tested was nominally a value in the order of 80 MN/m [43]. As the locked stiffness is also the limiting stiffness of the system, it must be incorporated into the wagon connection model. The limiting stiffness is the sum of all the stiffnesses of the structural

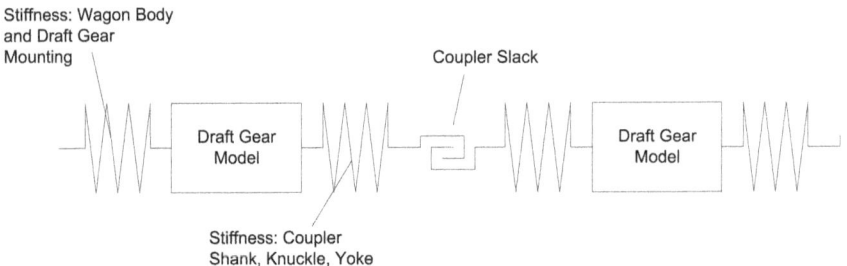

FIGURE 2.24 Components in a wagon connection model. (From [48], with permission.)

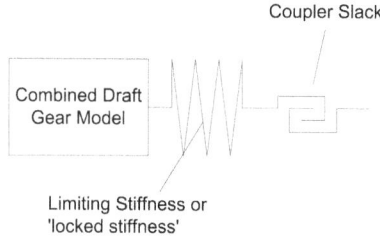

FIGURE 2.25   Simplified wagon connection model. (From [48], with permission.)

components and connections added in series, which includes the components such as the coupler shank, knuckle, yoke, draft gear structure, and the wagon body. It also includes any pseudo-linear stiffness due to gravity and bogie steering force components, whereby a longitudinal force is resisted by gravity as a wagon is lifted or forced higher on a curve. The limiting stiffness of a long train may therefore vary for different wagon loadings and on-track placement.

Wagon connection modeling can be simplified to a combined draft gear package model equivalent to two draft gear units and including one spring element representing locked or limiting stiffness (see Figure 2.25).

The friction-type draft gears are the most widely used draft gears [52]. An integrated draft gear model has to incorporate characteristics that can simulate the whole coupling system working under both draft and buff conditions as shown in Figure 2.26. In the case of elastomer and polymer draft gears, it is necessary for the development of a model to introduce strongly non-linear characteristics [42, 53]. An example of modeling of a polymer draft gear is shown in Figure 2.27.

There are three common approaches on how to model simplified wagon connections [48, 54–56]:

1. Simple spring-damper models
2. Lookup tables
3. Mathematical models

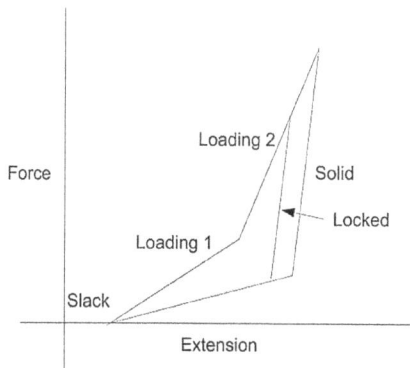

FIGURE 2.26   Integrated connection model. (From [42], with permission.)

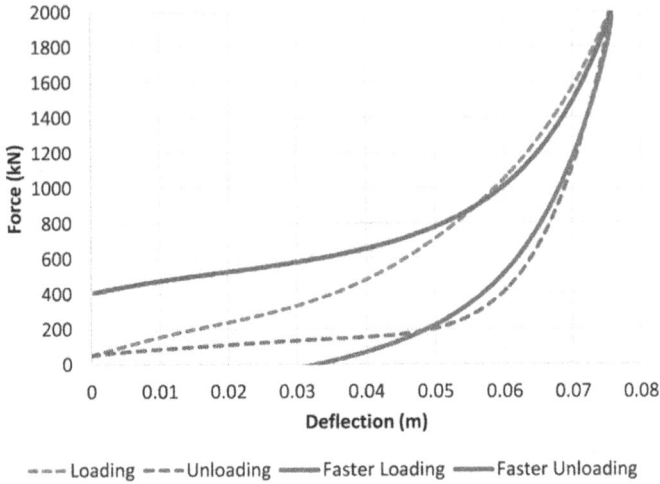

FIGURE 2.27 Example of outputs from a polymer draft gear model. (From [42], with permission.)

The difference in modeling of wagon connections between simple spring-damper and look-up table models is in the consideration of hysteresis. These two modeling approaches are good for fast calculation models. The mathematical model-based approach of modeling the friction and/or elastomeric damping with velocity-dependent functions require significant development and is only applicable if it is necessary to obtain more precise results.

The choice of the wagon connection modeling approach is a very important task for rail vehicle mechatronic systems that work with the control of longitudinal forces, accelerations, vehicle instability, and energy consumption. For example, if the train dynamics of passenger trains and short freight trains are involved, then the major concerns are connected with issues associated with passenger comfort or damage to sensitive freight. In addition, energy use might be also involved if the aim of the study is focused on the design and performance of hybrid rail vehicles.

REFERENCES

1. K. Knothe, S. Stichel, Rail Vehicle Dynamics, Springer, 2017.
2. A. H. Wickens, Fundamentals of Rail Vehicle Dynamics, CRC Press, 2003.
3. A. A. Shabana, K. E. Zaazaa, H. Sugiyama, Railroad Vehicle Dynamics: A Computational Approach, CRC Press, 2007.
4. A. Shabana, Dynamics of Multibody Systems, Cambridge University Press, Cambridge, UK, 1998.
5. J. S. Bendat, A. G. Piersol, Random Data: Analysis and Measurement Procedures, 4th Edition, John Wiley & Sons, 2010.
6. L. Marple, Digital Spectral Analysis With Applications, Prentice-Hall Series in Signal Processing, 1987.
7. J. J. Kalker, Three-Dimensional Elastic Bodies in Rolling Contact, Kluwer Academic Publishers, Dordrecht, The Netherlands, 1990.

8. J.-B. Ayasse, H. Chollet, Wheel-Rail Contact Mechanics, Chapter 7, In: S. Iwnicki, M Spiryagin, C. Cole, T. McSweeney (Eds), Handbook of Railway Vehicle Dynamics, 2nd Edition, CRC Press, Boca Raton, FL, 2020.

9. S. Bruni, J. P. Meijaard, G. Rill, A. L. Schwab, State-of-the-art and challenges of railway and road vehicle dynamics with multibody dynamics approaches, Multibody System Dynamics, 49, 1–32, 2020.

10. J. Piotrowski, W. Kik, A simplified model of wheel/rail contact mechanics for non-Hertzian problems and its application in rail vehicle dynamic simulations, Vehicle System Dynamics, 46(12), 27–48, 2008.

11. M. S. Sichani, R. Enblom, M. Berg, A novel method to model wheel-rail normal contact in vehicle dynamics simulation, Vehicle System Dynamics, 52(12), 1752–1764, 2014.

12. E. Vollebregt, A new solver for the elastic normal contact problem using conjugate gradients, deflation, and an FFT-based preconditioner, Journal of Computational Physics, 257(Part A), 333–351, 2014.

13. K. Knothe, R. Wille, B. W. Zastrau, Advanced contact mechanics – road and rail, Vehicle System Dynamics, 35(45), 361–407, 2001.

14. Z. Y. Shen, J. K. Hedrick, J. A. Elkins, A comparison of alternative creep force models for rail vehicle dynamic analysis, Vehicle System Dynamics, 12(1–3), 79–83, 1983.

15. J. J. Kalker, A fast algorithm for the simplified theory of rolling contact, Vehicle System Dynamics, 11, 1–13, 1982.

16. J. Piotrowski, B. Liu, S. Bruni, The Kalker book of tables for non-Hertzian contact of wheel and rail, Vehicle System Dynamics, 55(6), 875–901, 2017.

17. M. Spiryagin, O. Polach, C. Cole, Creep force modelling for rail traction vehicles based on the Fastsim algorithm, Vehicle System Dynamics, 51(11), 1765–1783, 2003.

18. M. Spiryagin, E. Vollebregt, M. Hayman, I. Persson, Q. Wu, C. Bosomworth, C. Cole, Development and computational performance improvement of the wheel-rail coupling for heavy haul locomotive traction studies, Vehicle System Dynamics, 2020, DOI:10.1080/00423114.2020.1803371.

19. O. Polach, Creep forces in simulations of traction vehicles running on adhesion limit, Wear, 258, 992–1000, 2005.

20. O. Polach, SBB 460 Adhäsionsversuche, Techn. Report No. 414, SLM Winterthur, 1992.

21. E. A. H. Vollebregt, S. D. Iwnicki, G. Xie, P. Shackleton, Assessing the accuracy of different simplified frictional rolling contact algorithms, Vehicle System Dynamics, 50(1), 1–17, 2012.

22. T. Dahlberg, Track Issues, Chapter 6, In: S. Iwnicki (Ed), Handbook of Railway Vehicle Dynamics, Taylor & Francis, London, 143–179, 2006.

23. F. Schmid (Ed), Wheel-Rail Best Practice Handbook, University of Birmingham Press, 2010.

24. R. Lewis, U. Oloffson (Eds), Wheel-Rail Interface Handbook, Woodhouse Publishing, 2009.

25. E. Di Gialleonardo, F. Braghin, S. Bruni, The influence of track modelling options on the simulation of rail vehicle dynamics, Journal of Sound and Vibration, 331(19), 4246–4258, 2012.

26. I. Kaiser, Refining the modelling of vehicle–track interaction, Vehicle System Dynamics, 50(Suppl. 1), 229–243, 2012.

27. W. Zhai, Z. Han, Z. Chen, L. Ling, S. Zhu, Train–track–bridge dynamic interaction: A state-of-the-art review, Vehicle System Dynamics, 57(7), 984–1027, 2019.

28. Y. Bezin, B. Pålsson, Multibody Simulation Benchmark for Dynamic Vehicle-Track Interaction in Switches and Crossings, University of Huddersfield, https://doi.org/10.34696/s60x-ay18.

29. S. Bruni, J. Vinolas, M. Berg, O. Polach, S. Stichel, Modelling of suspension components in a rail vehicle dynamics context, Vehicle System Dynamics, 49(7), 1021–1072, 2011.

30. X. Tan, J. Rogers, Equivalent viscous damping models of Coulomb friction in multi-degree-of-freedom vibration systems, Journal of Sound and Vibration, 185(1), 33–50, 1995.

31. F. Xia, Modelling of a two-dimensional Coulomb friction oscillator, Journal of Sound and Vibration, 265, 1063–1074, 2003.

32. N. Docquier, Multiphysics Modelling of Multibody Systems –Application to Railway Pneumatic Suspensions, PhD Thesis, Université Catholique de Louvain, 2010.

33. Y. Sugahara, A. Kazato, R. Koganei, M. Sampei, S. Nakaura, Suppression of vertical bending and rigid-body-mode vibration in railway vehicle car body by primary and secondary suspension control: Results of simulations and running tests using Shinkansen vehicle, Journal of Rail and Rapid Transit, 223(6), 517–531, 2009.

34. S. Alfi, S. Bruni, G. Diana, A. Facchinetti, L. Mazzola, Active control of airspring secondary suspension to improve ride quality and safety against crosswinds, Journal of Rail and Rapid Transit, 225(1), 84–98, 2011.

35. S. Bruni, G. Bucca, A. Collina, A. Facchinetti, Dynamics of the Pantograph-Catenary System, Chapter 16, In: S. Iwnicki, M Spiryagin, C. Cole, T. McSweeney (Eds), Handbook of Railway Vehicle Dynamics 2nd Edition, CRC Press, Boca Raton, FL, 2020.

36. A. Ambrósio, J. Pombo, P. Antunes, M. Pereira, PantoCat statement of method, Vehicle System Dynamics, 53(3), 314–328, 2015.

37. S. Bruni, J. Ambrosio, A. Carnicero, Y. H. Cho, L. Finner, M. Ikeda, S. Y. Kwon, J.-P. Massat, S. Stitchel, M. Tur, W. Zhang, The results of the pantograph-catenary interaction benchmark, Vehicle System Dynamics, 53(3), 412–435, 2015.

38. M. Spiryagin, P. Wolfs, C. Cole, V. Spiryagin, Y. Q. Sun, T. McSweeney, Design and Simulation of Heavy Haul Locomotives, Ground Vehicle Engineering Series, CRC Press, Boca Raton, FL, 2017.

39. M. Spiryagin, P. Wolfs, F. Szanto, C. Cole, Simplified and advanced modelling of traction control systems of heavy-haul locomotives, Vehicle System Dynamics, 53(5), 672–691, 2015.

40. S. Shrestha, Q. Wu, M. Spiryagin, Review of adhesion estimation approaches for rail vehicles, International Journal of Rail Transportation, 7(2), 79–102, 2019.

41. S. Shrestha, Q. Wu, M. Spiryagin, Friction condition characterization for rail vehicle advanced braking system, Mechanical Systems and Signal Processing, 134, 106324, 2019.

42. C. Cole, M. Spiryagin, Q. Wu, Y. Q. Sun, Modelling, simulation and applications of longitudinal train dynamics, Vehicle System Dynamics, 55(10), 1498–1571, 2017.

43. I. B. Duncan, P. A. Webb, The Longitudinal Behaviour of Heavy Haul Trains Using Remote Locomotives, Proceedings of the 4th International Heavy Haul Conference, 1989 11-15 September 1989, Brisbane, Australia, 587–590, 1989.

44. B. J. Jolly, B. G. Sismey, Doubling the Length of Coals Trains in the Hunter Valley, Proceedings of the 4th International Heavy Haul Conference, 11-15 September 1989, Brisbane, Australia, 579–583, 1989.

45. R. D. Van Der Meulen. Development of Train Handling Techniques for 200 Car Trains on the Ermelo-Richards Bay Line, Proceedings of the 4th International Heavy Haul Conference; 11-15 September 1989 Brisbane, Australia, 574–578, 1989.

46. C. Cole, Longitudinal Train Dynamics, Chapter 9, In: Iwnicki S (Ed), Handbook of Railway Vehicle Dynamics, Taylor & Francis, London, 239–277, 2006.

47. C. Cole, Longitudinal Train Dynamics and Vehicle Stability in Train Operations, Chapter 13, In S. Iwnicki, M. Spiryagin, C. Cole, T. McSweeney (Eds), Handbook of Railway Vehicle Dynamics, 2nd Edition, CRC Press, Boca Raton, FL, 457–520, 2020.

48. M. Spiryagin, C. Cole, Y. Q. Sun, M. McClanachan, V. Spiryagin, T. McSweeney, Design and Simulation of Rail Vehicles, CRC Press, Boca Raton FL, 2014.

49. V. K. Garg, R. V. Dukkipati, Dynamics of Railway Vehicle Systems, Academic Press, New York, NY, 1984.

50. S. Shrestha, M. Spiryagin, Q. Wu, Real-time multibody modeling and simulation of a scaled bogie test rig, Railway Engineering Science, 28, 146–159, 2020.
51. T. Piechowiak, Pneumatic train brake simulation method, Vehicle System Dynamics, 47(12), 1473–1492, 2009.
52. Q. Wu, C. Cole, S. Luo, M. Spiryagin, A review of dynamics modelling of friction draft gear, Vehicle System Dynamics, 52(6), 733–758, 2014.
53. Q. Wu, X. Yang, C. Cole, S. Luo, Modelling polymer draft gears, Vehicle System Dynamics, 54(9), 1208–1225, 2016.
54. C. Cole, M. Spiryagin, Q. Wu, C. Bosomworth, Modelling Issues in Passenger Draft Gear Connections in Proceedings of the 24th Symposium of the International Association for Vehicle System Dynamics (IAVSD 2015), Graz, Austria, 17–21 August 2015, In: M. Rosenberger, M. Plöchl, K. Six, J. Edelmann (Eds), The Dynamics of Vehicles on Roads and Tracks, CRC Press, London, 2016.
55. Q. Wu, M. Spiryagin, C. Cole, Advanced dynamic modelling for friction draft gears, Vehicle System Dynamics, 53(4), 475–492, 2015.
56. Q. Wu, M. Spiryagin, C. Cole, Longitudinal train dynamics: An overview, Vehicle System Dynamics, 54(12), 1688–1714, 2016.

3 Modeling of Electrical Systems for Rail Vehicles

An electric locomotive power train consists of the following elements:

- A source of electrical energy, either an electrical overhead power system or an on-board generator
- Optionally, energy storage devices which can perform functions such as the recovery of braking energy, the management of the loading imposed on the power source or the provision of vehicle operation while separated from its power source
- Power conditioning equipment including rectifiers, inverters, or dynamic braking equipment
- The traction motors

The topological arrangement of these elements and the technical choices available for each element are now explored.

3.1 ELECTRICAL TOPOLOGIES

3.1.1 DIESEL ELECTRIC LOCOMOTIVES

Diesel electric locomotives utilize a diesel prime mover to produce electrical energy which is then converted into a propulsive force by electrical traction motors. Figure 3.1 shows a typical arrangement for a freight locomotive. The diesel drives an alternator that produces AC power. This is rectified to produce DC which, via a DC bus, supplies traction to two inverters which in turn supply two groups of three traction motors which are directly mounted in the locomotive bogies or trucks. The induction motor is the dominant motor choice. The inverters will generally use field-oriented control (FOC) or direct torque control (DTC). Either combination allows the precise control of the tractive effort for both motoring and braking operations. A dynamic brake system allows electrical energy recovered from the motors during braking to be dissipated in braking resistors. The DC bus capacitor is required to carry the higher frequency currents produced by the inverters. It does not typically have enough energy storage capability to allow braking energy to be stored for later use.

3.1.2 ELECTRIC LOCOMOTIVES

Electric locomotives draw electrical energy from an overhead power system or in a few instances from a third rail [1]. Figure 3.2 shows an equipment arrangement that may be typical of a high-speed train. Power is drawn from an overhead AC system via

DOI: 10.1201/9781003028994-3

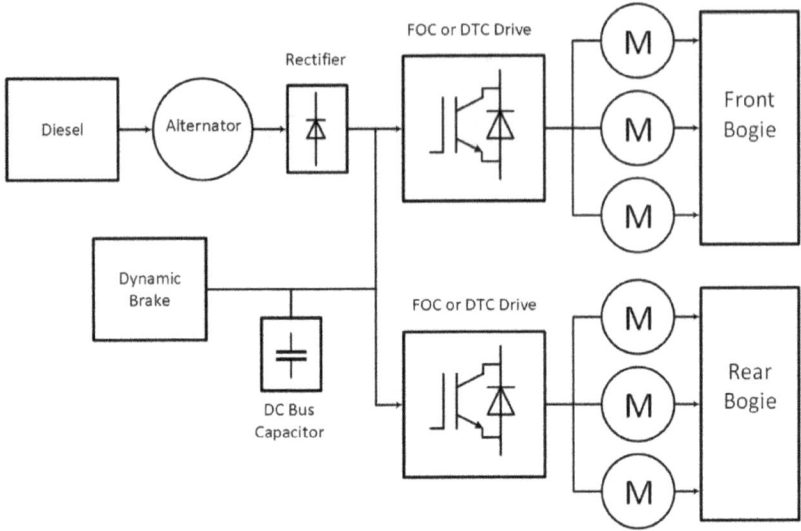

FIGURE 3.1   General equipment arrangement for a diesel AC electric locomotive.

a pantograph and transformer. The return current is via the rails. The transformer has two secondary windings that supply PWM rectifiers that in turn power a pair of independent DC buses. Each driven axle has a dedicated traction machine and inverter.

The overhead power system is frequently 25 kV at 50 or 60 Hz, though 15 kV and 16.7 Hz is common in parts of Europe [1]. DC overhead power systems do exist but are typically older legacy systems at much lower voltages such as 750, 1500, or 3000 Vdc.

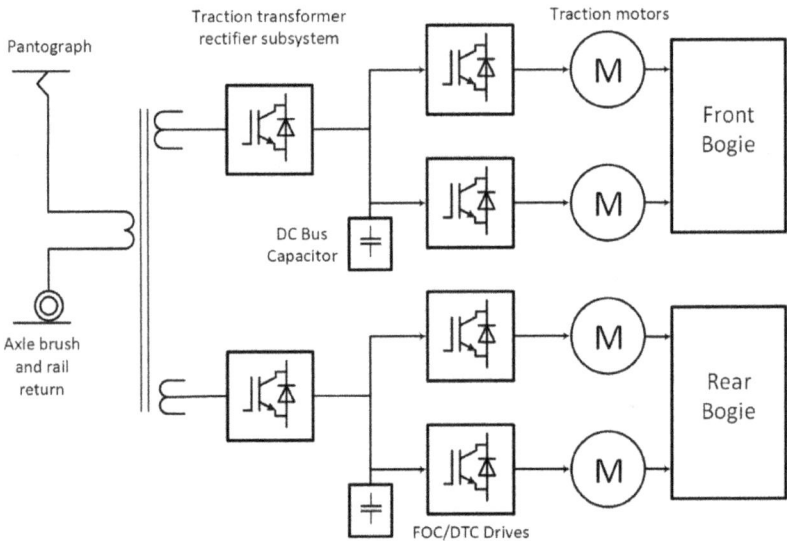

FIGURE 3.2   General equipment arrangement for an AC electric locomotive.

The lower voltage greatly limits the locomotive power and their common applications include tramways and light rail systems [1].

3.1.3 HYBRIDS

A hybrid system combines two or more energy sources or an energy source with at least one energy storage system. Hybrid systems are by definition more complex, but that complexity allows extra degrees of flexibility in operation that provides clear benefits. The benefit can be economic, environmental, or operational.

Hybrid traction systems using energy storage are the most common form of hybrid and can provide the following benefits:

- Fuel or electrical energy saving by storing braking energy for future use
- For electric passenger trains and trams, allowing operation on track sections without overhead power systems
- For diesel locomotives:
  - Management of the peak power demands required from the prime mover and the management of ramp rate limitations
  - The adjustment of prime mover loading to operate at optimum fuel or optimum emission points
  - Fuel savings if wayside recharging is available

Hybrid locomotive examples which are either in commercial production or have been in production include:

- Toshiba 500 kW HD300 shunting locomotive, which is a diesel, lithium-ion battery locomotive with permanent magnet (PM) synchronous motors [2, 3]
- Railpower Green Goat and Green Kid shunting locomotives, which are diesel, lead-acid battery with DC traction motors [4]
- Japan Transport Engineering Company (J-TREC) "sustina" diesel hybrid locomotives, which are diesel, lithium-ion battery locomotives with induction motors [5]
- J-TREC "sustina" battery hybrid locomotives, which can operate from an overhead catenary but can also operate on non-electrified lines [5]

3.1.3.1 Principles of Hybridization for Rail Vehicles

The design of a hybrid rail vehicle will be strongly determined by the intended operation. The first step in any design optimization will be to determine expected operational cycles in terms of both the required locomotive power and the requirement for energy storage [6–10]. Railway vehicles are frequently designed, or at least customized, for a specific duty. In a mineral export operation, a locomotive may be designed for operation on a specific export corridor and may spend its entire life in that service. Similarly, a commuter train may be designed for a specific line or a group of lines that have similar constructions. In these cases, train dynamics simulations can be conducted for the specific track sections, train consists, and driving strategies to determine the required locomotive power and energy in a time series form. These can be then used to rate the drive train components and assess fuel or energy usages.

A railway vehicle designer will seek to optimize a design according to some selected criteria. Energy or fuel savings are normally important contributors to the life cycle cost. Other important considerations may include the emission of pollutants, such as particulates, or changes to braking related maintenance costs.

Hybrid systems that seek to provide energy or fuel savings are likely to be more beneficial in railway systems where braking is frequent. Braking occurs frequently in metropolitan passenger railways but is much less common in heavy haul freight systems. Mineral export corridors are often designed, along with the trains themselves, to achieve a fuel-efficient operation. Braking will be infrequent and only necessitated by the crossing of major features such as mountain ranges. In these cases, the energy storage requirements can be many megawatt-hours and this may require batteries with weights of tens of tons. Batteries can be integrated into the locomotive structure or an alternative solution is to use a battery tender which is a purpose designed battery wagon that trails the locomotive [11].

Quite different design solutions result if the mission of a hybrid design is to integrate a power source that has specific operational features. For example, fuel cells may allow a locomotive to be fueled with hydrogen [7]. An operational constraint is that fuels cells can have quite long response times. An energy storage may be needed to support adequate power ramp rates. A gas turbine may allow the use of natural gas as a fuel and allow compact high-speed turbine alternator packages to be applied [11]. However, this combination may have a relatively narrow preferred operating range. A storage system allows the average turbine power and the peak traction power to be decoupled.

3.1.3.2 Hybrid Topologies

Hybrids are often characterized as series or parallel topologies. A series hybrid typically has an energy source and an energy storage and these power one motor system. A parallel hybrid has at least two methods of delivering driving power. Typically, an energy source and an energy storage system are mechanically connected to the drive train through two separate machines.

The rail vehicles shown in Figures 3.1 and 3.2 each feature DC busbars. These are natural locations where energy storage can be easily integrated as shown in Figure 3.3. The DC bus bar allows connections to batteries [7–9], supercapacitors [6], or flywheel storages [9–11]. For each storage, an interfacing converter is normally required, and this allows the energy storage to be controlled independently of the DC bus bar voltage. From the DC bus bar location, power can be both recovered from, or delivered to, the traction system. The drive packages nearly always allow a bidirectional flow of power between the mechanical and electrical subsystems. A hybrid produced in this way, adding energy storage at the DC bus bar, will be a series hybrid [12]. All of the hybrid locomotive examples above are series hybrids. Road vehicles frequently utilize a parallel hybrid topology where an internal combustion motor is assisted by an electrical motor. Both motors are mechanically coupled to the driving wheels [12]. As a result, a parallel hybrid has a far more complex mechanical transmission. A range of series and parallel hybrid approaches have been proposed, but not commercialized, for diesel locomotives. Diesel locomotives can be produced in diesel-hydraulic configuration where the diesel prime mover drives a hydraulic pump to produce high pressure oil to drive hydraulic motors. Parallel hybrid solutions might

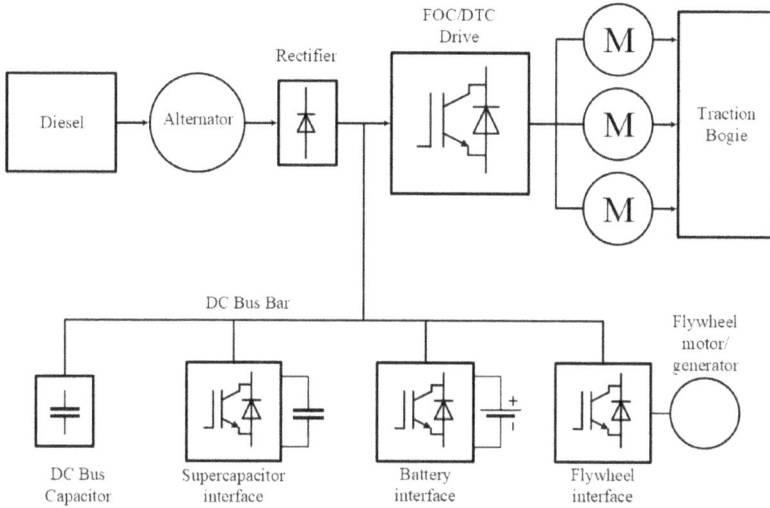

FIGURE 3.3 Series hybrid locomotive.

use a secondary electrical machine with electrical storage or a secondary hydraulic machine with hydraulic accumulators for energy storage [8].

3.2 TRACTION POWER SUPPLIES

3.2.1 ALTERNATORS AND GENERATORS

For a diesel locomotive, an alternator is used to produce power for the traction system. Strictly speaking, a generator is a DC machine with a commutator. Commutator machines are expensive, high maintenance items and impose significant operating voltage limitations. They are now largely a legacy technology. Modern locomotives make use of synchronous alternators which produce AC power. Rectifiers then convert the AC output into DC. The alternator output is readily adjusted by controlling the alternator field current provided to the rotor.

The alternator design must be matched to the diesel and these have typical operational speed ranges from 600 rpm to 1800 rpm. At lower loads the rotational speed is often reduced. Typically, a three-phase machine with four or six poles is used and the maximum alternator frequency is 25–60 Hz. Alternators are efficient machines. As an example, the ABB WGX500-560 family are 1200 Vdc, 600–1800 rpm, six-pole alternators in the 13.3 MW range. These claim a 96.9% peak efficiency [13].

3.2.2 RECTIFIERS

In an electric locomotive, AC current is collected from the overhead power system with a pantograph. Typically, the overhead system voltage is reduced to a lower level with a transformer and a rectifier is used to produce DC power for the supply of the traction package. Older locomotives will use thyristor rectifiers, but a modern locomotive will often use a PWM rectifier.

3.2.2.1 Thyristor Rectifiers

A simplified thyristor rectifier is shown in Figure 3.4 and its main waveforms are shown in Figure 3.5. The rectifier is half-controlled which offers power factor advantages, but this rectifier will not allow regeneration from the DC system into the AC network. Dynamic braking is possible with braking resistors. The voltage is controlled by the firing delay of thyristors $T1$ and $T2$. Thyristor $T1$ is fired at time t_1. Current increases in $T1$ at a rate determined by the commutating inductance L_C. Once fully established at t_2, current flows to the armature via $T1$ and $D2$. At time t_3, the zero crossing of the AC voltage, current flows to the flywheel path $T1$ and $D1$ until $T2$ is fired. The line current contains significant harmonics, especially third and fifth harmonics. The total harmonic distortion can easily be 20–30%. The line

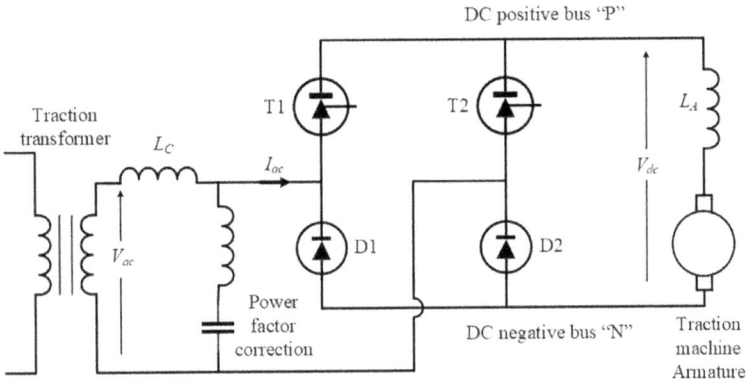

FIGURE 3.4   Thyristor rectifier.

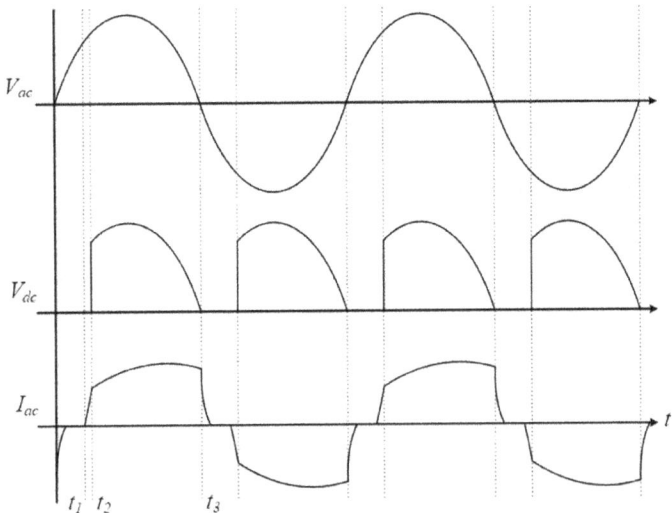

FIGURE 3.5   Thyristor rectifier waveforms.

current has a lagging power factor and power factor correction capacitors are also included. In many locomotives, several rectifier stages may be combined and operated with sequentially controlled firing angles to get power factor and harmonic advantages. Frequently, separate rectifiers are provided for both the armatures and field windings.

3.2.2.2 PWM Rectifiers

A PWM rectifier is shown in Figure 3.6 and its main waveforms are shown in Figure 3.7. The four switch devices, IGBTs in this case, are switched to produce a pulse width modulated voltage V_{PWM}. The PWM voltage can be controlled in magnitude and phase so that a sinusoidal current is drawn from the grid via the filter inductors L_F. The relative phase of the mains voltage and the centroid of the PWM

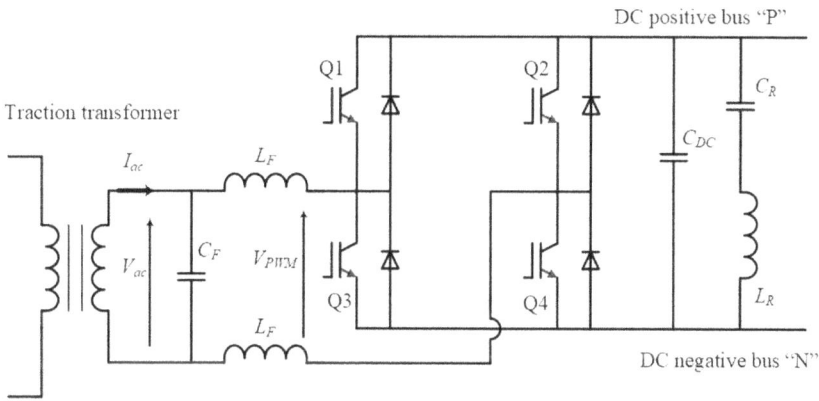

FIGURE 3.6 PWM rectifier.

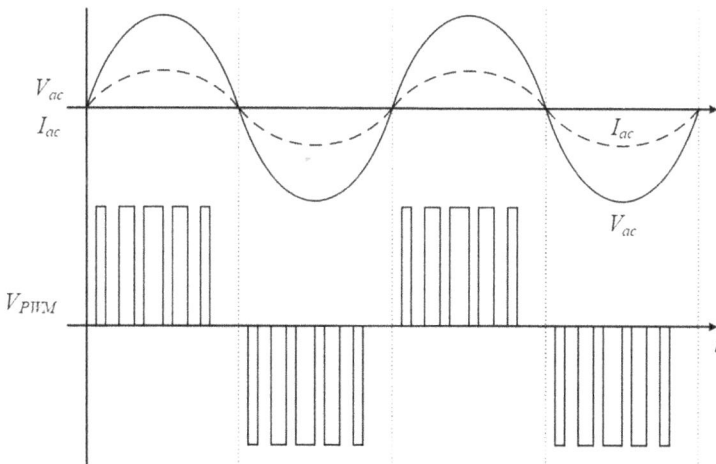

FIGURE 3.7 PWM rectifier waveforms.

waveform strongly affect power flow. In this case the current is controlled so that rectification occurs, that is energy flow from the AC system to the DC bus bar.

The PWM rectifiers are inherently reversible. That is, power flow can be forced from the DC system into the AC network. This allows energy to be returned to the grid during regeneration. The power factor of a PWM is controllable. The reactive power can be used to help control the overhead network voltage. The railway overhead network has higher levels of voltage fluctuation than are found in conventional power systems. Voltage rise will often limit the amount of power returned to the AC system during regenerative braking, and at least some braking resistors will still be required.

The PWM rectifier has a voltage output and is normally equipped with a DC bus capacitor. There is often a filter for the removal of ripple current at twice the mains frequency, i.e., 100 Hz or 120 Hz for 50 Hz and 60 Hz networks, respectively. In locomotives, multiple PWM rectifiers are often utilized. For megawatt scale rectifiers, the switching frequencies are in the range of hundreds of hertz to a few kilohertz. The switching instants of the PWM waveforms are staggered to cancel switching frequency harmonics. A PWM locomotive will have low levels of harmonic current, in the range of a few percent or less. The harmonic orders are odd and typically higher than 50.

3.2.3 ENERGY STORAGE

Energy storage systems have an energy rating. This has the dimension of Joules but is normally expressed in Watt-hours (Wh), kilowatt-hours (kWh), or megawatt-hours (MWh). They also have power ratings that apply during charging or discharging. These have the dimensions of Watts but kilowatts and megawatts are often used. In an ideal energy storage, the energy is the integral of the storage power. Conversely the power is the derivative of the stored energy. In simulation models, integrators are frequently used to model an energy storage element.

A key feature of any storage element is its cycle life. Cycle life will be affected by the stress levels imposed by a charge and discharge cycle. The cycle depth of discharge (DoD) refers to the energy, as a fraction of the battery capacity, removed during a cycle. A full discharge, or 100% DoD, is more stressful than a shallow discharge.

Railway applications are mobile. In this environment the energy and power densities, with respect to weight and volume, are critically important [14]. The current commercial storage technologies have energy densities that are orders of magnitude less than diesel fuel.

3.2.3.1 Batteries

Batteries have relatively high energy densities, hence are well suited to the recovery of braking energy. Lithium-ion batteries are the preferred battery in railway applications [15]. The battery technology has been intensively driven by electrical vehicle applications where battery packs of up to 100 kWh are now relatively commonplace. These battery packs have been shown to be robust with respect to shock and vibration, operating temperature ranges, and high charge and discharge rates. Modern battery packs are complex engineering systems that are quite different to older battery technologies. Lithium-ion battery packs incorporate a significant amount of electronics within their battery management systems (BMSs). It is critically important from a safety and battery life perspective that each cell is monitored and managed. The BMS will certainly include communications interfaces for remote monitoring and control. Battery systems often employ active cooling and warming methods. These may feature liquid coolants and heat pumps.

Battery systems into the tens and hundreds of megawatt-hour range using modular construction methods are increasingly deployed in grid scale batteries. Large batteries generally operate at DC voltages in the 400–1000 Vdc range. Many locomotive systems would prefer voltages up to a few kilovolts. These are technically possible but safety and regulatory requirements become increasingly complex above 1000 Vdc. If a higher DC bus bar operating voltage is required, then a battery may have to interface via an isolated DC–DC converter.

The term lithium-ion battery encompasses a wide family of battery technologies. By varying the construction and chemistry, considerable differences in battery performance can be obtained. Batteries can be optimized for energy density, cycle life, or power rating. Frequently, gains in one area come at the expense of performance in other areas. For railway applications, lithium iron phosphate (LFP) or lithium titanate oxide (LTO) chemistries are often preferred, especially on the grounds of safety [15]. These technologies offer long cycle and calendar lifetimes and can cater for high charge and discharge rates. They have energy densities in the range of 100–120 Wh/kg. Lithium nickel cobalt aluminum (NCA) or lithium nickel manganese cobalt (NMC) cells that are often found in electric vehicles will be in the range of 150 Wh/kg.

A battery charge or discharge power is often expressed in relation to its energy storage capacity by a "C" rate. A 1C discharge corresponds to discharging a battery in 1 hour, i.e., 1 MW for a 1 MWh battery. Lithium chemistries can achieve lifetimes of thousands of cycles at a 1C rate. Lithium batteries can support powers that significantly exceed this and 2C is often achievable. Extreme fast charging (XFC) is a topic or research in automotive applications. Here the goal is to deliver 80% charge in 10 min or less [16]. This is at least a 4.8C rate.

3.2.3.2 Flywheels

Modern flywheels have composite rotors using fiberglass, Kevlar, or carbon fiber materials [17, 18]. A high-speed rotor will be supported by a combination of magnetic and mechanical bearings. The highest energy/mass storage density is achieved with rotor materials that have a high specific strength [18]. The maximum energy density for a thin-walled cylindrical flywheel is:

$$U_{max} = S_{max}/2\rho \tag{7}$$

where S_{max} is the tensile strength of the material (Pa) and ρ is its density (kg/m^3) and U_{max} has the units of J/kg. The theoretical limit for graphite fibers, 545 Wh/kg, exceeds the storage capability of lithium-ion batteries which are typically 150 Wh/kg. These high energy densities typically require very high rotational speeds. Small flywheels have achieved speeds above 200,000 rpm. A large-scale flywheel for power system balancing flywheel rated at 100 kW, 25 kWh, rotates at 16,000 rpm [10]. The calculated energy density is much lower than a lithium-ion battery at 9.6 Wh/kg.

Composites are preferred as the stored energy increases with the square of velocity, and higher rotational speeds more than compensate for the lower materials density. The electrical drive is normally provided by a directly coupled inverter driven AC machine embedded within the flywheel. The electrical machines are compact due to the high rotational speeds and efficiencies can be very high. Flywheels can achieve very long cycle lifetimes and extremely high charge discharge rates if the flywheel is designed to allow the mechanical torques developed. The in-out energy efficiency for a flywheel storage will typically exceed 95% [18].

3.2.3.3 Super Capacitors

Super capacitors are a high-power high cycle life component that have rather limited energy storage capacity. Supercapacitors can achieve cycle lifetimes of 100,000 cycles or more. They can supply charge and discharge rates of 30–60C, i.e., charge and discharge times of 1–2 min. The energy density is in the range of 1–10 Wh/kg [15]. They are ideally suited to very high rate but lower energy applications. As an example, they pair well with an energy source such as a fuel cell that has a relatively low response rate [19].

Prior to the commercial availability of lithium-ion batteries with high C rates, supercapacitors were sometimes paired with battery technologies, such as lead acid batteries, to extend their cycle life. This approach might still be useful at very low temperatures. Supercapacitors will operate down to −40°C, while battery systems can have difficulties below −20°C [15].

3.2.4 DYNAMIC BRAKING ENERGY MANAGEMENT

Dynamic braking is frequently used to reduce the mechanical wear of train braking systems. Dynamic braking systems often have an advantage on long mountain range descents as they are able to sustain braking efforts indefinitely. Mechanical brakes may be limited by thermal masses of braking components. The braking requirements will determine the dynamic braking energy profile. This energy must be absorbed in an energy storage, returned to an overhead power system (if possible) or dissipated in braking resistors.

In modern locomotives, the traction motors and their controllers return the braking energy to the DC busbar. It is at this point where energy storages, braking resistors, and reversible PWM rectifiers can connect. There may be a technical need for an interfacing converter for each subsystem. A battery will generally interconnect via a DCDC converter that regulates the battery power. A braking resistor will often connect via a buck converter or DC chopper.

The action of these devices is often coordinated by using the DC bus bar voltage as a signaling mechanism. The return of braking energy to the DC bus bar will cause its voltage to rise at a rate determined by the DC capacitor. Any connected storage, or reversible PWM rectifier, will become active at threshold level and limit the voltage rise by absorbing energy. If the braking energy exceeds this capability, the voltage will continue to rise, albeit at a lower rate. At still higher DC bus voltage, the operation of the dynamic brake resistors is triggered. They must be designed to have the absorption capability to safely limit the DC bus voltage rise.

3.3 TRACTION MOTORS AND POWER ELECTRONICS

Railway traction motors have several features:

- A wide speed range including operation from zero speed
- High torque and power to volume ratios
- Intensive cooling, normally by forced air
- Tolerance to a high external contamination environment
- Durability in a high mechanical shock environment, as the motor is frequently an unsprung weight

The motors are always partnered with highly capable power electronics packages. Frequently, the motors are torque controlled to allow precise control over the tractive effort and dynamic braking. Electrical machines have two mechanisms for torque production. The Lorentz force, the force on a conductor in a magnetic field, is the primary torque production method in most high-power machines, especially DC motors, induction machines, and synchronous machines (SMs). The second method, saliency torque, relies on the magnetic alignment of a rotor that has a preferred magnetizing orientation. Saliency torque is important in some SMs and many PM machines such as brushless DC (BLDC) motors.

3.3.1 DC Motors

Many railway traction motors are described as series wound machines. While this is true, the connection and control of the field windings are modified during the different modes of the locomotive operation. It is helpful to model the machine as a separately excited machine with independent control of the armature and field windings. There are two clearly identifiable operating regions:

- The low-speed or torque limited region, which applies from standstill to approximately one-third of the vehicle's maximum operational speed. The field flux is maintained at the highest practical levels to maximize torque production.
- The constant power, field weakening, or high-speed region. Here the field flux must reduce because of limits on the armature voltage.

For a locomotive operating at low speed and high tractive effort, the series connected traction motor is operating in the low speed region. The armature current flows through the field windings. For higher torque ranges, certainly for the upper half of the machine rating, the field magnetic paths are saturated and the field flux is at the highest achievable level.

For higher locomotive speeds, the motor field strength must reduce as the machines enter the constant power region. This can be achieved by a reduction in the armature current and/or by the parallel connection of a field diverter resistor across the field winding. For a locomotive in dynamic braking, the machine will operate in a constant power mode. The field winding may be disconnected from the armature and separately excited to actively control the dynamic braking voltage produced by the armature and applied to the braking resistors.

3.3.1.1 Machine Models

The torque equation for a separately excited machine is:

$$T_{\text{mech}} = k_T \psi_f \left(i_f \right) i_a \tag{3.1}$$

where i_a is the armature current; ψ_f is the magnetizing flux and is a function of the field current i_f; and k_T is the torque constant. Figure 3.8 shows an equivalent circuit for the separately excited DC machine with the dependency of flux on the field current. The armature is to be controlled by adjusting the voltage, $v_a(t)$. The armature resistance and inductance are r_a and L_a, respectively. The armature voltage induced by the magnetizing field and the rotation of the armature is:

$$e_a = k_T \psi_f \left(i_f \right) \omega_m \tag{3.2}$$

where e_a is the armature back emf; ω_m is the rotational velocity; and k_T is the torque constant.

In Figure 3.8 the field resistance is r_f, L_{fi} is the effective inductance of the unsaturated iron path, and L_{fa} is the effective inductance of the air path. A linearized saturation characteristic, $\text{sat}(i_f, i_{\text{sat}})$, where i_{sat} is field current saturation value, can be used to model the flux dependency on the field current:

$$\psi_f = L_{fi}\text{sat}(i_f, i_{\text{sat}}) + L_{fa}i_f \tag{3.3}$$

$$\text{sat}(i_f, i_{\text{sat}}) = \begin{cases} i_{\text{sat}}, & i_f > i_{\text{sat}} \\ i_f, & -i_{\text{sat}} \le i_f \le i_{\text{sat}} \\ -i_{\text{sat}}, & i_f < -i_{\text{sat}} \end{cases} \tag{3.4}$$

Traction motors frequently have ratings of several hundred kilowatts and may range up to about 1 MW. Motors of this size have compensation windings or interpole windings to counteract the armature reaction and to maintain the best operating conditions for the commutator. A compensation winding is constructed by embedding current carrying conductors in the face of the field pole. An interpole is an additional magnetic pole and winding located between the main field poles. Both types

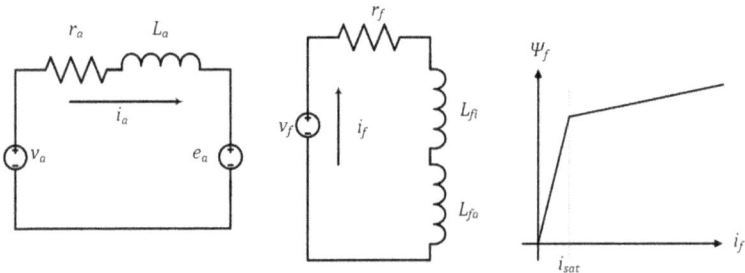

FIGURE 3.8 Separately excited DC machine.

TABLE 3.1

Motor Currents and Torque EMD D87 Series Connected Motor [20]

Armature Current (A)	Torque (kNm)	$k_T \psi_f (i_f)$ (Nm/A)
323	1.76	5.44
1644	13.4	8.15

of winding are series connected with the armature as they are intended to counteract the impact of the armature current upon the commutation process. The commutator diameter increases with voltage and, in a traction motor, the armature voltage will be limited to about 1400 V.

3.3.1.2 Case Studies

The D87 motors are a nominal 600 Vdc traction motor produced by General Motors Electromotive Division (EMD). These have a rating that exceeds 500 kW and depends on the cooling conditions. They are widely applied in heavy haul freight locomotives. For example, four D87 motors are applied in an EMD GP50 locomotive which is rated at 3500 Hp [20]. The motor parameters can be estimated from the published locomotive data. The D87 has an efficiency of 92% at 1000 A and 620 V [20]. This implies a total resistance, for the armature and series field of 0.039 Ω. Table 3.1 gives the motor current and torque at two operating points which can be used to estimate the field flux constants as: $i_{sat} = 323A$; $L_{fi} = 14.8 \times 10^{-3}$; $L_{fa} = 2.05 \times 10^{-3}$.

Figure 3.9 show some simulation results from a diesel-electric locomotive with six D87 motors. The simulation starts with the locomotive at standstill and the diesel

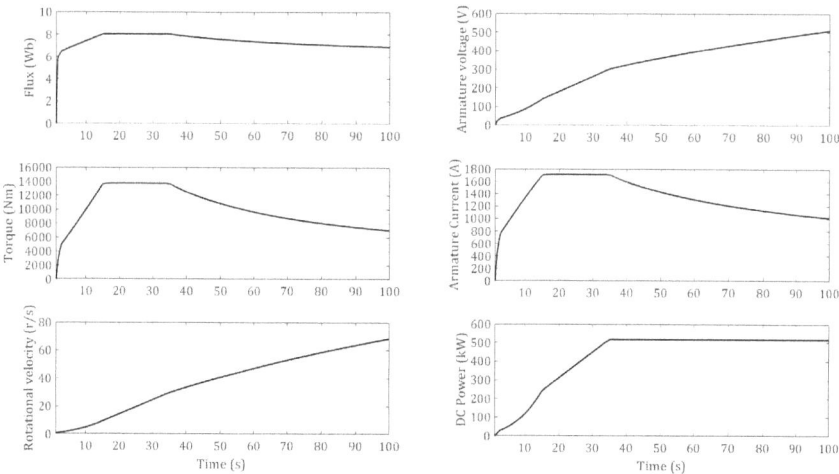

FIGURE 3.9 Diesel electric D87 locomotive traction motor response.

generator at idle. The maximum tractive effort is requested at $t = 0$ s. As the diesel generator response requires tens of seconds, a simulation of 100 s is presented. The generator reaches the current limit, 10 265 A or 1 710 A per traction motor, after approximately 15 s. The motors have their full magnetizing flux and their peak torque 13.9 kNm at this time. The locomotive accelerates with constant tractive effort until $t = 35$ s when the generator power limit of 3.095 MW or 516 kW per traction machine is reached. The locomotive then accelerates at constant power which implies reducing armature current, flux, and torque.

3.3.2 INDUCTION MACHINES

The squirrel cage induction machine is the most widely used railway traction machine. The rotor winding is constructed by die casting the rotor bars directly into a prepared stack of rotor laminations. The rotor, having an uninsulated construction, is extremely robust. Aluminum is frequently used in industrial applications but, in heavy haul, copper can be substituted. It has a lower electrical resistivity on a volume basis relative to aluminum so, in a space and cooling constrained motor design, the losses will be lower. Copper has a higher thermal specific heat and melting point, which allows higher short term overloads relative to an aluminum cage.

3.3.2.1 Machine Models

The phasor representation of an induction machine with a squirrel cage rotor is [21]:

$$\begin{bmatrix} \bar{v}_s(t) \\ 0 \end{bmatrix} = \begin{bmatrix} r_s & 0 \\ 0 & r_r \end{bmatrix} \begin{bmatrix} \bar{i}_s(t) \\ \bar{i}_r'(t) \end{bmatrix} + \frac{d}{dt} \begin{bmatrix} L_s & L_m \\ L_m & L_r \end{bmatrix} \begin{bmatrix} \bar{i}_s(t) \\ \bar{i}_r'(t) \end{bmatrix}$$
$$+ -j\omega_r \begin{bmatrix} 0 & 0 \\ L_m & L_r \end{bmatrix} \begin{bmatrix} \bar{i}_s(t) \\ \bar{i}_r'(t) \end{bmatrix}$$

(3.5)

An equivalent circuit diagram is shown in Figure 3.10. The stator circuit contains the applied stator space vector, $\bar{v}_s(t)$, the magnetizing inductance is L_m, and the stator leakage inductance is $L_{sl} = L_s - L_m$. The rotor circuit contains the rotor resistance, r_r, L_m, and the rotor leakage inductance is $L_{rl} = L_r - L_m$.

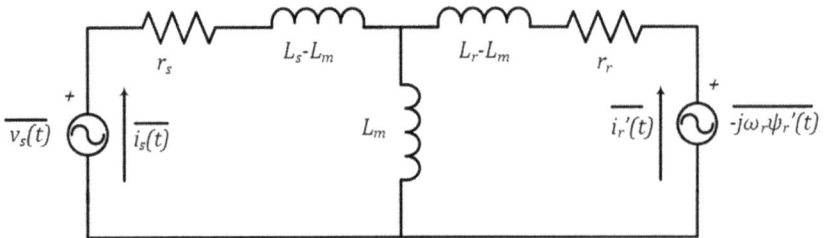

FIGURE 3.10 Equivalent circuit in phasor form.

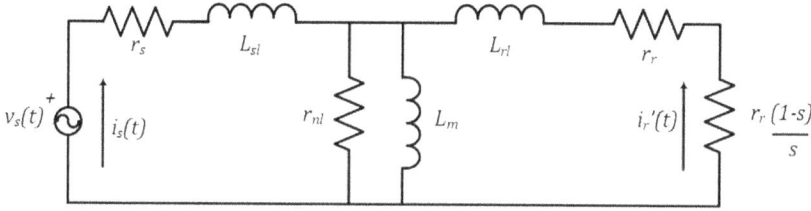

FIGURE 3.11 Equivalent circuit representation in phase variables.

The rotor voltage source is a speed dependent term and the electrical power transferred to this source is mechanical power delivered to the rotor. This voltage is:

$$-j\omega_r \left(L_m \overline{i}_s(t) + L_r \overline{i}_r'(t) \right) = -j\omega_r \overline{\psi}_r'(t) \tag{3.6}$$

where $\overline{\psi}_r'(t)$ is the rotor flux and $\overline{i}_r'(t)$ is the rotor current. Both quantities are referred to the stator side and scaled by the rotor stator turns ratio. The mechanical power delivered at the rotor is:

$$P_m = \frac{3}{2} Re\left(-j\omega_r \overline{\psi}_r(t)\overline{i}_r'(t)^* \right) = \omega_r\left[-\frac{3}{2}\overline{\psi}_r(t) \times \overline{i}_r'(t) \right] = \omega_r T_e \tag{3.7}$$

The equivalent circuit can be redrawn in phase variables, i.e., the phase voltages and currents, as shown in Figure 3.11. The circuit components can be physically measured with no-load and blocked rotor tests. The mechanical load element has been replaced by an equivalent load resistor. A *no load loss* resistor has been included to capture:

- Magnetic losses in the stator and rotor
- Parasitic mechanical losses such as windage and bearing friction

The equivalent mechanical load resistance is:

$$r_{mech} = r_r \left[\frac{1-s}{s} \right] \tag{3.8}$$

where s is the rotor slip, and

$$s = \frac{\omega_r}{\omega_e} \tag{3.9}$$

3.3.2.2 Field-Oriented Control

Modern induction machines are frequently controlled using field orientation or vector control principles that yield high quality dynamic performances. Consider the

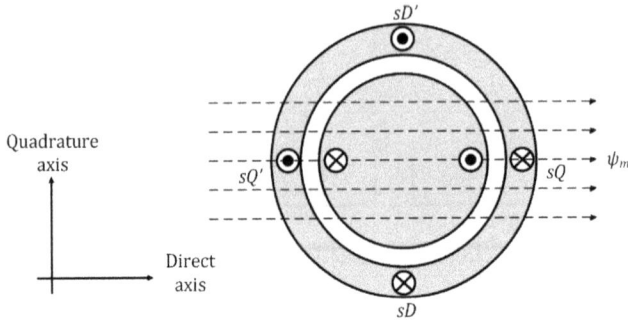

FIGURE 3.12   An equivalent two-phase induction machine.

generalized AC induction machine shown in Figure 3.12. A two winding representation is used. The stator windings for A, B, and C phases are replaced by two equivalent phase windings, direct (D) and quadrature (Q). Currents in these windings, i_{sD} and i_{sQ}, will provide an identical MMF as produced by the three phase currents i_{sA}, i_{sB}, and i_{sC}. These can be calculated using the Clarke Transform relationship:

$$\begin{bmatrix} i_{sD} \\ i_{sQ} \end{bmatrix} = \frac{2}{3} \begin{bmatrix} 1 & -\dfrac{1}{2} & -\dfrac{1}{2} \\ 0 & \dfrac{\sqrt{3}}{2} & -\dfrac{\sqrt{3}}{2} \end{bmatrix} \begin{bmatrix} i_{sA} \\ i_{sB} \\ i_{sC} \end{bmatrix} \tag{3.10}$$

Assume the direct axis winding D carries a fixed current that produces a magnetizing flux ψ_m in the direction of the direct axis. If, at $t = t_0$, a current is injected into the Q axis winding, an opposing current is induced into the mutually coupled rotor by Lenz's law. Assuming an equivalent turns ratio of 1:1, a current phase diagram can be drawn as shown in Figure 3.13.

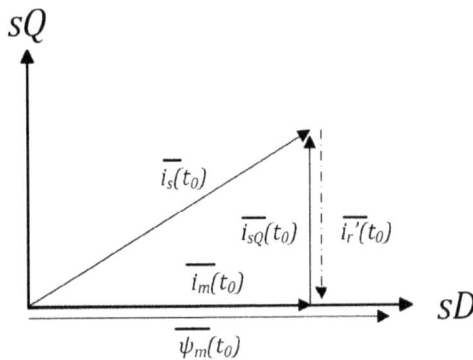

FIGURE 3.13   Rotor current and magnetizing flux at $t = t_0$.

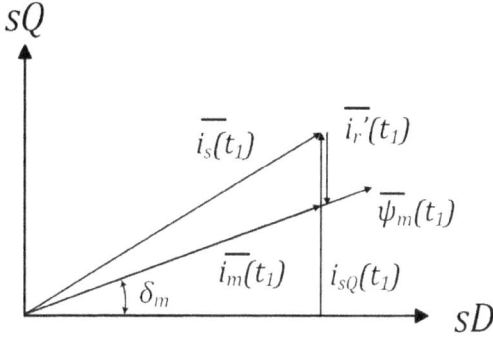

FIGURE 3.14 Realignment via mechanical rotation of the stator.

The magnetizing space phasor, $\overline{i_m}(t_0)$, is aligned with the D axis. It is the sum of the stator space phasor, $\overline{i_s}(t_0)$, and the rotor space phasor $\overline{i_r}'(t_0)$. The rotor current and magnetizing vectors are orthogonal, and the torque is maximized. The rotor current will exponentially decay over time. As the rotor current is reduced, the orthogonal relationship of the rotor current and the magnetizing flux is lost as shown in Figure 3.14

This alignment could be maintained by mechanically rotating the stator through an angle, δ_m, as shown in Figure 3.14. An alternative is adjusting the D and Q axis currents. A x,y reference frame that rotates so that the x axis remains aligned with the magnetizing current vector is [21]:

$$\begin{bmatrix} i_{sx} \\ i_{sy} \end{bmatrix} = \begin{bmatrix} \cos(\delta_m) & -\sin(\delta_m) \\ \sin(\delta_m) & \cos(\delta_m) \end{bmatrix} \begin{bmatrix} i_{sD} \\ i_{sQ} \end{bmatrix} \tag{3.11}$$

or

$$\overline{i_{sm}} = \overline{i_s}\, e^{-j\delta_m} \tag{3.12}$$

The torque equation becomes:

$$t_e(t) = -\frac{3}{2}\frac{P}{2}\,\overline{\psi_{sm}}(t) \times \overline{i_{sy}}(t) \tag{3.13}$$

Alternatively, as the magnetizing current is aligned to the x axis:

$$t_e(t) = -\frac{3}{2}\frac{P}{2}L_m\,\overline{i_{sx}}(t) \times \overline{i_{sy}}(t) \tag{3.14}$$

The machine torque is a product of the magnetizing flux and a quadrature stator current. All rotor flux field-oriented controls or vector control schemes seek to independently control the magnetizing flux and the quadrature current that is responsible for torque production. A major implementation issue is the need to align the stator current vector and the rotor magnetizing. There are two major implementation methods – the direct and indirect methods. The direct method directly measures the air gap magnetizing flux to produce D and Q axis signals. The rotor flux can be calculated from:

$$\overline{\psi}_r(t) = \frac{L_r}{L_m}\,\overline{\psi}_m(t) - L_{ls}\,\overline{i}_s(t) \tag{3.15}$$

The rotor flux magnitude, ψ_m, is controlled to track a reference flux, ψ_{ref}, according to operating speed. The machine is operated at rated flux at lower speeds. Field weakening is applied at higher speeds as the stator voltage is normally limited. Many controller models require that rated flux which separates the low- and high-speed regions is set. The rated flux of a traction machine is:

$$\lambda_{rated} = \frac{V_{ll}}{\sqrt{3}\,2\pi f_{rated}}\,(\text{Wb RMS}) \tag{3.16}$$

where V_{ll} is a rated machine line to line voltage and f_{rated} is the rated frequency.

Figure 3.15 shows the direct method of implementing a rotor-oriented field control. The flux controller adjusts the rotor x axis current to maintain the required flux. The torque estimator uses the stator current measurements and the rotor flux to calculate the machine electromagnetic torque. The torque controller adjusts the machine torque by managing the y axis current. The stator x and y axes current references are converted into D,Q components via a rotation by the rotor flux angle, and then into a,b,c phase currents via the inverse Clarke transform. The inverter is current controlled and impresses the required currents upon the stator windings.

Direct field-oriented control requires air gap flux sensors. This is practically difficult as the airgap environment is hot, and at elevated potentials. Rotor flux estimation based on state estimators or observers is an attractive alternative. The magnitude and location of the rotor flux vector is calculated from the stator currents and rotor

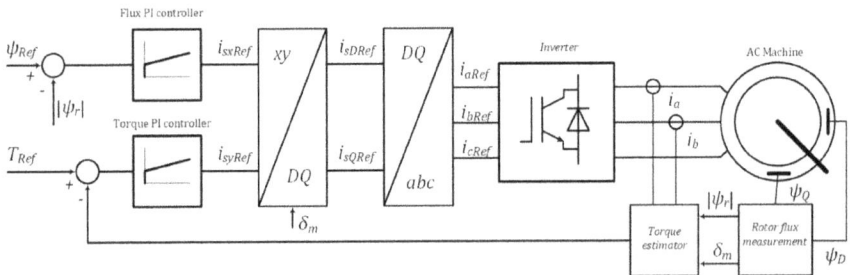

FIGURE 3.15   Direct rotor flux-oriented control.

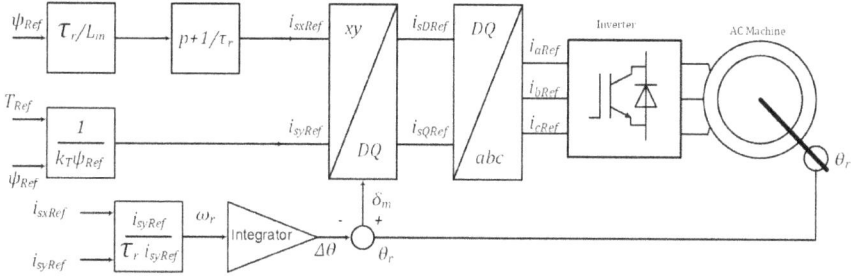

FIGURE 3.16 Indirect rotor flux field-oriented control.

speed as illustrated in Figure 3.16. The rotor flux angle can be calculated from the mechanical angle of the rotor and the integration of ω_r, the necessary rotor frequency for flux orientation:

$$\delta_m(t) = \theta_r(t) + \int_0^t \omega_r(t)\, dt \tag{3.17}$$

The rotor frequency can be found as follows. The stator frame rotor flux equation can be rewritten as:

$$\overline{i_r}'(t) = \frac{1}{L_r}\left(\overline{\psi_r}'(t) - L_m \overline{i_s}(t) \right) \tag{3.18}$$

Equation 3.6 contains an expression for the rotor voltage, substituting in Equation (3.18) yields:

$$L_m \overline{i_s}(t) = \overline{\psi_r}'(t) + \tau_r \left(p - j\omega_r \right)\overline{\psi_r}'(t) \tag{3.19}$$

where p is the derivative operator. The rotor flux axis, x, is aligned to the stator D axis. The real part gives:

$$L_m i_{sx}(t) = (1 + \tau_r p)\psi_{rx}'(t) \tag{3.20}$$

The imaginary part gives:

$$L_m i_{sy}(t) = -\tau_r \omega_r \psi_{rx}'(t) \tag{3.21}$$

$$L_m i_{sy}(t) = -\tau_r \omega_r L_m i_{rx}'(t) \tag{3.22}$$

$$\omega_r = -\frac{1}{\tau_r}\frac{i_{sy}(t)}{i_{rx}'(t)} \tag{3.23}$$

This term can now be integrated and added to the mechanical rotor angle to determine the rotor flux angle. The y axis current required to produce the necessary flux can be calculated from Equation (3.19). The torque producing current magnitude can be calculated from the machine torque constant:

$$i_{sx}(t) = \frac{T_{ref}}{k_T \Psi_{ref}} \qquad (3.24)$$

The computed angle depends on the rotor parameters, especially the rotor time constant. Some form of on-line adaption is normally present to track changes with rotor temperature.

The above models assume impressed machine currents. At powers in the megawatt range, the switching frequency of inverters is somewhat limited. Perfect control of the stator currents may not be completely possible. If so, the inverter must be treated as a voltage source. There are cross couplings between the D–Q axes that need to be compensated for in the inverter control. Modern drives are capable of producing torque step changes or full reversals equal to the drive rating in a few tens of milliseconds. However, such aggressive changes are undesirable and torque rate limiting will often be applied to reduce mechanical shocks.

3.3.2.3 Direct Torque Control

The DTC method [22] controls an induction machine by sequentially selecting a series of inverter output voltage phasors that cause the stator and rotor flux phasors to follow ideal trajectories in the DQ plane. As with all field-oriented controls, the torque production is maximized if these two fluxes are orthogonal.

A six-switch bridge (B6) inverter has three output terminals (a, b, c). Each terminal can connect to either the positive DC bus or the negative DC bus. The potential at the positive bus is designated as E, and that at the negative bus designated as 0. Table 3.2 shows the eight possible switch states for the B6 converter. Application of the Clarke Transform produces phasors that are represented diagrammatically in Figure 3.17.

TABLE 3.2

B6 Inverter Voltage Phasors

Switching State	V_a	V_b	V_c	V_D	V_Q
V_0	0	0	0	0	0
V_1	0	0	E	$-E/3$	$-E/\sqrt{3}$
V_2	0	E	0	$-E/3$	$E/\sqrt{3}$
V_3	0	E	E	$-2E/3$	0
V_4	E	0	0	$2E/3$	0
V_5	E	0	E	$E/3$	$-E/\sqrt{3}$
V_6	E	E	0	$E/3$	$E/\sqrt{3}$
V_7	E	E	E	0	0

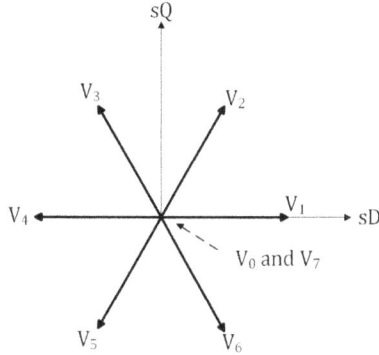

FIGURE 3.17 B6 inverter output voltage phasors.

At some instant, the stator and rotor fluxes in a machine may have relative positions as shown in Figure 3.18. The rotor and stator fluxes have a relative angle $\theta_{sr}(t)$. The stator flux vector, $\psi_s(t)$, neglecting the effect of the stator resistance, is the integration of the stator terminal voltages. The stator flux quickly responds to the short-term application of the stator voltage. The rotor flux vector, $\psi_r(t)$, responds more slowly to the stator voltage. The rotor and stator leakage inductances provide low pass filtering, and the rotor flux will change more slowly than the stator flux.

The change in the stator flux linkage is proportional to the volt-second product applied. A voltage vector v_l applied for Δt seconds results in a new stator flux $\psi_s(t+\Delta t)$. The stator flux vector magnitude has increased. This voltage vector application has increased the machine magnetization. However, the rotor-stator angle, the torque angle, has reduced. The six inverter voltage vectors each produce a different motion for the stator flux vector. An appropriate selection of the inverter states can force the stator flux vector to follow a circular trajectory. The magnetization flux is set by the trajectory radius. The torque can be adjusted by increasing or decreasing the stator frequency which controls the angle between the stator and rotor fluxes.

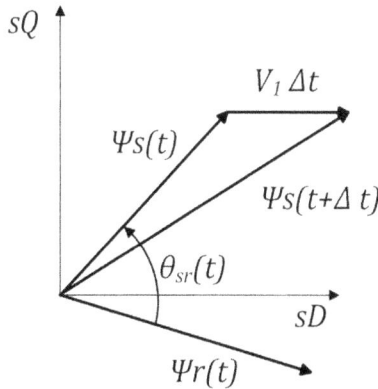

FIGURE 3.18 Influence of state V_1 upon the stator flux vector.

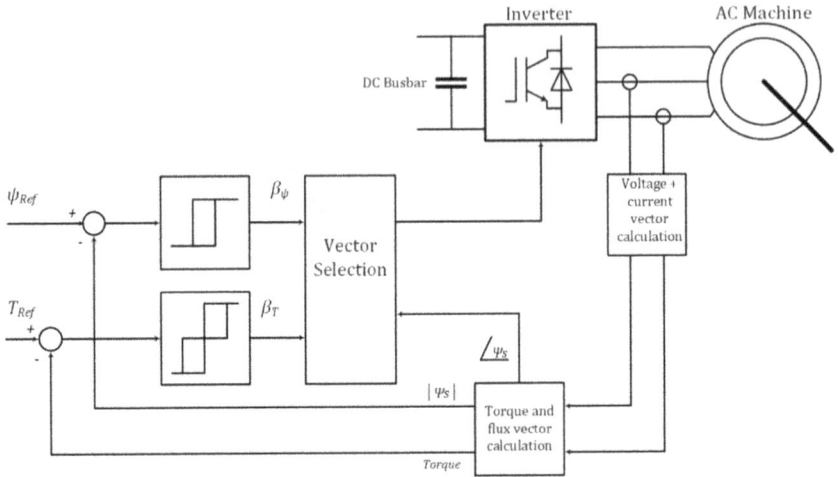

FIGURE 3.19 DTC drive.

The block diagram of a DTC controlled drive is shown in Figure 3.19. The stator voltage and current phasors, as well as electromagnetic torque and stator flux, are determined from measurement of the machine currents. The operating sector, as shown in Figure 3.20, is determined by the stator flux angle.

The electromagnetic torque is compared to the torque reference to produce a torque error. The decision variable, β_T, takes on values, +1, 0, or −1 if the torque is to be raised, held steady or reduced, respectively. The stator flux magnitude is compared to the flux reference. The flux error decision variable, β_ψ, takes on a Boolean value of +1 if the flux is to be raised or 0 if it is to be reduced. These decision variables, along with the angle of the stator flux, drive the vector selection lookup table that gives the best switching state choice for each sector and value of β_T and β_ψ.

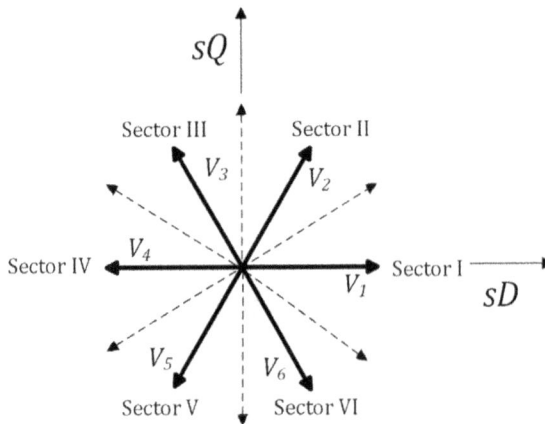

FIGURE 3.20 *DQ* plane sectors.

TABLE 3.3

Vector Selection for a DTC Drive (Counter-Clockwise Rotation) [22]

β_ψ	1			0		
β_T	1	0	−1	1	0	−1
Sector I	V_6	V_7	V_5	V_2	V_0	V_1
Sector II	V_2	V_0	V_4	V_3	V_7	V_5
Sector III	V_3	V_7	V_6	V_1	V_0	V_4
Sector IV	V_1	V_0	V_2	V_5	V_7	V_6
Sector V	V_5	V_7	V_3	V_4	V_0	V_2
Sector VI	V_4	V_0	V_1	V_6	V_7	V_3

The contents of the vector selection table, as presented in [22], are given in Table 3.3. The voltage sectors are shown in Figure 3.20. This is generically known as a *bang-bang* control. In each decision period, a specific vector is selected and applied. To achieve a smooth operation of the drive, the decision period should be short so that the discrete electrical states are held for periods that are much less than the machine time constants. One state decision can be made for each inverter switching period. In a modern inverter, switching can occur at a few kilohertz, so the decision period is typically a few hundred microseconds.

3.3.2.4 Case Studies

Induction machines have been the traction machine of choice in the past few decades. Table 3.4 gives several examples, which are drawn from references [7] and [23], of induction machines applied in modern heavy haul locomotives.

TABLE 3.4

Examples of Induction Machine Motors in Locomotives

Vehicle	Motors: Type & Rating	Vehicle	Motors: Type & Rating
German DB-Railion 189 electrical freight locomotive	4 × 1600 kW induction machines made by Siemens AG	German DB 152 electrical locomotive	4 × 1600 kW induction machines made by Siemens AG
Indian electric freight locomotive WAG-9	6 × 850 kW induction Motors, FRA6068, made by ABB	Chinese Railways DJ4 electrical locomotive	8 × 1200 kW induction machines made by Siemens AG and Zhuzhou Electric Locomotives Works
Swiss Railways SBB FLIRT RABe 521/523	4 × 500 kW Induction motors made by TSA Traktionssysteme	Swiss SBB Re 460 electrical locomotive	4 × 1525 kW induction machines made by ABB

TABLE 3.5
Traction Motor Parameters

Parameter	Value
Real power rating	500 kW
Line to line voltage	2027 Vrms
Number of phases	3
Base frequency	29 Hz
Number of poles	4
Stator resistance	132 mΩ
Stator reactance	3.14 mH
Rotor resistance	132 mΩ
Rotor reactance	3.14 mH
Magnetizing resistance	1240 Ω
Magnetizing reactance	117 mH

Table 3.5 gives the key parameters for a high-quality traction motor described in [24]. These machine parameters were derived from blocked rotor and no load tests. This machine achieves 500 kW at 1.83% slip at a shaft speed of 854 rpm and a mechanical torque of 5.59 kNm. The machine efficiency is 95.5% and the machine losses are 23.3 kW. In this condition the stator input power is 523 kW, apparent power is 585 kVA, the current is 167 A, and the power factor is 0.89. The motor torque peaks at 16.9 kN/m and a slip of 11.6%.

This machine is applied in a full locomotive simulation and some results are shown in Figure 3.21. The motor is torque controlled according to the locomotive tractive effort required. FOC is applied. In this simulation, the locomotive starts at

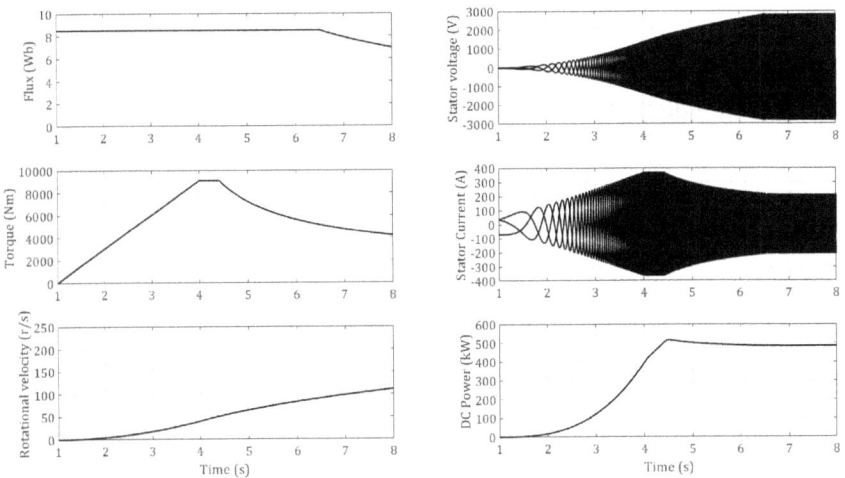

FIGURE 3.21 AC locomotive traction study.

zero velocity at $t = 1$ s and the maximum tractive effort is requested. The machine torque is restricted by a ramp rate limiter to increase linearly to 9 kNm at $t = 4$ s. The machine reaches its peak stator current at this time. Shortly after, at $t = 4.5$ s, the locomotive applies a power limit of 500 kW which can be seen in the DC Power trace which is the inverter bus bar power. Beyond this point the machine torque must fall as the locomotive accelerates in a constant power mode. As the locomotive accelerates, the stator voltage increases to maintain the machine rated flux. The limit is imposed by the DC bus bar voltage, 2650 Vdc in this case. The driving inverter uses state vector modulation (SVM) and the maximum line to line fundamental voltage is 2650 Vp or 1874 Vrms. The rated flux is 5.9 Wbrms or 8.4 Wbp. Field weakening starts at $t = 6.5$ s and at the rated machine frequency of 29 Hz or a rotational velocity of 91 r/s.

3.3.3 SYNCHRONOUS MACHINES

SMs appear in some rail applications including metropolitan passenger trains and some high-speed trains. The SM can use a PM rotor or a directly excited rotor. The SMs with PM rotor, PMSM machines, are brushless machines and offer higher electrical efficiency, power, and torque densities than induction machines. PMSM technology is becoming very popular for lighter vehicles such as electric cars and metropolitan passenger trains [25]. A recent PMSM developed by Alstom for high-speed passenger applications achieves 760 kW at 4500 rpm with a power density above 1 kW/kg of motor mass [23]. SMs of both the PM and wound rotor type have been applied in very fast trains [23].

PM machines have been prevalent at lower powers but there are examples of multi-megawatt motors applied in ship propulsion, especially in applications such as propulsion pods where volume is constrained [23]. For all the advantages of PMSMs, there are some operational factors to weigh. The directly excited SM has a clear advantage with respect to the ability to control the airgap flux. This allows the machine to have a broad field weakening range which increases the operating speed range. PMSM machines can be controlled to achieve field weakening but this is limited. Some magnets are damaged by quite moderate temperatures, especially if the machine is simultaneously expected to operate with field weakening. Rare earth materials are widely used in high-performance magnets. These have a supply chain that is vulnerable to disruption. This has produced a reluctance to use these magnets by some companies.

3.3.3.1 Machine Models

The SM has a stator which is similar to that found within an induction machine. As with the induction machines, the stator currents can be expressed as equivalent two-phase currents via the Clarke transform. The SM has a rotor that establishes the magnetizing flux ψ_m of the machine as shown in Figure 3.22. The rotor can have a wound field winding that is supplied with DC current, typically via slip rings. Alternatively, the rotor can carry PMs to establish the machine flux. The rotor magnets can be surface mounted on the rotor or embedded within the rotor. In this case the machine

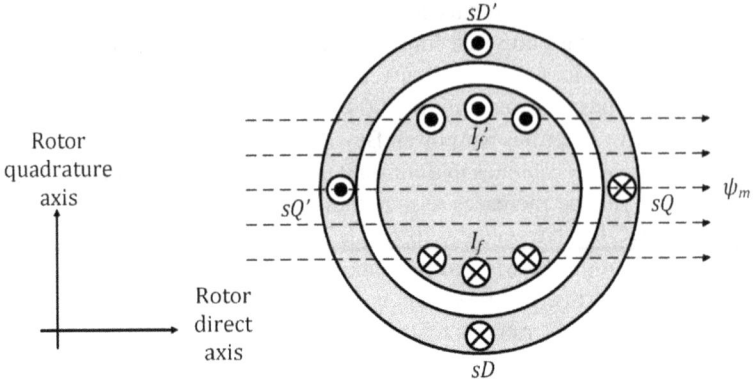

FIGURE 3.22   Two-phase SM.

is termed an interior PM machine (IPMSM). Interior magnet machines, often with liquid cooling, are emerging as a motor of choice in modern electric vehicles for powers to about 100 kW.

The magnetizing flux ψ_m is mechanically aligned to the rotor direct axis. The torque production in a SM depends on the magnitude of the magnetizing flux, the stator current vector, and the relative angle between the stator and rotor fluxes [21]:

$$t_e(t) = -\frac{3}{2}\frac{P}{2}\overline{\psi_m}(t) \times \left|\overline{i_s}(t)\right|\sin(\theta_s - \theta_r) \tag{3.25}$$

The stator voltage equations are:

$$\overline{u_s}(t) = R_s\overline{i_s}(t) + L_s\frac{d}{dt}\overline{i_s}(t) - j\omega\overline{\psi_m}(t) \tag{3.26}$$

In an inverter controlled machine, the torque production is optimized by keeping the stator and rotor fluxes orthogonal as shown in Figure 3.23. In the low speed region

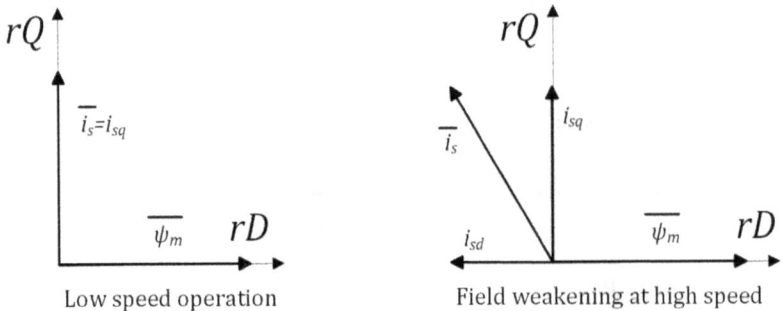

FIGURE 3.23   Steady state PMSM machine flux and current vectors.

this is readily achieved by aligning the stator current vector with the rotor quadrature axis. As the machine speed increases a point is reached where the stator voltage induced by the rotating rotor flux exceeds the available stator voltage. For a SM with a wound rotor, the back emf can be reduced by a reduction in the field winding current. For a PM machine, the rotor direct axis flux can be reduced by allowing the stator to carry an opposing current, aligned with the rotor direct axis, that causes field weakening.

The machine shown in Figure 3.22 has a cylindrical rotor. A machine with a smooth rotor does not have a preferred magnetic orientation and the torque production relies on the Lorentz force. Some SM rotors exhibit saliency, that is a significant difference between the direct and quadrature axis magnetizing inductance. This occurs in some large SMs, especially where the field poles are surface mounted. It can also occur in PM machines as the magnets themselves have a low relative permeability. Rotor saliency offers an opportunity to produce some additional torque that can be usefully exploited.

3.3.3.2 Machine-Commutated Converters

Load-commutated inverters (LCIs) are a class of inverters where induced stator winding voltages of a SM are used to commutate the inverter switches. As high power self-commutated inverters have become readily available, the range of potential applications has reduced. This approach still provides an important alternative to self-commutated inverters for high-speed drives in the high multi-megawatt range [26–27]. The LCI does have some important disadvantages related to starting and operation at low speeds where the machine generated commutation voltages are low.

LCIs have been applied at least once in a railway application. The Korean KTX high speed passenger train was equipped with SMs and LCIs in 2004 [23]. In a subsequent refurbishment these were replaced with induction machines with IGBTs self-commutated inverters.

3.3.3.3 Field-Oriented Control

Figure 3.24 shows a rotor-oriented torque control scheme in polar coordinates. Here the stator current vector is controlled in magnitude and angle terms. The rotor angle

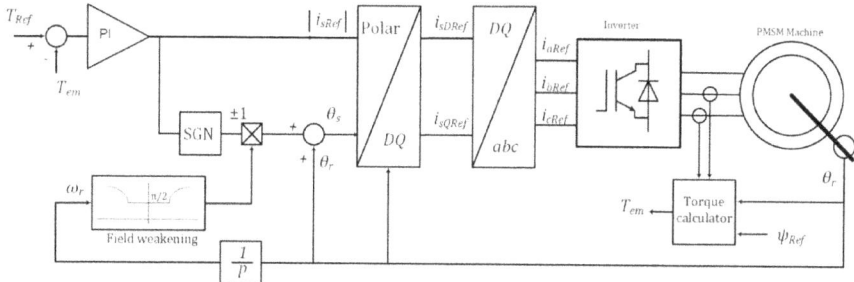

FIGURE 3.24 Rotor-oriented controlled PMSM drive in polar coordinates.

TABLE 3.6

Examples of SM Motors in Rail Vehicles and Locomotives

Vehicle	Motors: Type & Rating	Vehicle	Motors: Type & Rating
French SCNF REGOLIS EMUs REGIO 2N	Alstom PMSM, 12 pole 4053 rpm, 450 kW, 6 traction motors per carriage	Type 02 and Type 1000 Electric Train, Tokyo Metro	Toshiba PMSM
Korean SMRT C151	Toshiba PMSM 145kW, 8 traction motors per car	JR Kyushu 305 Series EMU	Toshiba PMSM
Korean KTX high-speed train, operational speed 300 km/h, in operation since 2004	12 × 1130 kW self-commutated SMs (later replaced by induction machines)	Alstom AGV very high speed train	760 kW enclosed self-ventilated, PM, Alstom 4500 rpm

is measured to provide the location of the rotor direct axis. In the low-speed region, the stator current is aligned with the rotor quadrature axis by maintaining the stator current angle $\pi/2$ radians in advance of the rotor angle. At high speeds, the stator current vector angle is advanced beyond $\pi/2$ radians to provide field weakening.

3.3.3.4   Case Studies

Table 3.6 lists a number of SM applications in trains drawn from references [23, 25, 27]. The Korean KTX is perhaps the only commercial LCI application. Given the availability of extremely capable megawatt-scale self-commutated inverter systems, it is unlikely that LCI technology will find many future rail applications.

PMSM drives are now appearing in some high speed trains and many electrical multiple unit (EMU) commuter trains. Part of the strong and growing interest in PMSM technology has been driven by energy efficiency and superior volumetric power density. The SCNF REGOLIS passenger trains feature machines with peak efficiencies of 96% [25]. In general terms, a PMSM machine will achieve an efficiency of 2–3% higher than a comparable wound rotor SM.

Toshiba [27] has undertaken comparative energy consumption studies for metropolitan trains, where some trains from an existing fleet have been retrofitted with PMSM traction systems. Clear reductions in energy consumption are achieved.

3.3.4   Brushless DC

The BLDC motor is an electronically commutated motor (ECM) that has a PM rotor and a stator winding that is driven by an inverter. They share many features with PM SMs to the extent that the dividing line is increasingly unclear. An accepted point of difference is that the BLDC has a trapezoidal back emf waveform, as shown in

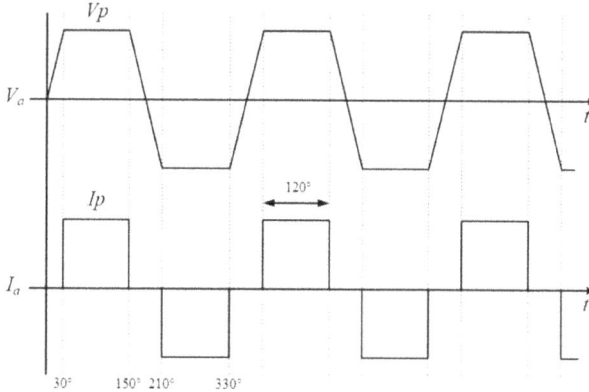

FIGURE 3.25 BLDC current and voltage waveforms.

Figure 3.25, whereas the PMSM has a sinusoidal back emf [28]. The BLDC was developed from DC motors through three key steps:

- The DC motor field is produced by PMs
- The mechanical arrangement is reversed – the motor field now rotates and the armature is stationary
- An electronic commutation system that sequentially directed the DC currents into the stationary armature windings

Most often the stationary armature has three windings that are connected in a wye arrangement. The electronic commutation system is a bridge inverter with three legs that operates in a six-step mode. Each armature coil carries 120° current pulses. The coils to be energized are selected depending upon the position of the rotor so that torque production is maximized. During this active period, the phase current if often controlled by a hysteresis based current regulator. If the back emf is constant during this active period, the phase power as well as the machine power and torque are constant. This ideally eliminates fluctuations in torque during the operating cycle. These fluctuations are termed *cogging torques* and are an undesirable feature in any drive system.

The BLDC and the PMSM both share a rotating PM rotor, a stationary armature or stator, and a driving inverter. The selection of the active armature coils in a BLDC is determined by the rotor position and this is effectively a form of rotor-oriented field control. Clearly the machines share deep similarities. Both are the subject of ongoing research and development activity that further blurs the distinctions. BLDC machines are frequently operated with sinusoidal currents and both machines routinely appear with phase numbers other than three.

BLDC machines that meet the criteria of having trapezoidal back emfs are typically found in applications of up to a few tens of kilowatts. Above this the PMSM offers benefits especially in regard to the control and minimization of cogging torques.

3.3.4.1 Machine Models

The BDLC stator currents and voltages are described by Equation (3.27):

$$\begin{bmatrix} v_{an} \\ v_{bn} \\ v_{cn} \end{bmatrix} = \begin{bmatrix} R & 0 & 0 \\ 0 & R & 0 \\ 0 & 0 & R \end{bmatrix} \begin{bmatrix} i_a \\ i_b \\ i_c \end{bmatrix} + \begin{bmatrix} L-M & 0 & 0 \\ 0 & L-M & 0 \\ 0 & 0 & L-M \end{bmatrix} p \begin{bmatrix} i_a \\ i_b \\ i_c \end{bmatrix} + \begin{bmatrix} e_{an} \\ e_{bn} \\ e_{cn} \end{bmatrix} \quad (3.27)$$

where v_{an}, v_{bn}, v_{cn} are the applied stator phase voltages, e_{an}, e_{bn}, e_{cn} are the stator back emf voltages, R is the stator resistance, L is the stator winding self-inductance, M is the stator winding mutual inductance, and p is the derivative operator.

The electromagnetic torque can be derived from the equivalence of the electrical and mechanical power as:

$$t_e(t) = \left(e_{an}i_a + e_{bn}i_b + e_{cn}i_c \right)/\omega_r \quad (3.28)$$

The machine torque production for both the PMSM and the BDLC are also determined by the magnetizing flux, the torque producing components of the stator flux, and the relative angle as described in Equation (3.25). The vector diagrams shown in Figure 3.23 also apply.

3.3.4.2 Field-Oriented Control

A rotor-oriented control scheme is shown in Figure 3.26. The rotor position determines which armature coils are excited. The armature current magnitude is used to control the machine torque. The armature currents are controlled by current feedback and a hysteresis switching method. For the purposes of torque feedback control, the electromagnetic torque is estimated from the power delivered to the machine armature by the inverter. As the inverter is a nearly lossless converter, the machine power can be estimated from the power at the inverter DC bus bar. For high-speed operation, field weakening can be achieved by advancing the stator current angle past $\pi/2$ radians.

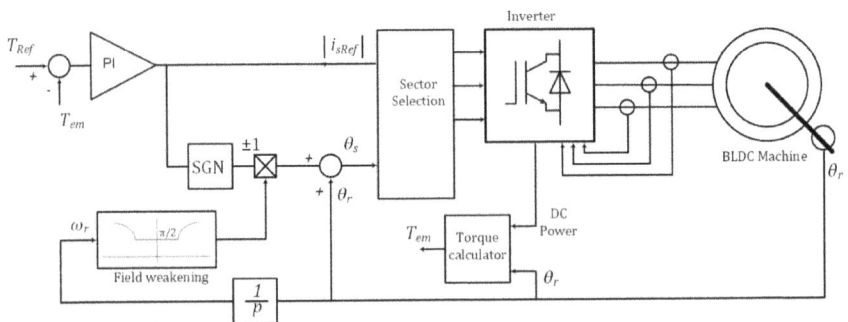

FIGURE 3.26 BLDC with rotor-oriented control.

3.3.4.3 Case Studies

BLDC machines, defined as those with trapezoidal back emf voltages, do not significantly feature in railway traction applications. If a permanent machine with a rating of tens of kilowatts or more is required, the PMSM is often a better technical choice. As noted above, the dividing line between the technologies is quite indistinct.

There are developments that may exploit saliency properties of the BLDC. Reluctance motors (RMs) feature a rotor with a high degree of saliency and rely on the preferential alignment of the rotor to produce torque. A BLDC motor can be combined with a RM to produce a machine without magnets. The rotor is extremely robust and capable of operation at high temperatures and speeds. Tutelea et al., presents the FEM analysis of a 1.2 MW BLDC, multiphase reluctance machine for railway traction [29]. It is important to note that this chapter describes a simulated rather than a physical machine.

3.3.5 SLIP CONTROL

Modern locomotives and trains normally operate with torque controlled drives that seek to maximize both the tractive and dynamic braking efforts. Rail vehicles are commonly equipped with one of four traction control schemes [30–32]:

- **A whole of vehicle traction control system:** where the same torque value is given to all the traction machines of a rail traction vehicle;
- **Bogie traction control:** where vehicles have one inverter per bogie (also called group traction control) and it is appropriate to set a torque value for that group alone (see Figure 3.27);
- **Individual wheelset traction control:** where each wheelset has its own inverter (also called axle traction control);
- **Independently rotating wheel traction control:** where the torque value is adjusted for each wheel's traction motor.

Traction controls schemes are commonly implemented using creep or wheel slip control principles. The terms *slip* and *creep* are often used interchangeably. For the estimation of wheel slip or creep, a good estimate of the vehicle speed is essential.

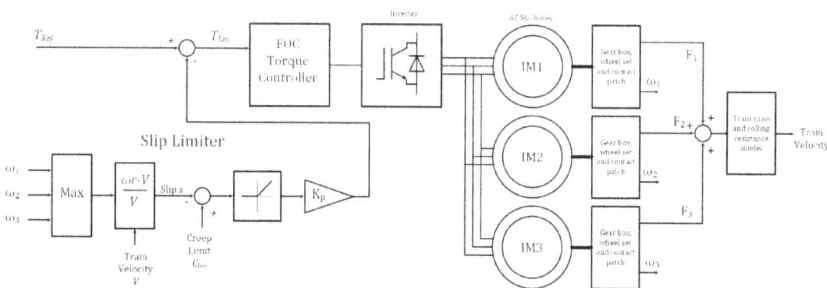

FIGURE 3.27 Bogie level slip control system.

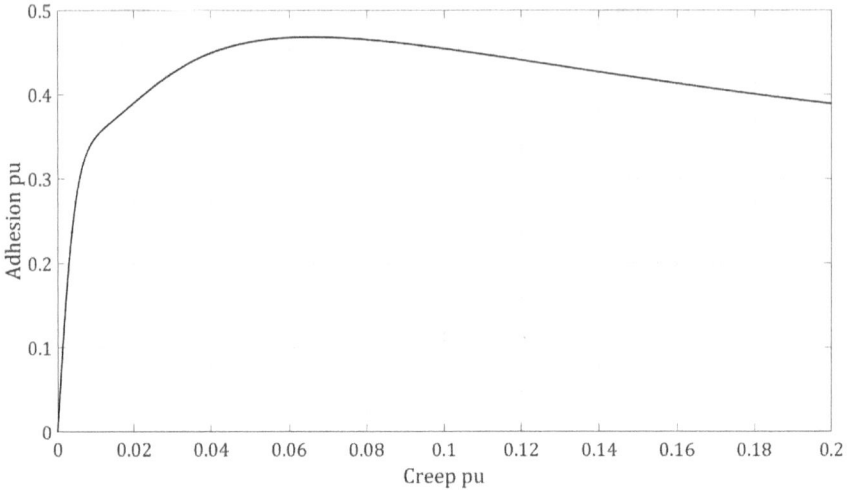

FIGURE 3.28 Adhesion versus creep.

The ground speed can be measured directly by devices such as Doppler radar speed sensors. If a vehicle has multiple axles, the lowest rotational speed may be used as an indication of the vehicle speed under traction. During braking, the highest rotational speed is used. The slip or creep is estimated based on the following relation:

$$s = \frac{\omega r - V}{V} \tag{3.29}$$

where ω is the wheelset angular velocity; V is the rail vehicle velocity; and r is the rolling radius of the wheels. Modern traction control systems can use conventional or extended slip control techniques. The basic traction control principle for rail traction vehicles is that the drivetrain torque should be quickly adjusted so that the driven wheel or wheelset remains in the stable wheel slip zone, i.e., to the left the peak tractive effort as shown in Figure 3.28.

The conventional adhesion/traction control systems usually operate to limit wheel slip to below a selected threshold. If the slip limit is approached, the torque produced by the traction machine is reduced. Often a high gain proportional controller is applied. Extended slip control techniques attempt to operate at the peak of the adhesion-slip curve. This is very important for poor adhesion conditions. Various algorithms can be used to detect the initiation of the wheel slip. For example, the initiation of wheel slip is often marked by the sudden acceleration of the wheelset as adhesion is lost. The detection of excessive acceleration can be used to temporarily reduce the traction machine torque which can subsequently be re-applied once the slip condition is arrested.

3.3.5.1   Case Studies

Figure 3.27 shows a bogie level slip control system. In this case the bogie has three traction machines supplied from a common inverter. For this configuration the

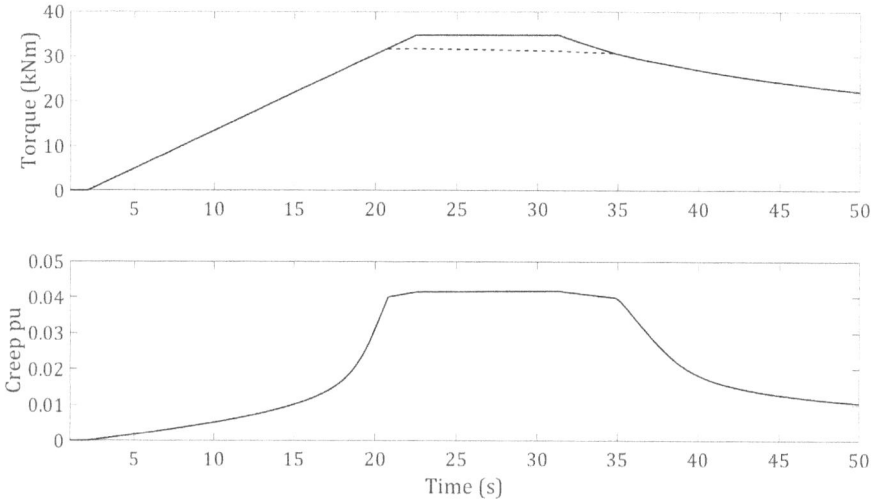

FIGURE 3.29 Bogie toque commands and slip.

wheelsets should have identical rolling radii to assist in torque sharing between the induction machines. An advantage of the arrangement is that it is unlikely that just one machine can experience extreme levels of slip. A relative wheel slip event will quickly reduce the induction machine torque relative to its peers. Wheel slip is more likely to be a group behavior.

This slip limiter has access to the train velocity, V, via an instrument independent of the wheel sets. The vehicle velocity is used to determine the highest value of wheel slip amongst the bogie wheelsets. If the slip exceeds a threshold value, the slip limit C_{lim}, a proportional control with gain K_p, becomes active. This reduces the torque reference from T_{ref} to a slip limited torque signal Tl_{im}.

In this case study, the traction machines are as described in Table 3.5. The wheelsets have an axle load of 220 kN and the adhesion characteristics are as shown in Figure 3.28. the locomotive is initially at idle with a train velocity of 1 m/s. At $t = 2$ s, the full tractive effort is requested and the bogie torque command ramps linearly to 35 kN/m at $t = 22$ s. In this case the creep limit has been set to 0.04. The slip limited becomes active at approximately 21 s. In Figure 3.29, the top traces are the torque traces T_{ref} and Tl_{im}. The limited torque signal is shown as a dashed trace. The locomotive is slip limited until $t = 35$ s where a traction power limit becomes active and independently reduces the traction demand signal.

REFERENCES

1. R. J. Hill, Electric railway traction. Part 3. Traction power supplies, Power Engineering Journal, 8(6), 275–286, 1994.
2. Hybrid Locomotives, https://www.toshiba.co.jp/sis/railwaysystem/en/products/locomotive/hybrid.htm.

3. Toshiba Locomotives, https://www.toshiba.co.jp/sis/railwaysystem/en/products/catalog/pdf/c6ToshibaLocomotive.pdf.
4. R. Cousineau, Development of a hybrid switcher locomotive the Railpower Green Goat, IEEE Instrumentation & Measurement Magazine, 9(1), 25–29, 2006.
5. J-TREC sustina Hybrid, https://www.j-trec.co.jp/eng/sustina/060/index.html
6. C. R. Akli, X. Roboam, B. Sareni, A. Jeunesse, Energy Management and Sizing of a Hybrid Locomotive, 2007 European Conference on Power Electronics and Applications, IEEE, 1–10,, 2007, September.
7. A. R. Miller, K. S. Hess, D. L. Barnes, T. L. Erickson, System design of a large fuel cell hybrid locomotive, Journal of Power Sources, 173(2), 935–942, 2007.
8. M. Meinert, P. Prenleloup, S. Schmid, R. Palacin, Energy storage technologies and hybrid architectures for specific diesel-driven rail duty cycles: Design and system integration aspects, Applied energy, 157, 619–629, 2015.
9. A. Jaafar, C. R. Akli, B. Sareni, X. Roboam, A. Jeunesse, Sizing and energy management of a hybrid locomotive based on flywheel and accumulators, IEEE Transactions on Vehicular Technology, 58(8), 3947–3958, 2009.
10. M. Spiryagin, P. Wolfs, F. Szanto, Y. Q. Sun, C. Cole, D. Nielsen, Application of flywheel energy storage for heavy haul locomotives, Applied energy, 157, 607–618, 2015.
11. R. F. Thelen, J. D. Herbst, M. T. Caprio, A 2 MW Flywheel for Hybrid Locomotive Power. 2003 IEEE 58th Vehicular Technology Conference. VTC 2003-Fall (IEEE Cat. No. 03CH37484), IEEE, 5(October), 3231–3235, 2003.
12. A. Emadi, Y. J. Lee, K. Rajashekara, Power electronics and motor drives in electric, hybrid electric, and plug-in hybrid electric vehicles, IEEE Transactions on Industrial Electronics, 55(6), 2237–2245, 2008.
13. http://new.abb.com/motors-generators/traction-motors-and-generators/traction-generators.
14. M. Spiryagin, Q. Wu, P. Wolfs, Y. Sun, C. Cole, Comparison of locomotive energy storage systems for heavy-haul operation, International Journal of Rail Transportation, 6(1), 1–15, 2018.
15. M. Brady, Assessment of Battery Technology for Rail Propulsion Application (No. DOT/FRA/ORD-17/12), Federal Railroad Administration, United States, 2017.
16. S. Ahmed, I. Bloom, A. N. Jansen, T. Tanim, E. J. Dufek, A. Pesaran, A. Burnham, R. B. Carlson, F. Dias, K. Hardy, M. Keyser, Enabling fast charging–A battery technology gap assessment, Journal of Power Sources, 367, 250–262, 2017.
17. S. Wen, Analysis of maximum radial stress location of composite energy storage flywheel rotor, Archive of Applied Mechanics, 84(7), 1007–1013, 2014.
18. R. F. Post, T. K. Fowler, S. F. Post, A high-efficiency electrochemical battery, Proceedings of the IEEE, 81(3), 462–474, 1993.
19. Z. Hong, Y. Han, Q. Li, W. Chen, October, Design of Energy Management System for Fuel Cell/Supercapacitor Hybrid Locomotive, 2016 IEEE Vehicle Power and Propulsion Conference (VPPC), 1–5, IEEE, 2016.
20. W. W. Hay, Railroad Engineering, 1, John Wiley & Sons, New York, 1982.
21. P. Vas, Vector Control of AC Machines, Oxford Science Publications, Oxford, 1990.
22. M. Depenbrock, JuneDirect Self-Control (DSC) of Inverter Fed Induktion Machine, 1987 IEEE Power Electronics Specialists Conference, 632–641, IEEE, 1987.
23. A. M. El-Refaie, Motors/generators for traction/propulsion applications: A review, IEEE Vehicular Technology Magazine, 8(1), 90–99, March 2013, DOI: 10.1109/MVT.2012.2218438.
24. M. Spiryagin, P. Wolfs, C. Cole, V. Spiryagin, Y. Q. Sun, T. McSweeney, Design and Simulation of Heavy Haul Locomotives and Trains, CRC Press, Taylor & Francis Group, Boca Raton, FL, 2016, 2017.

25. C. André-Philippe, M. Pascal, B. Odile, NovemberThe Permanent Magnet Synchronous Motor From a Customer Point of View: REGIO 2N vs REGIOLIS, 2016 International Conference on Electrical Systems for Aircraft, Railway, Ship Propulsion and Road Vehicles & International Transportation Electrification Conference (ESARS-ITEC), 1–4, IEEE, 2016.

26. A. Tessarolo, C. Bassi, G. Ferrari, D. Giulivo, R. Macuglia, R. Menis, Investigation into the high-frequency limits and performance of load commutated inverters for high-speed synchronous motor drives, IEEE Transactions on Industrial Electronics, 60(6), 2147–2157, June 2013, DOI: 10.1109/TIE.2012.2192897.

27. Permanent Magnet Synchronous Motor (PMSM) https://asia.toshiba.com/permanent-magnet-synchronous-motor/.

28. P. Pillay, R. Krishnan, Application characteristics of permanent magnet synchronous and brushless DC motors for servo drives, IEEE Transactions on Industry Applications, 27(5), 986–996, Sept–Oct. 1991, DOI: 10.1109/28.90357.

29. L. N. Tutelea, I. Torac, G. Giovinco, F. Marignetti, I. Boldea, 1.2 MW, 2-6Krpm BLDC-MRM Traction Drive: Preliminary Design With Key FEM Inquiries, 2018 XIII International Conference on Electrical Machines (ICEM), Alexandroupoli, 2018, 1989–1995, DOI: 10.1109/ICELMACH.2018.8507239.

30. L. Liudvinavičius, L. P. Lingaitis, G. Bureika, Investigation on wheel-sets slip and slide control problems of locomotives with AC traction motors, Maintenance and Reliability, 4, 21–28, 2011.

31. M. Spiryagin, P. Wolfs, F. Szanto, C. Cole, Simplified and advanced modelling of traction control systems of heavy-haul locomotives, Vehicle System Dynamics, 53(5), 672–691, 2015.

32. M. Spiryagin, P. Wolfs, C. Cole, S. Stichel, M. Berg, M. Plöch, Influence of AC system design on the realisation of tractive efforts by high adhesion locomotives, Vehicle System Dynamics, 55(8), 1241–1264, 2017.

4 Control Systems

This chapter is focused on the application of theoretical aspects of control applicable in mechatronic system designs used in rail vehicles. The application of the two common control systems of open-loop and closed-loop control systems for mechatronic rail vehicle designs is discussed. The background and fundamentals are provided as high-level guidance for classical and modern control approaches applicable in rail vehicle mechatronic systems. Details of the major areas of control method applications in the railway field are also addressed in this chapter through the provision of an extensive list of relevant references.

4.1 INTRODUCTION

A mechatronic system is required for the application of many powerful control methods that are supported by control engineering to provide robust solutions considering the great number of uncertainties presented in rail vehicle/track systems. Generally, the control of mechatronic systems for rail vehicles is divided into four categories:

- Control of traction and braking systems
- Control of vehicle dynamics
- Control of train dynamics
- Control of other auxiliary equipment and systems

It is necessary to mention that every controlled mechanical system of a rail vehicle is a mechatronic system because the design of a mechatronic system by definition is commonly performed based on the specific design procedure that requires not only optimization of the controller but also the optimization of the whole system including sensors and actuators. However, the implementation of appropriate control principles is considered as one of key design elements of controlled mechatronic systems. Such systems should contain a controller that is built based on various control functions. When these functions are performed inside of a controller, the system (also called a plant or a process) achieves the desired outcomes. The control systems used in rail vehicles are commonly designed as closed-loop control systems. However, open-loop control systems are still in use or might be used if required in a specific rail vehicle design. In the next section the difference between these two control systems will be explained, and examples will be shown on how to use such systems for the design of a mechatronic system of a train or a rail vehicle.

4.2 OPEN-LOOP AND CLOSED-LOOP CONTROL SYSTEMS

An open-loop control system is a system where the system output has no influence on the control action provided to the system through the input reference signal. In this case, the open control system has no knowledge about the output condition and such a

DOI: 10.1201/9781003028994-4

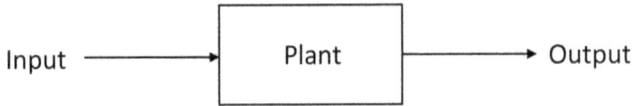

FIGURE 4.1 Generalized concept of an open-loop control system.

system has no possibility to self-control any errors and to adjust a set input value. The generalized concept of an open-loop control system is shown in Figure 4.1. A good example of an open-loop control system is a manual train driving control process. For example, in a train traction mode, the input is defined through notch position and the input signal is interpreted as the power to be produced by the power plant which is acting on the processes defined as the train/track interaction system. However, the output (e.g., a variation of power applied at the rail resulting in a change of the speed of the train) is not self-correcting and this leads to large deviations in the output value because the open-loop system cannot reduce or overcome train power operational variations and disturbances coming from the external loads, factors, and conditions.

If the system uses the output of the control system to adjust the input signal, then such a feature is called feedback of a closed-loop control system. It also forms the definition of a closed-loop control system that can be stated as: the closed-loop control system is the type of a control system as shown in Figure 4.2 that allows to achieve the expected results by means of observing and comparing the output with the input and then taking appropriate control action by adjusting the input signal. This type of control system is also called a feedback system. The general block diagram of a single-loop feedback system is shown in Figure 4.3. The accuracy of the output is

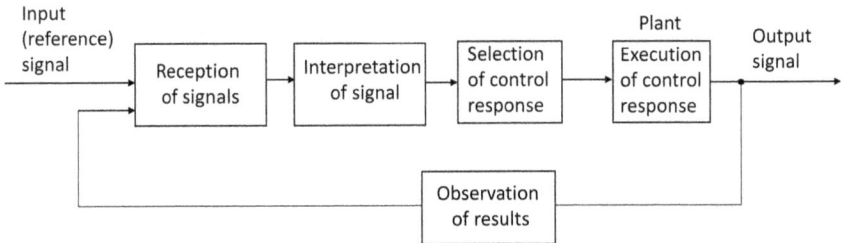

FIGURE 4.2 Generalized concept of a closed-loop control system.

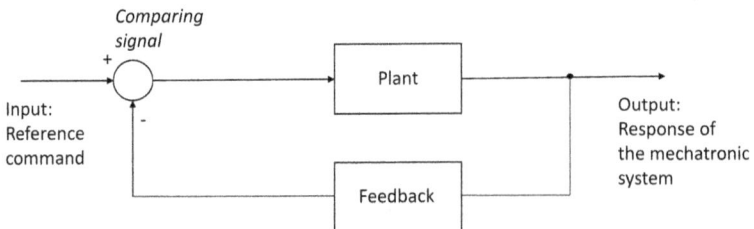

FIGURE 4.3 General block diagram of a single-loop feedback system.

strongly reliant on the feedback which in general is dependent on the control methods embedded in the system design. A good example of a closed-loop control system is an automatic train driving control where all output parameters (e.g., train speed, etc.) are controlled by a system based on the reference signal (e.g., measured from a speed sensor, etc.) produces a new input signal to increase the accuracy of the output.

Considering both control system concepts and their outputs in terms of achieving the desired results, it makes clear why the closed-loop control systems have found wide application in the field of rail vehicle mechatronics.

4.3 CLASSICAL CONTROL

Considering their systems engineering contents, the general design requirements of control loops in a mechatronic system should be formulated as [1]:

- Robust stability of the controller under parameter variations and uncertainties
- Rejection of undesirable inputs caused by unexpected parameter variations and uncertainties
- Control command following relative to the reference command given to the system

In order to achieve these requirements, it common to use a classical control method that avoids high-order differential equations to represent a physical system (plant) by means of the usage of the Laplace transform that allows replacing a plant presented in the form of ordinary-differential equation in the time domain with a plant presented in the form of a regular algebraic polynomial in the frequency domain. Such a manipulation allows faster and easier processing with less computational effort applied to solve a control problem. In most applications, the transfer function and proportional integral derivative (PID) feedback control described in the next two sections can be used to satisfy the generalized design requirements mentioned above.

4.3.1 CLOSED-LOOP TRANSFER FUNCTION

A standard feedback control architecture is shown in Figure 4.4 and it represents a single-input-single-output (SISO) control system. The application of multi-input-multi-output (MIMO) systems that have more than one input/output is also common

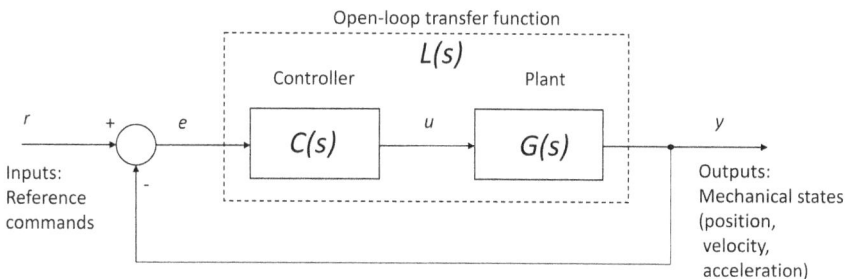

FIGURE 4.4 A standard control loop for a SISO control system.

but, in order to understand principles for the design of closed-loop transfer functions, it is always better to consider a simplistic system as shown in Figure 4.4. The input considered is the reference input $r(t)$ and output considered is $y(t)$. Assuming that Controller and Plant are linear and invariant, the Laplace transform for variables is used for the analysis of the whole system. It means that the Controller can be presented with transfer function $C(s)$ and the Plant with transfer function $G(s)$. The design task for the configuration shown in Figure 4.4 is to find suitable structures and parameters for the dynamic transfer function $C(s)$.

All system properties important for the design of the closed-loop transfer function can be written as:

- Open-loop transfer function

$$L(s) := C(s)G(s) \tag{4.1}$$

- Sensitivity function

$$S(s) := \frac{1}{1+L(s)} \tag{4.2}$$

- Complementary sensitivity function

$$T(s) := \frac{L(s)}{1+L(s)} \tag{4.3}$$

Closed-loop transfer function, $T_{cl}(s)$, as a relation between input and output can then be delivered as:

$$Y(s) := C(s)G(s)E(s) \tag{4.4}$$

$$E(s) := R(s) - Y(s) \tag{4.5}$$

$$Y(s) := C(s)G(s)(R(s) - Y(s)) \tag{4.6}$$

$$Y(s)(1 + C(s)G(s)) := C(s)G(s)R(s) \tag{4.7}$$

$$Y(s) := R(s)\frac{C(s)G(s)}{1+C(s)G(s)} \tag{4.8}$$

$$T_{cl}(s) := \frac{Y(s)}{R(s)} = \frac{C(s)G(s)}{1+C(s)G(s)} \tag{4.9}$$

The closed-loop transfer function described by Equation (4.9) presents a fundamental idea for the design of the system shown in Figure 4.3, which can be described as the feedback control for this system that should address parameter uncertainties that might exist in the plant. In case of a complex system, it is necessary to eliminate intermediate components that represents subsystems in order to not to complicate the system. Such a design approach will lead to easy delivery of a relation between input and output of the whole complex system.

More description on how to deliver plant transfer functions can be found in [2–5] and railway specific examples in the design of a rail vehicle active suspension system can be found in [6, 7].

4.3.2 PID Feedback Control

There is a common opinion that the first real PID-type controller was developed by Elmer Sperry in 1911. However, a theoretical PID control law that describes theoretical analysis for three control actions was presented in 1922 by Nicolas Minorsky. An actual boom of PID control application started in pneumatic controllers from 1932. Its development has subsequently been progressed including in industrial and computerized control [8].

The three types of control actions used in PID control are a proportional action, an integral action, and derivative action.

The proportional control (P) action represents a proportional gain, K_p, to the control error and, as shown in Figure 4.4, the control action of the system can be expressed in the time domain as:

$$u(t) = K_p e(t) = K_p \big(r(t) - y(t) \big) \tag{4.10}$$

The transfer function for a proportional controller can then be defined as:

$$C(s) = K_p \tag{4.11}$$

The integral control (I) is represented proportionally to the integral of the control error and the control action of the system can be expressed, considering an integral gain, K_i, in the time domain, as:

$$u(t) = K_i \int_0^t e(t) dt \tag{4.12}$$

The transfer function for an integral controller can then be defined as:

$$C(s) = \frac{K_i}{s} \tag{4.13}$$

The derivative action (D) represented by a derivative gain, K_d, is based on the past values of the control error. The ideal derivative control action in the time domain can be written as:

$$u(t) = K_d \frac{de(t)}{dt} \tag{4.14}$$

The transfer function for an integral controller can then be defined as:

$$C(s) = K_d s \qquad (4.15)$$

The output of a PID controller, being equal to the combination of all three control actions in the time domain, can be written as follows:

$$u(t) = K_p e(t) + K_i \int_0^t e(t) dt + K_d \frac{de(t)}{dt} \qquad (4.16)$$

The transfer function of a PID controller can be defined as:

$$C(s) = K_p + \frac{K_i}{s} + K_d s = \frac{K_d s^2 + K_p s + K_i}{s} \qquad (4.17)$$

The first part of Equation (4.11) represents the series form of a PID controller, while the second part is the parallel form which is also widely used for the implementation of a PID controller. The choice of controller type based on control actions (P, PI, PD, PID) is dependent on the areas of applications and the decision should be made considering the process dynamics. More description on how to deliver control actions can be found in [2–5, 8] and railway specific examples in the PID controller design for rail vehicle active suspension, braking, and traction systems can be found in [9–18].

4.4   MODERN CONTROL APPROACH

Unlike the classical control approach that is based upon converting a system's differential equations into a transfer function, the modern control approach (also called state-space approach) is based on the usage of linear, time-invariant differential equations. The modern approach allows to avoid the major disadvantage of the classical approach, namely its applicability only for linear, time-invariant systems or systems that can be approximated. In other words, the modern control approach can be used for the non-linear systems with non-zero initial conditions, backlash, saturation, and zones [3]. The time-varying system is a system where one or more parameters of the system vary as a function of time. As a result, the system is represented in a time-domain approach that allows easy modeling of the control systems in computer simulations. Some basic principles of state-space techniques are covered in the following sections.

4.4.1   STATE SPACE REPRESENTATION

The time-domain analysis and the design of a mechatronic system very often uses the techniques which are called the state of a system. The term *the state of a system* can be defined as a set of variables which, in cooperation with the input functions, is used to define and describe the dynamic behavior of the system and to provide the future state and output(s). For the dynamic system, the state of a system is defined by means of a set of state variables, also called the state variables, that describes the future response of the system based on the present state, the inputs, and the equations that describe the dynamics. A set of variables should be linear, should be linearly independent and cannot be written as a linear combination. In other words, the state variables are variables whose values change over time in a way that depends on the values they have at any given time and on the given values of excitation inputs. In this case, the values of output variables are strongly dependent on the values of the state variables. The state variables should not be mixed with the system variables. In order to avoid any misinterpretation, the following definitions should be used [3]:

- **System variable:** Any variable that responds to an input or initial condition in a system.
- **State variables:** The smallest set of state variables.
- **State vector:** A vector formed with state variables as its elements.
- **State space:** The n-dimensional space whose axes are the state variables.
- **State equations:** A set of n simultaneous, first-order differential equations with n variables, where the n variables to be solved are the state variables.
- **Output equations:** The algebraic equation that express the output variables of a system as a linear combination of the state variables.

For example, a state-space representation of a linear system with p inputs, q outputs, and n state variables is represented by the following state and output equations, respectively:

$$\dot{x} = Ax + Bu \tag{4.18}$$

$$y = Cx + Du \tag{4.19}$$

where x is the state vector, $\dot{x}$ is the derivative of the state vector with the respect to time, y is the output vector, u *is* the input (or control) vector, A is the state (or system) matrix with a $n \times n$ dimension, B is the input coupling matrix with a $n \times p$ dimension, C is the output matrix with a $q \times n$ dimension, and D is the feedthrough (or feedforward) matrix with a $q \times p$ dimension.

More description regarding the state-space representation can be found in [2–4] and railway specific examples for applying the state-space representation for rail vehicle mechatronic systems and their control can be found in [19–24].

4.4.2 POLE PLACEMENT METHOD

The pole placement method, also known as the full-state feedback, is a method used in the feedback controller design to place the closed-loop poles of a plant in predetermined locations in a complex plane that dictates a stability of the system [3]. For understanding the basics of this method, we need to consider an open-loop system that can report every state in the state vector, x, which is used as a feedback signal rather than use an output value y as shown in Figure 4.5.

Plant

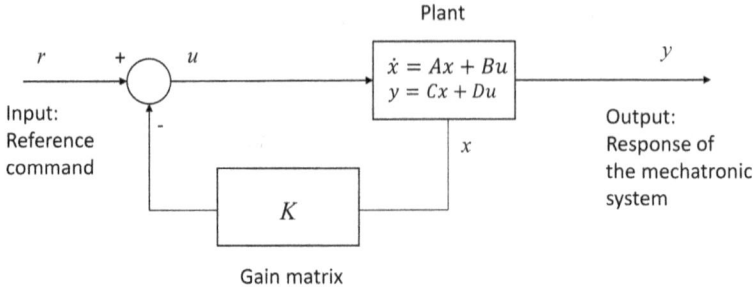

FIGURE 4.5 Full-state feedback control system.

Considering Equation (4.18), it is clear that a controller has to modify the state matrix A to change the dynamics of a system. In this case, the poles of the system are eigenvalues of the matrix A. The location of the poles dictates the stability of the system which means that any movement of the poles changes the system stability. If eigenvalues are at undesirable locations, it is necessary to use the pole placement method to move them in appropriate locations. In this case, the gain matrix (also called feedback matrix) that is used as a feedback control should be calculated using the pole placement in order to get a stable behavior of a whole system. The system must be considered controllable in order to implement this method.

For example, considering the schematic of a full-state feedback system shown in Figure 4.5, it is possible to build a controller using the pole placement method. First, it is necessary to assume that the reference command is set to zero, i.e., $r = 0$. For simplicity, we also assume that the D matrix is a zero matrix. Then, the control input can be defined as:

$$u = -Kx \tag{4.20}$$

The state-space equations for the system shown in Figure 4.5 can be written as:

$$\dot{x} = Ax + B(-Kx) = (A - BK)x \tag{4.21}$$

$$y = Cx \tag{4.22}$$

This equation contains the closed-loop A_{CL} matrix, $A_{CL} = A - BK$, that gives us the ability to move eigenvalues, which are equal to the closed-loop poles, by choosing an appropriate value of K. This method is good for a system that has one or two states. For the systems that have more than two states, the mathematical apparatus starts to be overwhelmed, and it is better to consider other control methods.

4.4.3 OBSERVER DESIGN TECHNIQUE

In control theory, when there is no chance in practice to measure all state variables of a system it is better to build an observer that allows estimating them. The observer requires to have all required control signals to satisfy the controllability condition

Plant

r   +   u $\dot{x} = Ax + Bu$ y
                            $y = Cx + Du$

Input: Output:
Reference   Response of
command   the mechatronic
                                                            system

            K $\hat{x}$ State
                                        observer

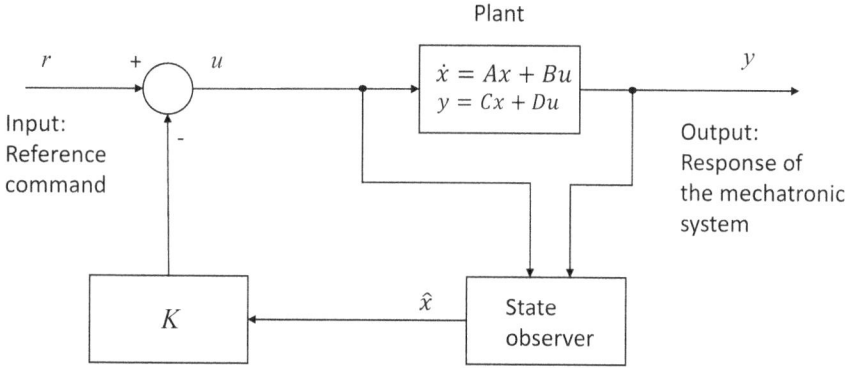

FIGURE 4.6   Observer-based design for a feedback control system.

of the system. The term of *controllability* usually assumes that the system has all control signals which allows the system to reach any state in a finite amount of time. The observer also requires satisfying the observability condition. The term *observability* means that all critical states (see state vector x in Equation (4.18), which can also be called practical states) of a mechatronic system can be obtained from system outputs. In this case, the controller system using an observer can be designed as shown in Figure 4.6.

As shown in Figure 4.6, the state observer, also known as a state estimator, is an estimation system that provides an estimate of the internal state of a given plant from measurements of the input and output of a whole mechatronic system.

For example, considering the schematic of the feedback system shown in Figure 4.6, it is possible to build a controller using an observer technique. Assuming that the D matrix is a zero matrix and the observer is represented by gain L, and also that the observer is basically a digital twin copy of the real plant, it is possible to write the estimated output from the observer as:

$$\hat{y} = C\hat{x} \tag{4.23}$$

The state-space equations for the observer can be written as:

$$\dot{\hat{x}} = A\hat{x} + Bu + L\left(y - \hat{y}\right) \tag{4.24}$$

$$\hat{y} = C\hat{x} \tag{4.25}$$

The error dynamics of the observer can be defined as:

$$\dot{e} = \dot{x} - \dot{\hat{x}} = \left(A - LC\right)e \tag{4.26}$$

where $(A - LC)$ are the poles.

The feedback control is then expressed for such a system as:

$$u = -K\hat{x} \tag{4.27}$$

This shows that, in the case of the estimated state of feedback, the measurements of all state variables of the system are not required to be performed. However, in order to make the performance of this design technique at the stable system level, it requires that the dynamics of the observer have to be much faster than the real plant itself. As a result, the implementation of such designs of a feedback control system requires the usage of a microprocessor-based system in practical mechatronic system applications.

More description on the observer design can be found in [3, 25] and specific examples for its railway applications can be found in [26–30].

4.4.4 OPTIMAL CONTROL

The optimal control is designated to control a plant with the optimized objective function that is built based on a certain optimal criterion. The optimal control design of the system can be done using a performance index that should be selected by an engineer as a performance measure and the optimal control is adjusted to achieve the minimum index value. This means that an optimal control described by a set of differential equations delivers the paths of the control variables that minimizes the objective function.

For example, in the design of linear control systems, there are three performance indices in use [2]:

- The integral of the absolute magnitude of the error (IAE) criterion:

$$J_1 = \int_0^\infty |e(t)| \, dt \qquad (4.28)$$

- The integral of the squared error (ISE) criterion:

$$J_2 = \int_0^\infty e^2(t) \, dt \qquad (4.29)$$

- The integral of time multiplied by the absolute value of error (ITAE) criterion:

$$J_3 = \int_0^\infty t |e(t)| \, dt \qquad (4.30)$$

However, considering practical aspects of control applications that commonly require permitting some deviation from the optimum, it was found that the ISE criterion is the most desirable index in most practical applications. In addition, the ISE criterion does not require a complicated mathematical apparatus which thus makes it easy to deal with mathematically [2]. Considering the full-state feedback control system presented in Figure 4.5, the specific form of performance index based on the ISE criterion can be defined in terms of the time-domain as:

$$J = \int_0^{t_f} \left(x^T x \right) dt \tag{4.31}$$

where x is the state vector, x^T is the transpose of the y matrix, and t_f is the final time.

More information about the optimal control theory aspects can be found in [25, 31–33]. In the sections below we will consider some ideas about how to achieve optimal control techniques in rail vehicle mechatronic systems by using different control and estimation techniques.

4.4.4.1 Linear–Quadratic Regulator

The aim of this controller to operate a dynamic system at minimum processing cost. It allows to define a Linear–Quadratic (LQ) problem as a quadratic function that describes the cost to operate the system dynamics which are represented by a set of linear differential equations. In order to solve this problem, it is necessary to use the LQ regulator (LQR) which is defined as a feedback controller. For the full-state feedback control system presented in Figure 4.5, the cost function in terms of the time-domain can be represented by the equation:

$$J = \int_0^{t_f} \left(x^T Q \; x + u^T R \; u \right) dt \tag{4.32}$$

This equation is commonly known as the Riccati equation. The first term reflects the expected performance of the controller and the second term reflects the control effort requirement. By tuning the weighting matrices Q and R, a feedback control gain K_{LQR} is calculated to minimize the cost function presented by Equation (4.32). Selection of appropriate weighting factors is very important to ensure a good design of the optimal controller. The ideal feedback control for such a dynamic system is:

$$u = -K_{LQR} \hat{x} \tag{4.33}$$

More description on the LQ control theory can be found in [25, 33] and specific examples for the LOR's railway applications can be found in [11, 34–36].

4.4.4.2 Kalman Filter

The Kalman Filter (KF) represents an estimator used in the LQ problem which is focused on the estimation of the instantaneous state of a linear dynamic system with adding the white Gaussian system noise generated from the measurements performed on the dynamic system. The KF was developed by Rudolf E. Kalman in 1960 [37] as "a technique for filtering of and prediction in linear Gaussian systems" [38]. The KF is also called a LQ state estimator (LQE). Assuming that there is no feedback term D present in the system, then the linear dynamics of the plant is described with the following equations considering stochastic disturbances w_d, also known as process noise, and measurement noise w_n, also known as sensor noise:

$$\dot{x} = Ax + Bu + w_d \tag{4.34}$$

$$y = Cx + w_n \tag{4.35}$$

The equation of the KF can then be written as:

$$\hat{x}_k = K_k y_k + (1 - K_k)\hat{x}_{k-1} \tag{4.36}$$

where k is the subscript of the state for the current estimate, $k-1$ is the subscript of the state for the previous estimate, and K_k is the Kalman gain. In Equation (4.36), all measurement values are known, and the only unknown component is the Kalman gain, which requires to be calculated for each consequent state.

Considering that the KF is based on the recursive state estimation technique and it works with two sets of equations (Time Update for a prediction and Measurement Update for a correction), the KF algorithm for linear Gaussian state transitions and measurements can be presented as shown in Figure 4.7. In this figure: $\hat{x}_k^-$ is the prior estimate (also called the mean) that represents the rough estimate before the measurement update correction; P_k^- is the prior error covariance that defines the posterior state transition probability; $\hat{x}_k$ is the estimate of the state x at time k; $\mathbb{Q}$ is process noise covariance; R is covariance of measurement noise; I is unit matrix; C is the correction term; and P_k is the updated error covariance that is needed for the future estimate $(k+1)$. The updated estimate with the measurement equation (Equation 2 in the Corrector Block in Figure 4.7) shows the discrepancy between the predicted measurement $C\hat{x}_k^-$ and the actual measurement y_k.

This algorithm makes the KF computationally quite efficient. However, the state transitions and measurements are commonly non-linear in real practice. In order to handle the non-linearities, it is necessary to use the extended KF which relaxes the

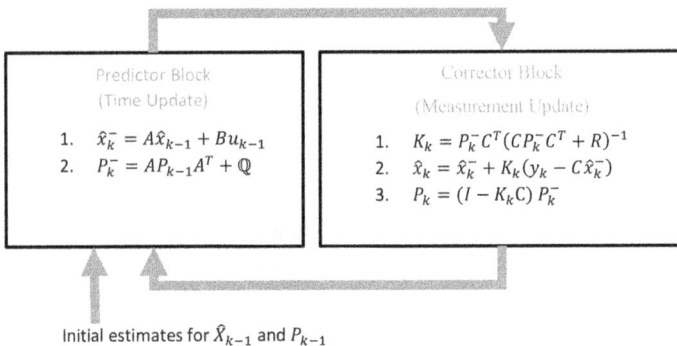

Predictor Block
(Time Update)

1. $\hat{x}_k^- = A\hat{x}_{k-1} + Bu_{k-1}$
2. $P_k^- = AP_{k-1}A^T + \mathbb{Q}$

Corrector Block
(Measurement Update)

1. $K_k = P_k^- C^T (CP_k^- C^T + R)^{-1}$
2. $\hat{x}_k = \hat{x}_k^- + K_k(y_k - C\hat{x}_k^-)$
3. $P_k = (I - K_k C) P_k^-$

Initial estimates for $\hat{x}_{k-1}$ and P_{k-1}

FIGURE 4.7 KF algorithm for linear Gaussian state transitions and measurements. (Adapted from [39].)

linearity assumption. More description on the KF theories for linear and non-linear dynamic systems can be found in [31, 37, 40, 41] and specific examples for the KF's railway applications can be found in [42–51].

4.4.4.3   Linear–Quadratic–Gaussian Control

The Linear–Quadratic–Gaussian Control (LQG) is a control approach that is based on the integration of a LQ regulator and state estimation based on the KF [52, 53]. As stated in [54], the LQG controller is "aimed at optimizing the performance based on specific penalty values assigned to the state vector x and the input vector u." In addition, [54] also states that "the Kalman–Bucy filter or its digital equivalent is used rather than the more general of Kalman filter." The Kalman–Bucy filter (named after Richard Snowden Bucy) is a continuous time version of the KF, where the Kalman gain for the dynamic system described with Equations (4.34) and (4.35) is calculated as:

$$K_k = P_k^- C^T R^{-1} \tag{4.37}$$

Assuming the observation of the full state of the system is not feasible or practical, a KF or Kalman–Bucy filter is commonly used to estimate the system's state variables based on a reduced set of measurements. In this case, the schematic diagram of LQG control is shown in Figure 4.8.

Specific examples for the railway applications of LQG control can be found in [22, 54–57].

4.4.4.4 H_2 and H_∞ Methods

Two popular performance measures used in optimal and robust control theory are H_2 and H_∞ norms [58]. While the H_2 norm works with the variance of the output given

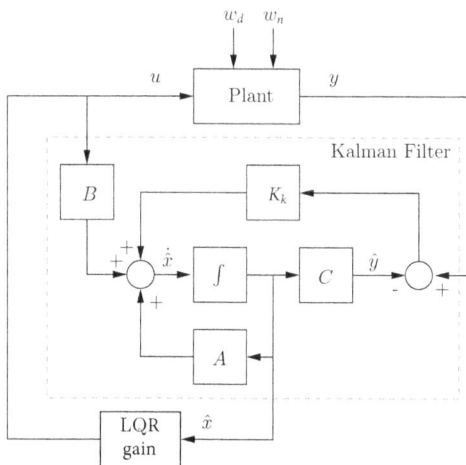

FIGURE 4.8   Schematic diagram of LQG control.

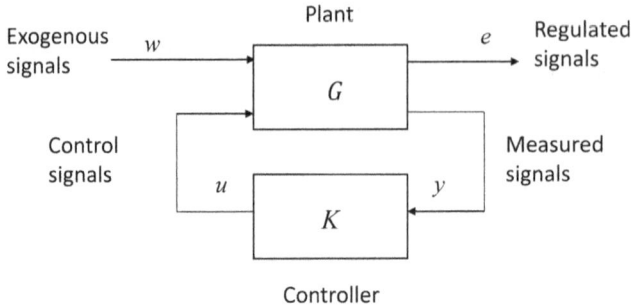

FIGURE 4.9 Generalized feedback system for optimal and robust control studies.

white noise disturbances, the H_∞ norm is found to be more appropriate for specifying both the level of plant uncertainty and the signal gain from the disturbance inputs to error outputs in the controlled system [59].

The generalized feedback system based on the optimal and robust control method is shown in Figure 4.9. The control and measured signals presented in Figure 4.9 were covered in previous sections of this chapter, but the regulated signals and exogenous signals should be explained. The regulated signals commonly are very small in value and are an error signal obtained from signal estimation and control inputs. The exogenous signals may generally include reference commands issued to the controller and disturbances acting on the system (e.g., sensor noise, etc.).

The control system design via H_2 (quadratic) optimization is one of further development in terms of LQG control design methodology [60–62]. This optimization combines all the commonly used quadratic control and filter methodologies: LQR, Kalman–Bucy filter, and LQG. It is necessary to mention that the LQG is considered as a stochastic problem while H_2 is defined as a deterministic problem. Considering that, in practical applications, some difficulties quite often exist to establish the precise stochastic properties of disturbances and noise signals, these signals should be considered very carefully because these parameters play a role of design parameters. The idea of H_2 optimization is to develop a minimizing norm on the output feedback problem from the noise signals. If we consider a closed-loop transfer function T_{ew} from w to e as for the control system shown in Figure 4.9, then we can write the standard control problem as: to find the stabilizing value of K so it meets the following condition:

$$\min \|T_{ew}\|_2 \tag{4.38}$$

In this control problem, it is possible to say that the minimization of an LQ integral cost (see Equation (4.32)) is reformulated as the H_2 system norm minimization as per Equation (4.38).

The control method via the H_∞ optimization is called robust control and it deals with the ability of a mechatronic system to work under uncertainty and it requires the application of advanced mathematical methods to solve the control problem. If we consider a closed-loop transfer function T_{ew} from w to e as for the control system shown in Figure 4.9, then we can write the standard control problem as: to find the stabilizing value of K so it meets the following condition:

$$\|T_{ew}\|_\infty \leq \gamma \tag{4.39}$$

where γ is the threshold value ($\gamma > 0$) to get the robust control.

Specific examples for the railway applications of H_2 and H_∞ control methods can be found in [22, 63–68].

4.4.4.5 Model Predictive Control

Model Predictive Control (MPC) is an advanced method that is slightly different from the LQR (infinite horizon) method because the MPC uses a finite horizon at each time step, and the control algorithm should keep future time steps in account for the control purpose. As a result, LQR can achieve global stability properties, but MPC has a complex performance that allows it to achieve an optimal for local control applications. The basic structure of the MPC-based control system is shown in Figure 4.10. In this figure, the estimator requires the current plant measurements as inputs to produce the current dynamic state of the process. The latter, as well as the reference command (target), are used as inputs in the MPC controller. The MPC controller consists of a plant model and an optimizer. The plant model is used to predict future plant output based on past and current values in the dependent variables. The plant model is controlled with the optimizer that solves an optimization problem and delivers optimal future control actions. The optimizer is generally built considering constraints and the pre-defined cost function. The plant model is, in consequence, a decision maker in the MPC controller.

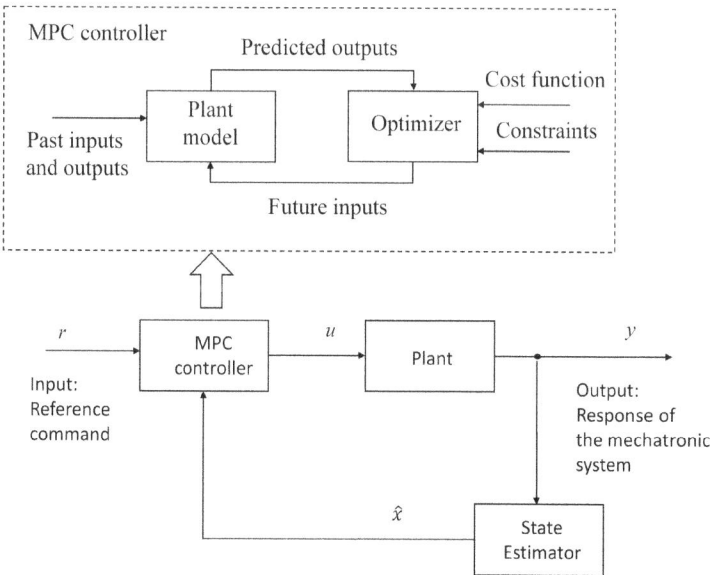

FIGURE 4.10   Generalized feedback system for MPC control studies.

In real practice, the models which are required to be used in MPC are designed to represent the dynamic behavior of complex systems that might include large time delays and high-order dynamics.

More description on the MPC control theory for linear and non-linear dynamic systems can be found in [69–71] and specific examples for the MPC control system's railway applications can be found in [63, 72–75].

4.5   NON-CLASSICAL CONTROL METHODS

The non-classical control methods include fuzzy and neural network-based control techniques that are considered as a part of intelligent machine design. The intelligent machine from the machine design perspective is a machine that is endowed with the ability to reason. Therefore, the advanced machine control methods should provide an ability to a control system to analyze analog input values in terms of numerical or logical variables and to achieve control system outputs that are already successfully achievable by humans or to form a new knowledge that can improve the performance of the control system further. The main disadvantage with these methods is that they rely on numeric or measured data to form system models.

In this section, a brief look at basic concepts associated with fuzzy and neural network-based control methods are provided.

4.5.1   Fuzzy Control

The fuzzy logic control is built on a set of rules. One major advantage with fuzzy logic is that a set of rules can include experienced human experts' linguistic rules, describing how to design the mechatronic control system. The linguistic rules become very important when there is some limitation in measured data presented in the control system. The mathematical modeling of fuzzy concepts was presented in 1965 by Lotfi Zadeh. In the publication [76], Prof. Zadeh described how to present classes of objects mathematically considering that those do not have precisely defined criteria of membership function in the fuzzy set such as when, for example, some information cannot be described in the numerical values.

The rules used in fuzzy logic are translated into *if-then* rules and in this form, they are included in the fuzzy logic algorithm. A fuzzy logic control structure can be tuned simply by changing the weight of some rule/s.

Fuzzy logic controllers serve the same function as the more conventional controllers, but they are simple in design and take no regard for dynamic, stability, and mathematical modeling. In other words, the fuzzy logic control works based on heuristics and mathematical models provided by fuzzy logic, rather than via mathematical models provided by differential equations. This is particularly useful for controlling systems whose mathematical models are nonlinear or for which standard mathematical models are simply not available. The implementations of fuzzy controls are, in some sense, imitations of the control laws that humans use. The example of a fuzzy logic control system is shown in Figure 4.11. As it is possible to see, a fuzzy controller works with data sets. The term *set* means a collection of elements.

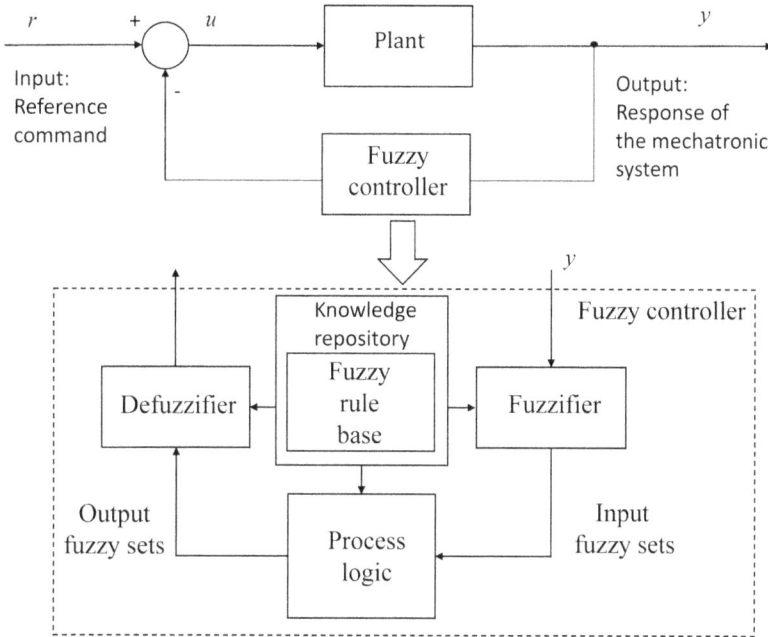

FIGURE 4.11   Example of simplistic fuzzy logic control system.

In this case, an element can be defined as an element of a set or a member of a set. In fuzzy logic, the membership can occur in varying degrees. Therefore, it is necessary to define a degree of membership by means of a membership function which is specified for each set. An example of membership function is shown in Figure 4.12. The fuzzy sets are often defined as triangle or trapezoid-shaped curves and a degree of membership has a peak value equal to 1.

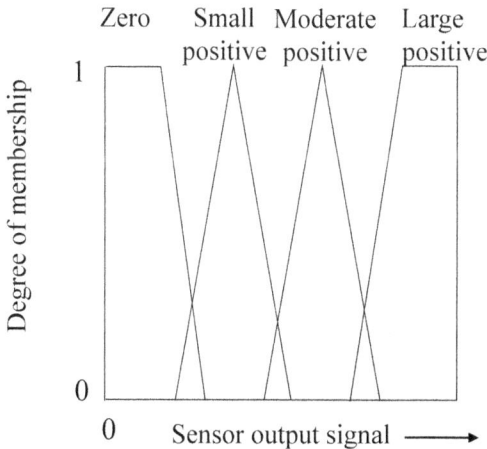

FIGURE 4.12   Example of degree of membership assigned for each fuzzy set.

The fuzzy logic algorithm includes three processes [77, 78]:

- **Fuzzification:** This process is processed by the Fuzzifier block (see Figure 4.11) which compares a signal from each sensor with a set of possible linguistic variables and defines a membership.
- **Execution:** This process is performed by the Process logic block as shown in Figure 4.11. In this block, all applicable rules in the knowledge repository are used to compute the degree of memberships, i.e., to receive fuzzy output functions.
- **Defuzzification:** This process is performed by the Defuzzifier block (see Figure 4.11) which defuzzifies the fuzzy output functions and converts an output fuzzy set to a single control value.

The simplicity of the rules makes it easier to use fuzzy logic in many control problems where the mathematical precision can be neglected. However, this control should be used very carefully because the emulation of the decision-making ability is not always equal to a human expert behavior considering changes in the dynamic behavior of a system and its responses.

In railway mechatronic design, fuzzy control systems are commonly used in train control, suspension control, and slip control systems. Some practical examples can be found in [26, 50, 79–85].

4.5.2 NEURAL NETWORK-BASED CONTROL

In 1943, McCulloch and Pitts proposed a model of the artificial neuron that has started to be a basis for neural network and neural control engineering. The neural control has been developed with a similar goal as in fuzzy control to avoid excessive mathematical approaches and formalism used in classical control. The neural control is commonly built using feedback control topologies with various implementations of neural networks [86]. The neural network uses predetermined numbers of neurons which are connected to each other to accomplish defined actions [87]. The neurons are processing elements which accept a certain number of inputs and apply weighting factors to them. The weighted inputs are then added or multiplied to calculate the net weighted input and forwarded to the output function. The algorithm is described mathematically in Figure 4.13. The architecture, system and method of artificial neural network implementation are explained in patent [88]. For practical tasks, the neural networks can be trained through simulations with the consideration of error back propagation [26].

In some cases, a neural control can be used in combination with classical or fuzzy control methods to improve the dynamic response of control methods. Such a control combination is commonly referred to as a hybrid control method.

In railway mechatronic designs, the neural control systems are proposed for use in train control, suspension control, adhesion control, and monitoring systems. Some practical examples can be found in [26, 89–93].

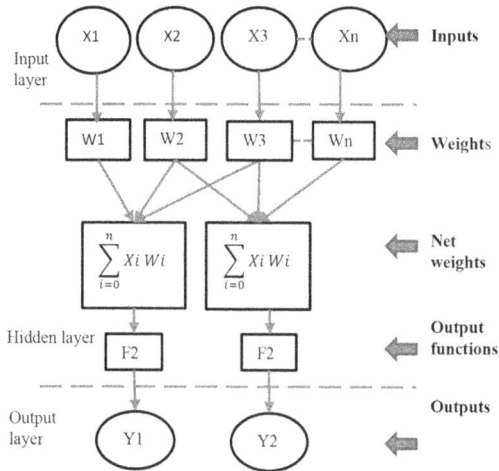

FIGURE 4.13 Neural network process. (Adapted from [39].)

REFERENCES

1. K. Janschek, Control Theoretical Aspects, Chapter 10, Mechatronic Systems Design – Methods, Models, Concepts, Springer-Verlag, Heidelberg, Germany, 2012.
2. S. M. Shinners, Modern Control System Theory and Design, 2nd Edition, Wiley, New York, 1998.
3. N. S. Nise, Control Systems Engineering, 8th Edition, Wiley, NY, 2019.
4. W. J. Palm III, Modeling, Analysis, and Control of Dynamic Systems, 2nd Edition, Wiley, New York, 1999.
5. W. Bolton, Mechatronics: Electronic Control Systems in Mechanical and Electrical Engineering, 6th Edition, Pearson, London, UK, 2015.
6. T. X. Mei, H. Li, Control design for the active stabilization of rail wheelsets, ASME Journal of Dynamic Systems, Measurement, and Control, 130(1), 011002, 2008.
7. A. Qazizadeh, S. Stichel, H. R. Feyzmahdavian, Wheelset curving guidance using H_∞ control, Vehicle System Dynamics, 56(3), 461–484, 2018.
8. A. Visioli, Practical PID Control, Series – Advances in Industrial Control, Springer-Verlag, London, UK, 2006.
9. J. J. Choi, S. H. Park, J. S. Kim, Dynamic adhesion model and adaptive sliding mode brake control system for the railway rolling stocks, Journal of Rail and Rapid Transit, 221(3), 313–320, 2007.
10. A. D. Cheok, S. Shiomi, Combined heuristic knowledge and limited measurement based fuzzy logic antiskid control for railway applications, IEEE Transactions on Systems, Man, and Cybernetics, Part C, Applications and Reviews, 30(4), 557–568, 2000.
11. J. T. Pearson, R. M. Goodall, T. X. Mei, G. Himmelstein. Active stability control strategies for a high speed bogie. Control Engineering Practice, 12, 1381–1391, 2004.
12. G. Jacazio, D. Risso, M. Sorli, L. Tomassini, Adaptive control for improved efficiency of hydraulic systems for high-speed tilting trains, Journal of Rail and Rapid Transit, 226(3), 272–283, 2012.
13. H. Zamzuri, A. C. Zolotas, R. M. Goodall, Intelligent control approaches for tilting railway vehicles, Vehicle Systems Dynamics, 44(supl), 834–842, 2006.

14. F. Hassan, A. C. Zolotas, R. M. Margetts, Optimised PID control for tilting trains, Systems Science & Control Engineering, 5(1), 25–41, 2017.
15. F. Hassan, A. C. Zolotas, T. Smith, Optimized Ziegler–Nichols based PID control design for tilt suspensions, Journal of Engineering Science and Technology, 10(5), 17–24, 2017.
16. Y. Tian, S. Liu, W. J. T. Daniel, P. A. Meehan, Investigation of the impact of locomotive creep control on wear under changing contact conditions, Vehicle System Dynamics, 53(5), 692–709, 2015.
17. M. Spiryagin, P. Wolfs, C. Cole, S. Stichel, M. Berg, M. Plöchl, Influence of AC system design on the realisation of tractive efforts by high adhesion locomotives, Vehicle System Dynamics, 55(8), 1241–1264, 2017.
18. M. Spiryagin, P. Wolfs, C. Cole, V. Spiryagin, Y. Q. Sun, T. McSweeney, Design and Simulation of Heavy Haul Locomotives and Trains, CRC Press, Boca Raton, FL, 2017.
19. I. Pratt, R. Goodall, Controlling the ride quality of the central portion of a high-speed railway vehicle, Proceedings of the 1997 American Control Conference, (Cat. No.97CH36041), 1, 719–723, 1997, DOI: 10.1109/ACC.1997.611895.
20. H. Dai, X. Hao, H. Zhang, The research on robust control design railway of vehicle active suspension, Proceedings of the IEEE International Vehicle Electronics Conference (IVEC'99) (Cat. No.99EX257), 1, 58–61, 1999, DOI: 10.1109/IVEC.1999.830622.
21. P. Li, R. Goodall, V. Kadirkamanathan, Parameter estimation of railway vehicle dynamic model using Rao-Blackwellised particle filter, Proceedings of 2003 European Control Conference (ECC), 2384–2389, 2003, DOI: 10.23919/ECC.2003.7085323.
22. A. C. Zolotas, R. M. Goodall, Modelling and Control of Railway Vehicle Suspensions, In: M. C. Turner, D. G. Bates (Eds), Mathematical Methods for Robust and Nonlinear Control. Lecture Notes in Control and Information Sciences, 367, 373–412, Springer, London, UK, 2007.
23. H. Yamazaki, Y. Karino, T. Kamada, M. Nagai, T. Kimura, Effect of wheel-slip prevention based on sliding mode control theory for railway vehicles, Vehicle System Dynamics, 46(4), 255–270, 2008.
24. E. D. Gialleonardo, M. Santelia, S. Bruni, A. Zolotas, A Simple Active Carbody Roll Scheme for Hydraulically Actuated Railway Vehicles Using Internal Model Control, ISA Transactions, Elsevier, 2021 May 08. https://doi.org/10.1016/j.isatra.2021.03.003.
25. R. C. Dorf, R. H. Bishop. Modern Control Systems, 13th Edition, Pearson, London, UK, 2017.
26. D. Frylmark, S. Johnsson, Automatic Slip Control for Railway Vehicles, Master's Thesis, Linköpings Universitet, Vehicular Systems, Department of Electrical Engineering, Linköping, Sweden, 2003.
27. X. Wei, S. Lin, H. Liu, Distributed fault detection observer for rail vehicle suspension systems, 24th Chinese Control and Decision Conference (CCDC), 33963401, 2012, DOI: 10.1109/CCDC.2012.6244541.
28. L. Weerasooriya, T. X. Mei, H. Li, Y. Luo, Application of State Estimators in Active Control of Railway Wheelsets, In: M. Klomp, F. Bruzelius, J. Nielsen, A. Hillemyr (Eds), Advances in Dynamics of Vehicles on Roads and Tracks (IAVSD 2019), Lecture Notes in Mechanical Engineering, Springer, Cham, Switzerland, 2020.
29. L. Weerasooriya, Analysis on Actuator Dynamics in Active Wheelset Control, PhD Thesis, School of Computing, Science & Engineering, University of Salford, Salford, UK, 2019.
30. S. Shrestha, M. Spiryagin, Q. Wu, Friction condition characterization for rail vehicle advanced braking system, Mechanical Systems and Signal Processing, 134, 1–16, 2019.

31. R. F. Stengel, Optimal Control and Estimation, Dover Publications, New York, 1994.
32. D. E. Kirk, Optimal Control Theory: An Introduction, Dover Publications, New York, 2004.
33. F. L. Lewis, D. L. Vrabie, V. L. Syrmos, Optimal Control, 3rd Edition, Wiley, Hoboken, NJ, 2012.
34. H. Selamat, S. D. A. Bilong, Optimal controller design for a railway vehicle suspension system using Particle Swarm Optimization, 9th Asian Control Conference (ASCC), 1–5, 2013, DOI: 10.1109/ASCC.2013.6606076.
35. M. Graa, M. Nejlaoui, A. Houidi, Z. Affi, L. Romdhane, Modeling and control of rail vehicle suspensions: A comparative study based on the passenger comfort, Proceedings of the Institution of Mechanical Engineers, Part C: Journal of Mechanical Engineering Science, 232(2), 260–274, 2018.
36. G. Schandl, P. Lugner, C. Benatzky, M. Kozek, A. Stribersky, Comfort enhancement by an active vibration reduction system for a flexible railway car body, Vehicle System Dynamics, 45(9), 835–847, 2007.
37. R. E. Kalman. A new approach to linear filtering and prediction problems, Journal of Basic Engineering, 82, 35–45, 1960.
38. S. Thrun, W. Burgard, D. Fox, Probabilistic Robotics, The MIT Press, Cambridge, MA, 2006.
39. S. Shrestha, Q. Wu, M. Spiryagin, Review of adhesion estimation approaches for rail vehicles, International Journal of Rail Transportation, 7(2), 79–102, 2019.
40. M. S. Grewal, A. P. Andrews, Kalman Filtering Theory and Practice Using MATLAB®, Wiley, NY, 2001.
41. G. Welch, G. Bishop, An Introduction to the Kalman Filter, TR 95-041, Department of Computer Science, University of North Carolina at Chapel Hill, 1–16, 2006.
42. T. X. Mei, R. M. Goodall, H. Li, Kalman filter for the state estimation of a 2-AXLE railway vehicle, Proceedings of 1999 European Control Conference (ECC), 2431–2435, 1999, DOI: 10.23919/ECC.1999.7099687.
43. T. X. Mei, H. Li, R. M. Goodall, Kalman filters applied to actively controlled railway vehicle suspensions, Transactions of the Institute of Measurement and Control, 23(3), 163–181, 2001.
44. C. P. Ward, R. M. Goodall, R. Dixon, G. A. Charles, Adhesion estimation at the wheel–rail interface using advanced model-based filtering, Vehicle System Dynamics, 50(12), 1797–1816, 2012.
45. G. Charles, R. M. Goodall, Low adhesion estimation, The Institution of Engineering and Technology International Conference on Railway Condition Monitoring, Birmingham, UK; 2006 Nov., 96101, 2006.
46. G. Charles, R. Goodall, R. Dixon, Model-based condition monitoring at the wheel–rail interface, Vehicle System Dynamics, 46, 415–430, 2008.
47. S. Strano, M. Terzo, On the real-time estimation of the wheel-rail contact force by means of a new nonlinear estimator design model, Mechanical Systems and Signal Processing, 105, 391–403, 2018.
48. M. Jesussek, K. Ellermann, Fault detection and isolation for a nonlinear railway vehicle suspension with a hybrid extended Kalman filter, Vehicle System Dynamics, 51, 1489–1501, 2013.
49. S. Alfi, S. Bruni, L. Mazzola L, et al., Model – Based Fault Detection in Railway Bogies, Proceedings of 22nd International Symposium on Dynamics of Vehicles on Roads and Tracks, Manchester, UK, 1–7, 2011.
50. I. Hussain, T. X. Mei, R. T. Ritchings, Estimation of wheel–rail contact conditions and adhesion using the multiple model approach, Vehicle System Dynamics, 51, 32–53, 2013.

51. P. Pichlík, J. Zděnek, Comparison of Locomotive Adhesion Force Estimation Methods for a Wheel Slip Control Purpose, Proceedings of 9th International Conference on Electronics, Computers and Artificial Intelligence (ECAI-2017), Targoviste, Romania, 1–4, 2017.

52. J. Doyle, Guaranteed margins for LQG regulators, IEEE Transactions on Automatic Control, 23(4), 756–757, 1978, DOI: 10.1109/TAC.1978.1101812.

53. M. Athans, A tutorial on the LQG/LTR method, Proceedings of 1986 American Control Conference, 1289–1296, 1986, DOI: 10.23919/ACC.1986.4789131.

54. B. Fu, R. L. Giossi, R. Persson, et al., Active suspension in railway vehicles: A literature survey, Railway Engineering Science, 28, 3–35, 2020.

55. A. C. Zolotas, R. M. Goodall, Improving the tilt control performance of high-speed railway vehicles: An LQG approach, IFAC Proceedings Volumes, 38(1), 25–30, 2015.

56. A. C. Zolotas, R. M. Goodall, Advanced Control Strategies for Tilting Railway Vehicles, Proceedings of UK Automatic Control Council Control 2000 Conference, Cambridge, UK, 2000.

57. R. Zhou, A. Zolotas, R. Goodall, LQG control for the integrated tilt and active lateral secondary suspension in high speed railway vehicles, Proceedings of 8th IEEE International Conference on Control and Automation, 1621, 2010, DOI: 10.1109/ICCA.2010.5524239.

58. J. C. Doyle, K. Glover, P. P. Khargonekar, B. A. Francis, State-space solutions to standard $\mathcal{H}2$ and $\mathcal{H}\infty$ control problems, IEEE Transactions on Automatic Control, 34(8), 831–847, 1989.

59. J. Doyle, Robust and Optimal Control, Proceedings of the 35th Conference on Decision and Control, Kobe, Japan, 1595–1596, 1996.

60. K. Zhou, J. C. Doyle, K. Glover, Robust and Optimal Control, Prentice Hall, Englewood Cliffs, NJ, 1996.

61. A. Sinha, Linear Systems Optimal and Robust Control, CRC Press, Boca Raton FL, 2007.

62. L. Fortuna, M. Frasca, Optimal and Robust Control: Advanced Topics With MATLAB®, CRC Press, Boca Raton FL, 2012.

63. P. E. Orukpe, X. Zheng, I. M. Jaimoukha, A. C. Zolotas, R. M. Goodall, Model predictive control based on mixed $\mathcal{H}_2/\mathcal{H}_\infty$ control approach for active vibration control of railway vehicles, Vehicle System Dynamics, 46, 151–160, 2008.

64. A. C. Zolotas, G. D. Halikiasy, R. M. Goodall, A comparison of tilt control approaches for high speed railway vehicles, Proceedings of 14th International Conference on Systems Engineering ICSE, 632636, 2000.

65. T. Kamada, K. Hiraizumi, M. Nagai, Active vibration suppression of lightweight railway vehicle body by combined use of piezoelectric actuators and linear actuators, Vehicle System Dynamics, 48, 73–87, 2010.

66. T. X. Mei, R. M. Goodall, Robust control for independently rotating wheelsets on a railway vehicle using practical sensors, IEEE Transactions on Control Systems Technology, 9(4), 599–607, 2001, DOI:10.1109/87.930970.

67. T. Gajdar, P. Korondi, I. Rudas, H. Hashimoto, Robust and sliding mode control for railway wheelset, Proceedings of 1996 IEEE IECON, 22nd International Conference on Industrial Electronics, Control, and Instrumentation, 1, 250–255, 1996, DOI: 10.1109/IECON.1996.570961.

68. T. Hirata, S. Koizumi, R. Takahashi, H∞ control of railroad vehicle active suspension, Automatica, 31(1), 13–24, 1995.

69. E. F. Camacho, C. Bordons, Model Predictive Control, 2nd Edition, Springer, London, UK, 2007.

70. L. Wang, Model Predictive Control System Design and Implementation Using MATLAB®, Springer, London, UK, 2009.

71. J. B. Rawlings, D. Q. Mayne, M. M. Diehl, Model Predictive Control: Theory, Computation, and Design, 2nd Edition, Nob Hill Publishing, Santa Barbara, CA, 2019.

72. L. Zhang, X. Zhuan, Optimal operation of heavy-haul trains equipped with electronically controlled pneumatic brake systems using model predictive control methodology, IEEE Transactions on Control Systems Technology, 22(1), 13–22, 2014.

73. L. Cheng, P. Acuna, R. P. Aguilera, J. Jiang, J. Fletcher, C. Baier, Model Predictive Control for Energy Management of a Hybrid Energy Storage System in Light Rail Vehicles, Proceedings of 11th IEEE International Conference on Compatibility, Power Electronics and Power Engineering, Cadiz, Spain, 683–688, 2017, DOI:10.1109/CPE. 2017.7915255.

74. X. Liu, J. Xun, B. Ning, L. Yuan, An Approach for Accurate Stopping of High-Speed Train by Using Model Predictive Control, 2019 IEEE Intelligent Transportation Systems Conference (ITSC), Auckland, New Zealand, 846–851, 2019, DOI:10.1109/ITSC.2019.8917237.

75. X. Xu, J. Peng, R. Zhang, B. Chen, F. Zhou, Y. Yang, K. Gao, Z. Huang, Adaptive model predictive control for cruise control of high-speed trains with time-varying parameters. Journal of Advanced Transportation, Article ID 7261726, 1–11, 2019.

76. L. A. Zadeh, Fuzzy sets, Information and Control, 8(3), 338–353, 1965.

77. L.-X. Wang, A Course in Fuzzy Systems and Control, Prentice-Hall, NJ, 1996.

78. H. T. Nguyen, N. R. Prasad, C. L. Walker, E. A. Walker, A First Course in Fuzzy and Neural Control, CRC Press, Boca Raton, FL, 2003.

79. C. H. Chang, L. M. Jia, N. S. Xu, X. D. Zhang, The application of fuzzy control to automatic train operation, IFAC Proceedings Volumes, 29(1), 7674–7679, 1996.

80. R. J. Stonier, S. Kuppa, P. J. Thomas, C. Cole, Fuzzy Modelling of Wagon Wheel Unloading Due to Longitudinal Impact Forces, ASME/IEEE 2005 Joint Rail Conference, Pueblo, CO, Paper No: RTD2005-70034, 59–64, 2005.

81. S. Sezer, A. E. Atalay, Application of fuzzy logic based control algorithms on a railway vehicle considering random track irregularities, Journal of Vibration and Control, 18(8), 1177–1198, 2012.

82. M. Spiryagin, V. Spiryagin, V. Ulshin, Active Steering Control of a Rail Vehicle: A New Approach, Annals of DAAAM for 2008 & Proceedings of the 19th International DAAAM Symposium "Intelligent Manufacturing & Automation: Focus on Next Generation of Intelligent Systems and Solutions", Trnava, Slovakia, 640–641, 2008.

83. L. B. Jordan Jr., Locomotive Traction Control System Using Fuzzy Logic, United States Patent US5424948A, 1995.

84. M. Garcia-Rivera, R. Sanz, J. A. Perez-Rodriguez, An antislipping fuzzy logic controller for a railway traction system, Proceedings of the 6th IEEE International Conference on Fuzzy Systems, Barcelona, Spain, 1, 119–124, 1997, DOI: 10.1109/FUZZY.1997.616355.

85. M. Spiryagin, K. S. Lee, H. Yoo, Control system for maximum use of adhesive forces of a railway vehicle in a tractive mode, Mechanical Systems and Signal Processing, 22, 709–720, 2008.

86. F. L. Lewis, S. S. Ge, Neural Networks in Feedback Control Systems, In: M. Kutz (Ed), Mechanical Engineers' Handbook: Instrumentation, Systems, Controls, and MEMS, 3rd Edition, John Wiley & Sons, Hoboken, NJ, 2, 791–825, 2006.

87. K. Gurney, An Introduction to Neural Networks, UCL Press, London, UK, 1997.

88. M. Moussa, A. Savich, S. Areibi, Architecture, System and Method for Artificial Neural Network Implementation, United States Patent US8103606B2, 2012.

89. T. Gajdar, I. Rudas, Y. Suda, Neural Network Based Estimation of Friction Coefficient of Wheel and Rail, IEEE International Conference on Intelligent Engineering Systems, Budapest, Hungary, 315–318, 1997.

90. M. Nagai, A. Moran, Y. Tamura, S. Koizumi, Identification and control of nonlinear active pneumatic suspension for railway vehicles using neural networks, Control Engineering Practice, 5(8), 1137–1144, 1997.

91. N. Hossein-Zadeh, M. Dhanasekar, Y. Q. Sun, ANN Application to Predict the Wheel–Rail Impact Forces Due to the Short Wavelength Defects in Rail, Proceedings of 2nd International Conference on Artificial Intelligence in Science and Technology (AISAT 2004), Hobart, Australia, 212–217, 2004.

92. C. R. Cole, Vehicle Dynamics Production System and Method, United States Patent US6853889, 2005.

93. S. Falomi, M. Malvezzi, E. Meli, et al., Determination of wheel-rail contact points: Comparison between classical and neural network based procedures, Meccanica, 44, 661–686, 2009.

5 Actuators

5.1 INTRODUCTION

Actuators and sensors are a fundamental component of mechatronic systems. Actuators are used to drive the mechatronic system according to a control action elaborated by the controller which can be either a specified movement (position control, speed control) or a force or torque (force control). Sensors instead provide the controller with measured signals describing the state of the system being controlled (the *plant*), enabling the implementation of feedback control strategies, see Chapter 4. It should be noted that actuators themselves are feedback-controlled systems, so in a mechatronic railway vehicle there are usually multiple feedback control loops nested one into another.

Actuators are typically complex devices involving the use of power made available in some form (e.g., electrical power from batteries, fluid power from pressurized air or oil) which is transformed to mechanical power and used to apply the control action to the plant. Therefore, the use of actuators involves significant additional complexity in a railway vehicle, although it is recognized that active control may result in substantial simplifications of the vehicle's mechanical structure [1]. Different types of actuators are suitable for mechatronic railway vehicles and the choice of the principle of actuation impacts significantly on the overall design of the vehicle. It is one main aim of this chapter to present the different principles of actuation that can be used in mechatronic railway vehicles, outlining their advantages and drawbacks.

For instance, electro-mechanical actuators are fast, relatively compact, and do not require the use of fluid, but some of their failure modes, particularly jamming due to a failure in the mechanical transmission, may lead to safety issues. Hydraulic actuators provide large forces and large power density, are inherently free from jamming and, in the case of failure, can be switched to passive mode but these actuators require the use of pressurized oil, placing an additional complexity into the design of the vehicle, and can be quite expensive if a large bandwidth of actuation is required. Pneumatic actuators are simple, relatively inexpensive, use pressurized air which is already available on the vehicle for braking and other functions, but their bandwidth is insufficient for implementations of active control requiring fast actuation, e.g., active stabilization of the bogie or active control of car body flexible modes.

It is also important to note that actuators and sensors are dynamic systems, so a deviation of the actual control action produced by the actuator from the desired one shall be expected. It follows that actuator dynamics must be considered carefully in the design of a mechatronic system using suitable mathematical models. A second main aim of this chapter is therefore to introduce mathematical models for actuators mainly used in railway vehicles.

This chapter is organized as follows: Sections 5.2, 5.3, and 5.4 cover electro-mechanical, hydraulic, and pneumatic actuators, respectively.

5.2 ELECTRO-MECHANICAL ACTUATORS

Electro-mechanical actuators use the principles of electro-magnetism to apply control forces or torques that can be used to control the motion of a mechanical system (the plant). In general, an electro-mechanic actuator consists of an electric motor and of a mechanical transmission providing the interface of the motor to the plant. The electric motor is controlled using a feedback control loop in which the controlled variable can be either a force or a measured displacement/rotation representative of the system's position. A controlled electric motor is termed a *servomotor*. Recent developments of power electronics have made the control of servomotors much simpler than previously, removing some of the limitations associated with the different typologies of electric motors.

Electro-mechanical actuators are typically composed by a servomotor coupled to a mechanical transmission which provides a suitable interface of the servomotor to the plant. In particular, the mechanical transmission may include:

- A ball screw and nut device to convert the rotation of the output shaft of the servomotor into a linear movement.
- A gear train to amplify the torque generated by the servomotor by means of a suitable gear ratio.

Advantages of electro-mechanical actuators are the compact design with high force density, accurate and fast actuation, high efficiency, no risk of fluid leakage/contamination which is otherwise possible for hydraulic actuators. However, electromechanical actuators also have significant drawbacks, namely: risk of jamming due to failures in the mechanical transmission, relatively high cost, high stiffness at high frequency. In particular, the risk of jamming has so far hindered the application of electromechanical actuators to mechatronic railway vehicles in those cases where the actuation system is meant to provide a safety-relevant function, such as active steering or active stabilization.

In this section, we provide first an overview of different types of electric motors used in electro-mechanical actuators, focusing on direct current (DC) motors and alternating current (AC) permanent magnet synchronous motors (PMSMs) which are the ones most relevant to the scopes of this book. Then, we introduce the main features of mechanical transmissions and finally we provide an example of the complete model of an electro-mechanical actuator formed by a PMSM with ball screw transmission.

5.2.1 DIRECT CURRENT MOTORS

DC motors are rotary motors fed by a mono-phase DC current power supply. Like all electric rotary motors, they are formed by a rotating body called the *rotor* and a non-rotating part called the *stator*. Electric windings are hosted in slots in the surface of the rotor and form the *armature* of the rotor. A DC electric voltage v_a is applied to the armature so that a current i_a flows in the armature circuit. The stator produces a magnetic field, having a fixed direction in space, which interacts with the electric current flowing in the armature circuit producing a mechanical torque acting on the rotor. Since the magnetic field generated by the stator is fixed in space, a constant direction of the current i_a flowing in the armature circuit would lead to a reversal of the mechanical torque at each half-turn of the rotor, resulting in an average zero torque over one turn. To avoid this, the armature windings are electrically connected to a collector which is formed by a sequence of conductive segments electrically insulated one from the other. The armature voltage is applied to the collector and then to the windings through sliding contacts called brushes. In this way, the current flowing in each winding is reversed at each half-turn of the rotor and the resulting torque maintains a fixed direction over the entire revolution of the rotor.

The stator may consist either of a permanent magnet, in which case the intensity of the magnetic field cannot be varied, or of a set of non-rotating electric windings, called *field* windings, forming an electro-magnet. In either case, the stator is designed so that the lines of the magnetic field are as much as possible pointing in radial direction as this reduces the fluctuations of the torque with the angular position of the rotor. In some cases, more than one couple of magnetic poles are used to obtain a better radial alignment of the magnetic field.

In the case of the stator being formed by an electro-magnet, there are two electric circuits in the DC motor, called respectively the *armature circuit* (formed by the wirings of the rotor) and the *field circuit* (formed by the wirings of the stator). See Figure 5.1 where the back electromotive force e_a caused in the armature circuit by the rotation of the rotor is also shown. The two circuits can be fed by independent voltages v_f and v_a or can be connected in parallel (shunt-wound DC motor), in series (series-wound motor) or partly in series and partly in parallel (compound-wound motor).

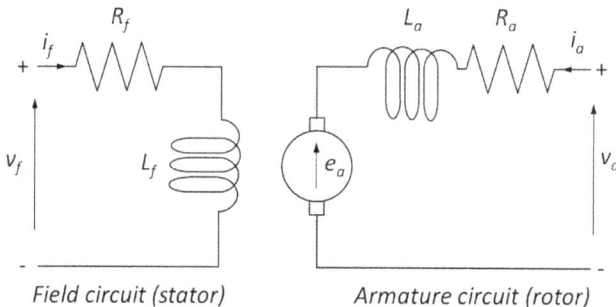

Field circuit (stator)   Armature circuit (rotor)

FIGURE 5.1   Field circuit and armature circuit in a DC motor.

In this chapter, we will confine our analysis to the case of a motor with separately excited field and armature circuits, but more details on the other cases can be found in [2].

The commutation provided by the collector and brushes is one weak point of the DC motor, as the brushes tend to wear out, requiring maintenance and may be subject to contact losses and to sparks which are undesired. The sliding contact of the brushes to the collector also causes energy loss, reducing the efficiency of the motor and increasing the need for motor cooling. Therefore, a different design of the DC motor is sometimes used, called a *brushless* motor. In the brushless motor, the rotor is formed by permanent magnets and the stator is formed by electro-magnets. In order to provide a constant direction for the torque applied to the rotor, the tension applied to the electro-magnets is varied periodically over one revolution of the rotor, this being accomplished by a power semi-conductor device. The correct synchronization between the switching of the tension applied to the field windings and the angular position of the rotor must be ensured by measuring the rotation of the rotor using a suitable sensor.

DC motors are characterized by features that make them one of the preferred choices for servomotors which include high torque versus mass ratio, ease of controllability, and relatively simple power electronics required for driving the motor.

The equations describing the behavior of a DC motor are presented below, considering the case of a motor with separately excited field and armature circuits. The mechanical torque T_m applied on the stator is proportional through a constant k_T to the product of the flux linkage ψ_f times the armature current i_a, see also Section 3.3.1.1:

$$T_m = k_T \psi_f i_a \qquad (5.1)$$

whereas the back electromotive force e_a is proportional to the product of the flux linkage ψ_f times the angular speed of the rotor ω_m through a constant k_e:

$$e_a = k_e \psi_f \omega_m \qquad (5.2)$$

The mechanical power produced by the motor is $W_m = T_m \omega_m$ whereas the electric power involved in the conversion from electric to mechanical energy is $W_e = e_a \omega_m i_a$. Assuming ideal energy conversion, it follows that $W_m = W_e$ which, based on Equations (5.1) and (5.2) implies $k_T = k_e$.

The model in the time domain of the DC motor with separately excited field and armature circuits can be written considering two first-order differential equations to relate the tensions and currents in the field and armature circuits shown in Figure 5.1 and one first-order differential equation to describe the motion of the rotor subject to the torque T_m produced by the motor, to a resisting torque T_L produced by the load driven by the motor and to a viscous resisting torque proportional to the angular speed of the rotor ω_m through a constant coefficient c_r, representing the effect of dissipative forces in the bearings and of aerodynamic resisting forces acting on the rotor. In this last equation, $J_m + J_L$ is the total mass moment of inertia of the rotor plus the load (see Section 5.2.3 for more information about term J_L). The resulting third-order system of equations for the simple case considered here is:

$$\begin{cases} v_a = R_a i_a + L_a \dfrac{di_a}{dt} + e_a \\[2mm] v_f = R_f i_f + L_f \dfrac{di_f}{dt} \\[2mm] T_m - T_L - c_r \omega_m = \left(J_m + J_L \right) \dfrac{d\omega_m}{dt} \end{cases} \qquad (5.3)$$

Using Equations (5.1) and (5.2), Equation (5.3) can be rewritten as:

$$\begin{cases} L_a \dfrac{di_a}{dt} + R_a i_a = v_a - k_e \psi_f \omega_m \\[2mm] L_f \dfrac{di_f}{dt} + R_f i_f = v_f \\[2mm] \left(J_m + J_L \right) \dfrac{d\omega_m}{dt} + c_r \omega_m = k_T \psi_f i_a - T_L \end{cases} \qquad (5.4)$$

where the coupling of the electrical equation describing the armature circuit and the mechanical equation of the motor is made explicit. Additionally, it must be noted that ψ_f is a function of the current i_f flowing in the field circuit, so there is also a coupling of these two equations to the one describing the field circuit.

Equation (5.4) provides a third-order model of the DC motor. In the case of the motor with separately excited circuits being operated at constant field tension v_f, or in the case of the stator consisting of permanent magnets, the flux linkage takes a constant value so that products $k_e \psi_f = k_T \psi_f$ can be replaced by a constant value k_ψ. At the same time, the second line of Equation (5.4) provides a time-invariant relationship between the field voltage and current, so the motor can be modeled as a second-order system.

We conclude this section by deriving the steady-state torque-speed curve for the DC motor. In a steady-state condition, the armature current takes a constant value given by:

$$i_a = \frac{v_a - k_e \Psi_f \omega_m}{R_a} \qquad (5.5)$$

and the steady-state torque $T_{m,ss}$ produced by the motor is:

$$T_{m,ss} = k_T \Psi_f i_a = k_T \Psi_f \frac{v_a}{R_a} - \frac{k_T k_e}{R_a} \psi_f^2 \omega_m \qquad (5.6)$$

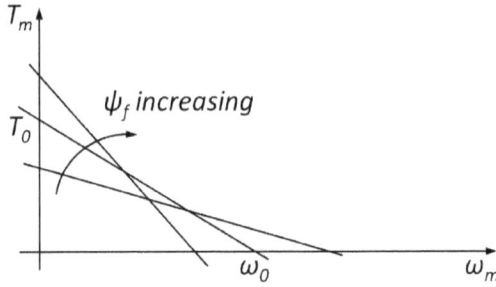

FIGURE 5.2   Steady-state torque-speed curve for a DC motor.

which can also be rewritten as:

$$T_{m,ss} = T_0 \left(1 - \frac{\omega_m}{\omega_{m0}} \right) \tag{5.7}$$

with T_0 the starting torque of the motor and ω_{m0} the speed at which a zero value of the torque is produced by the motor, as shown in Figure 5.2. We can see that, by reducing the flux linkage ψ_f, the value of ω_{m0} is increased, which means the range of speeds in which the motor produces a positive torque is extended, but this is traded for a lower starting torque.

5.2.2 ALTERNATING CURRENT MOTORS

There are several types of AC motors but in this chapter we will restrict ourselves to two main types: induction motors and synchronous motors. Thanks to the use of power electronics, both these types of motors can be used as servomotors.

In both induction motors and synchronous motors, the stator consists of three windings named a, b, and c having a phase shift of $\frac{2\pi}{3}$ one from each other. When the three windings are fed by a balanced three-phase AC system, it can be shown [3] that:

$$i_a = I\cos(\omega t)$$

$$i_b = I\cos\left(\omega t - \frac{2\pi}{3} \right) \tag{5.8}$$

$$i_c = I\cos\left(\omega t - \frac{4\pi}{3} \right)$$

with ω and I, respectively the frequency and amplitude of each of these ACs, a rotating magnetic field is generated having constant magnitude and angular speed equal to:

$$\omega_f = \frac{\omega}{p} \tag{5.9}$$

with p the number of pole pairs in the stator windings. The rotating magnetic field is used in different ways by induction motors and synchronous motors to generate a mechanical torque.

5.2.2.1 Induction Motors

In an induction motor, the rotor winding is formed by a cage consisting of conductive bars and end-rings or alternatively of short-circuited conductive wires. In both cases, the rotor winding is not energized, thus there is no need of using sliding contacts as in a brushed DC motor. When the rotor is still or rotates at an angular speed ω_r different from the angular speed of the rotating magnetic field generated by the stator, the flux linkage of the rotating magnetic field through the rotor winding changes in time. Therefore, an electric current is induced in the rotor windings and hence electromagnetic forces are applied on the rotor, resulting in a mechanical torque.

From the above description of the working principle of an induction motor, it follows that, in order to produce a non-zero torque, the angular speed of the rotor must be different from that of the rotating magnetic field. Therefore, an important parameter describing the working condition of this type of motor is the *slip s*, defined as:

$$s = \frac{\omega_f - \omega_r}{\omega_f} \tag{5.10}$$

It can be shown [2] that the speed-torque characteristic curve of the asynchronous motor for a given magnitude and frequency of the voltage applied to the stator windings takes the shape shown in Figure 5.3. This curve exhibits a non-zero starting torque T_s, a maximum torque T_{max}, and a synchronous speed ω_f corresponding to zero torque. At speeds higher than the synchronous speed the torque becomes negative, i.e., the motor behaves like a brake.

The control of an induction motor can be performed using an *inverter*, i.e., a circuit using high-frequency switched semiconductor devices called *thyristors* to transform a DC power supply into an AC three-phase system with variable frequency. Additionally, pulse-width modulation (PWM) is used to change the voltage of the

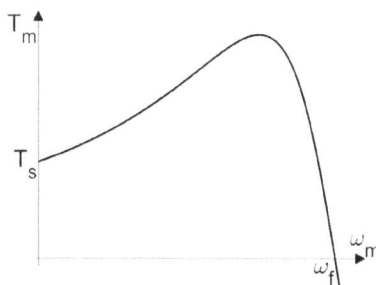

FIGURE 5.3 Speed-torque characteristic curve for an induction motor.

three-phase AC system produced by the inverter. The PWM inverter feeds the stator windings of the motor. By varying the frequency and amplitude of the AC system fed to the stator windings, it is possible to set the torque produced by the motor to a desired value. Typical strategies for asynchronous motors include *field-oriented control* (FOC) and *direct torque control* (DTC). Both these techniques are described in Section 3.3.2 of this book.

5.2.2.2 Synchronous Motors

In a synchronous motor, the rotating magnetic field produced by the stator interacts with a magnetic field produced by the rotor, generating a mechanical torque on the rotor that tends to align the rotor's magnetic field to the rotating field produced by the three-phase stator windings. The rotor's magnetic field can be produced either by conductive windings powered through sliding electric contacts or by permanent magnets. In this chapter, we focus on PMSMs, which are more suitable for use as servomotors. PMSMs offer higher power and torque densities compared to induction motors and hence are generally preferred for high-end servomotor applications, as an alternative to DC motors.

The equations representing the three stator windings of the motor are:

$$v_a = R_s i_a + \frac{d\psi_a}{dt}$$

$$v_b = R_s i_b + \frac{d\psi_b}{dt} \tag{5.11}$$

$$v_c = R_s i_c + \frac{d\psi_c}{dt}$$

where v_a, v_b, and v_c are the voltages applied to the three windings, i_a, i_b, and i_c are the currents in the three stator windings, R_s is the electrical resistance of the windings and ψ_a, ψ_b, and ψ_c are the stator flux linkages. The three scalar equations can be represented as the projection along the axes of the stator's windings, denoted as a, b, and c of a single vector equation using space vectors $\bar{v}_s$, $\bar{i}_s$, and $\bar{\psi}_s$ [3]:

$$\bar{v}_s = R_s \bar{i}_s + \frac{d\bar{\psi}_s}{dt} \tag{5.12}$$

The flux linkage space vector $\bar{\psi}_s$ is given by the sum of two terms, the first being related to the magnetic field generated by the stator and takes the form $L_s \bar{i}_s$ with L_s the self-inductance of the stator winding. The second term is produced by the permanent magnets in the rotor and therefore is aligned with the magnetic axis of the rotor. Hence, the following complex-valued expression of the flux linkage space vector is obtained:

$$\bar{\psi}_s = L_s \bar{i}_s + \psi_m e^{j\theta_r} \tag{5.13}$$

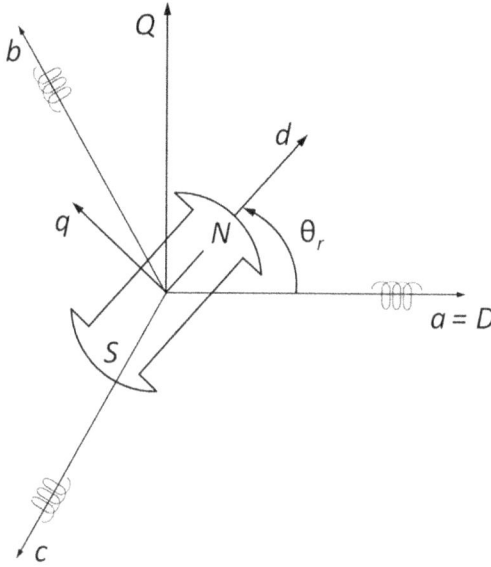

FIGURE 5.4 Directions of the axes of the three-phase windings (a, b, c), direct and quadrature axes in the fixed stator reference (D, Q) and in the moving rotor reference (d, q). N-S is the magnetic axis of the rotor.

with j the imaginary unit and θ_r the angle formed by the magnetic axis of the rotor with the axis of the a winding in the stator (see Figure 5.4).

The voltage and current space vectors can be expressed with respect to a reference rotating with the rotor having axes d (direct) and q (quadrature), also shown in Figure 5.4:

$$\bar{v}_s = \left(v_{sd} + j v_{sq} \right) e^{j\theta_r}$$
$$\bar{i}_s = \left(i_{sd} + j i_{sq} \right) e^{j\theta_r} \qquad (5.14)$$

In Equation (5.14), v_{sd}, v_{sq}, i_{sd}, and i_{sq} are the components of the voltage and current space vectors along the axes of the d-q reference and the direction of the d axis is chosen as the magnetic axis of the rotor.

Introducing the expressions from Equations (5.13) and (5.14) into Equation (5.12) and considering that L_s and ψ_m are constant, the following complex-valued equation is obtained:

$$\left(v_{sd} + j v_{sq} \right) e^{j\theta_r} = \left[R_s \left(i_{sd} + j i_{sq} \right) + L_s \left(\frac{di_{sd}}{dt} + j \frac{di_{sq}}{dt} \right) + j \frac{d\theta_r}{dt} L_s \left(i_{sd} + j i_{sq} \right) \right.$$
$$\left. + j \frac{d\theta_r}{dt} \psi_m \right] e^{j\theta_r} \qquad (5.15)$$

Removing the common term $e^{j\theta_r}$ from both sides of Equation (5.15), separating the real and imaginary parts of the equation and rewriting the time derivative of angle θ_r as the angular speed of the rotor ω_r we get:

$$L_s \frac{di_{sd}}{dt} + R_s i_{sd} = v_{sd} + \omega_r L_s i_{sq}$$

$$L_s \frac{di_{sq}}{dt} + R_s i_{sq} = v_{sq} - \omega_r L_s i_{sd} - \omega_r \psi_m$$

(5.16)

Furthermore, it can be shown [3] that torque T_m produced by the motor is:

$$T_m = \frac{3}{2} p \psi_m i_{sq}$$

(5.17)

with p the number of pole pairs in the stator.

Equation (5.17) shows that the torque produced by the motor is proportional to the quadrature-axis component of the stator current i_{sq}, thus providing the theoretical basis of vector control of the asynchronous motor. The currents i_a, i_b, and i_c fed to the three stator windings are measured and transformed to their two-phase equivalents in a D-Q reference fixed to the stator i_{sD} and i_{sQ}. These two currents are then transformed in the two-phase currents i_{sd} and i_{sq} in the reference fixed to the rotor, using the angular position of the rotor θ_r measured by a position sensor (e.g., an encoder, see Chapter 6). The torque produced by the motor obtained according to Equation (5.17) is compared to the reference torque T_{ref} and a PI controller is used to define the desired value for the quadrature current i_{sqref} which is then back-transformed into the references for the three stator currents i_{aref}, i_{bref}, and i_{cref}. These references are used to control the currents fed to the stator windings using a PWM inverter containing current control loops to provide the desired current output [3].

The dynamic model of a PMSM is finally obtained considering Equation (5.16) together with the equation of motion describing the motion of the rotor subject to the torque T_m produced by the motor, to a resisting torque T_L due to the load and to a viscous resisting torque $c_r\omega_m$ representing speed-dependent dissipative forces acting on the rotor. This equation is the same as the third line in Equation (5.4), but the expression of the motor torque is provided for the PMSM by Equation (5.17). The resulting dynamic model for the PMSM motor is therefore:

$$\begin{cases} L_s \dfrac{di_{sd}}{dt} + R_s i_{sd} = v_{sd} + \omega_r L_s\, i_{sq} \\[2mm] L_s \dfrac{di_{sq}}{dt} + R_s i_{sq} = v_{sq} - \omega_r L_s\, i_{sd} - \omega_r \psi_m \\[2mm] (J_m + J_L)\dfrac{d\omega_m}{dt} + c_r\omega_m = \dfrac{3}{2} p \psi_m i_{sq} - T_L \end{cases}$$

(5.18)

5.2.3 Mechanical Transmission

To form an electro-mechanical actuator, the servomotor is coupled to a mechanical transmission providing a suitable interface to the mechanical system being actuated. The role of the mechanical transmission is to provide speed reduction and/or to convert the rotary motion of the motor shaft into a linear movement of the actuator.

5.2.3.1 Gear Trains

Actuators of interest for mechatronic railway vehicles are typically required to produce a large actuation torque or force at relatively low speed, whereas electrical motors work at relatively large speed. Therefore, a gear train can be used to have different angular speeds ω_m and ω_L for the motor and load shafts, respectively, with $\omega_L < \omega_m$ (see Figure 5.5).

The gear ratio of a gear train is defined as the ratio of the speed of the motor shaft to the speed of the load shaft:

$$n = \frac{\omega_m}{\omega_L} \tag{5.19}$$

For a simple pair of gears, the gear ratio is given by the ratio of the number of teeth in the output gear to the number of teeth in the pinion, i.e., the gear mounted on the motor shaft. In the case where a high gear ratio is needed, more than one pair of gears can be used, but this is seldom done in electro-mechanical actuators as it would result in significant increase in the mass and size of the actuator.

Neglecting compliance effects in the transmission, the equations of motion previously established for the DC motor, third line of Equation (5.4), and for the PMSM, third line of Equation (5.18), are still valid provided the resisting torque due to the load T_L and the moment of inertia of the load J_L are replaced by equivalent expressions T_L' and J_L' which are derived below.

Figure 5.6 shows the torques acting on the rotor and on the input and output shafts of the transmission. The torques acting on the motor are the motor torque T_m, the viscous resisting torque $c_r\omega_m$, and the resisting torque T_L'. In accordance with Newton's third law, the torque applied to the input shaft of the transmission is opposite to T_L', so the mechanical power input to the transmission is $W_1 = T_L'\omega_m$, whereas the

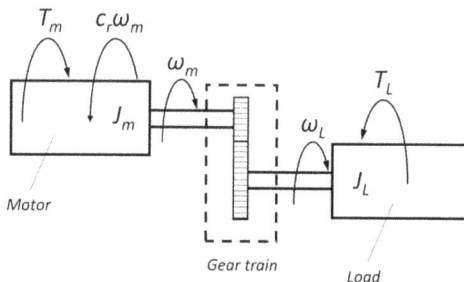

FIGURE 5.5 Interface between the motor and load established via a gear train.

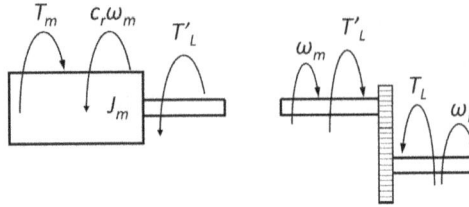

FIGURE 5.6   Torques acting on the rotor and on the input and output shafts of the transmission.

mechanical power output from the transmission is $W_2 = T_L \omega_L$. Considering the efficiency η of the transmission defined as $\eta = W_2/W_1$ we have:

$$\eta = W_2/W_1 = (T_L \omega_L)/(T'_L \omega_m) \qquad (5.20)$$

From Equation (5.20) and considering Equation (5.19), the expression for the equivalent load torque considered acting on the rotor is:

$$T'_L = \frac{T_L \omega_L}{\eta \omega_m} = \frac{T_L}{\eta n} \qquad (5.21)$$

The equivalent moment of inertia of the load J'_L is defined as the moment of inertia of a rigid body rotating at rotor speed ω_m and providing the same kinetic energy as the actual inertia of the load J_L rotating at speed ω_L. Hence, we consider:

$$\frac{1}{2} J'_L \omega_m^2 = \frac{1}{2} J_L \omega_L^2 \qquad (5.22)$$

from which we obtain:

$$J'_L = J_L \left(\frac{\omega_L}{\omega_m} \right)^2 = \frac{J_L}{n^2} \qquad (5.23)$$

5.2.3.2   Ball Screw Transmission

Conversion from rotary to linear motion is usually obtained using a ball screw mechanism where the shaft of the servomotor is connected either directly or through a gear train to a threaded shaft (the *screw*) which is coupled to a multi-race ball bearing (the *nut*). During the rotation of the screw, the balls are flowing through the races of the nut so that, once they reach the final race, they are moved back to the initial race by a recirculation channel. Ball screws provide the same rotary to linear conversion of motion as lead screws but remove dry friction effects which are undesired in electro-mechanical actuators since they cause low efficiency, but especially because dry friction is accompanied by wear of the screw thread reducing the service life of the transmission. However, the races in the nut and screw of a ball screw mechanism are also subject to rolling contact fatigue phenomena which may cause a failure of the transmission and possible jamming of the actuator.

A rotation by one turn of the screw produces a linear movement of the nut by one lead s of the screw, hence the relationship of the linear speed of the nut v_n to the angular speed of the screw ω_s is:

$$v_s = \frac{s}{2\pi}\omega_s \tag{5.24}$$

Denoting the efficiency of the ball screw by η_s, the following relationship is obtained between the load torque T_L applied to the screw and the resisting force F_L from the load:

$$F_L v_s = \eta_s T_L \omega_s \tag{5.25}$$

and, considering Equation (5.24), the following expression is found for torque T_L:

$$T_L = \frac{F_r v_s}{\eta_s \omega_s} = \frac{s}{2\pi\eta_s} F_L \tag{5.26}$$

Denoting the total mass undergoing the linear motion of the nut by m_L and the moment of inertia of the screw by J_s (which is often negligible), the equivalent moment of inertia J_L associated with the ball screw and load is:

$$J_L = J_s + m_L \left(\frac{s}{2\pi}\right)^2 \tag{5.27}$$

If the screw is directly connected with the motor shaft, the load torque T_L provided by Equation (5.26) and the moment of inertia J_L provided by Equation (5.27) are directly the ones appearing in the equation of motion of the rotor, see Equation (5.4) for the DC motor and Equation (5.18) for the PMSM. Otherwise, if the motor is connected to the ball screw through a gear train, Equations (5.21) and (5.23) should be used to obtain the equivalent torque T_L' and the equivalent moment of inertia J_L' of the load.

5.2.4 MODEL OF AN ELECTROMECHANICAL ACTUATOR WITH BRUSHLESS AC MOTOR

An example of the complete model of an electro-mechanical actuator is developed in this section, considering a PMSM driving a load through a ball screw directly attached to the motor shaft. This model is used in [4] to analyze the behavior of a system for active stabilization and active curving of a high-speed railway vehicle but note, however, a different notation is adopted here compared to the one used in the reference.

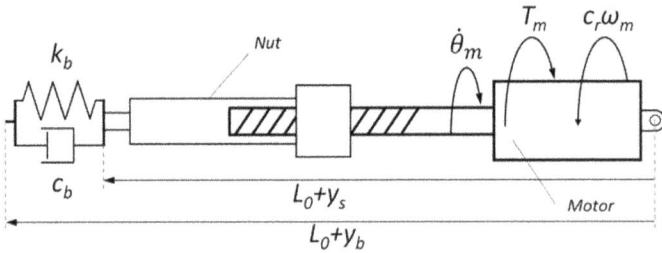

FIGURE 5.7 Electro-mechanical actuator formed by a PMSM and a ball screw; top: view of the system, bottom: schematic representation of the mathematical model. (Photo from [5].)

As shown in Figure 5.7, the actuator is mounted on the side of the bogie, with the motor attached to the bogie frame through a spherical joint and the nut connected to the car body through a rubber bushing. The bogie is assumed to undergo a longitudinal movement y_b relative to the car body.

Denoting by k_b and c_b the stiffness and damping coefficients of the bushing, the force applied to the nut is:

$$F_L = k_b \left(y_s - y_b \right) + c_b \left(\dot{y}_s - \dot{y}_b \right) \tag{5.28}$$

with:

$$y_s = \frac{s}{2\pi} \theta_m ; \dot{y}_s = \frac{s}{2\pi} \dot{\theta}_m \tag{5.29}$$

where θ_m is the rotation of the motor shaft. Considering Equation (5.26) and assuming $\eta_s = 1$, we have:

$$T_L = \frac{s}{2\pi} \left[k_b \left(\frac{s}{2\pi} \theta_m - y_b \right) + c_b \left(\frac{s}{2\pi} \dot{\theta}_m - \dot{y}_b \right) \right] \tag{5.30}$$

Bringing Equation (5.30) into Equation (5.18) provides the final model of the actuator, which is of a fourth order, involving two electrical states (currents i_{sd} and i_{sq}) and two mechanical states (rotation and angular speed of the rotor, θ_m, $\dot{\theta}_m$):

$$
\begin{cases}
v_{sd} = R_s i_{sd} + L_s \dfrac{di_{sd}}{dt} - \omega_r L_s i_{sq} \\[2mm]
v_{sq} = R_s i_{sq} + L_s \dfrac{di_{sq}}{dt} + \omega_r L_s i_{sd} + \omega_r \psi_m \\[2mm]
(J_m + J_L)\ddot{\theta}_m + \left[c_r + c_b \left(\dfrac{s}{2\pi} \right)^2 \right] \dot{\theta}_m + k_b \left(\dfrac{s}{2\pi} \right)^2 \theta_m = \dfrac{3}{2} p \psi_m i_{sq} + \dfrac{s}{2\pi} c_b \dot{y}_b + \dfrac{s}{2\pi} k_b y_b
\end{cases}
$$

$$(5.31)$$

Models for other types of electro-mechanical actuators, e.g., using a DC motor in place of the PMSM and/or including a gear train can be obtained following the example presented above. For instance, a model of an actuator using a DC motor and ball screw is introduced in [6] considering the case of an active vertical suspension for a two-axle railway vehicle.

5.3   HYDRAULIC ACTUATORS

5.3.1   FLUID POWER SYSTEM BASICS

Industrial hydraulics is the engineering discipline of transferring energy via fluids (usually mineral oils). Pressures typically range from 70 to 700 bar with 140–350 bar popular in mobile applications. Hydraulic supply is generally via positive displacement pumps and the issue of pressure control is of utmost importance for both the equipment and the safety of personnel. This leads to a vast array of pressure controls and pump flow and pressure control systems. Fundamentally, all hydraulic equipment functions are based on the transfer of force, being proportional to pressure and area, and the pressure drop that occurs when oil flows through a restriction. Force or torque can be increased by increasing either the pressure or size of the actuator. Another fundamental of industrial hydraulic systems is that the fluid is relatively stiff (although not incompressible) and hydraulic pumps are positive displacement. This means that control of hydraulic systems and the pressures generated is fundamentally about controlling flow. A pressure control such as a relief valve controls pressure by releasing flow when a certain pressure is reached. Applying these features to actuators means that hydraulic cylinders and motors are basically *position* control devices, unlike pneumatic actuators that tend to maintain almost constant force over a range of displacement due the compressibility of the air. Conversely, to achieve force control, hydraulic systems need fine control of the fluid flow to the actuator.

A further important feature of hydraulic systems is that the working fluid recirculates as seen in Figure 5.8. This means that both cooling and filtration are extremely important. Heat is generated in every area where leakage or venting from a high pressure to a low pressure occurs. A positive displacement pump may have an efficiency of 90%. In such a case, the 10% power loss ends up in the oil as heat.

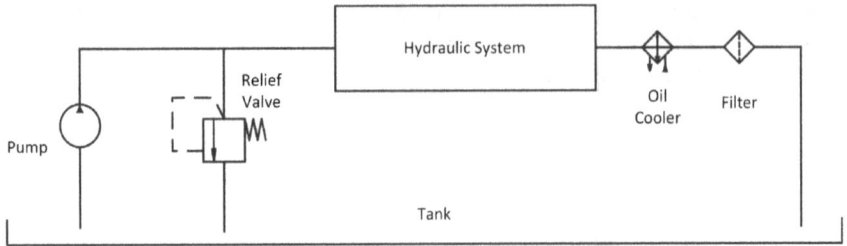

FIGURE 5.8   Basic components of a hydraulic circuit.

Hydraulic circuits therefore always have a cooling or a heat dissipative system adequate to maintain oil temperature in the correct range for the environmental operating conditions. Pressure controls are also a large source of energy loss and heat generation and circuits should be designed so that venting of oil across high pressure differences is minimized. Several approaches are used. Older technology with fixed displacement used either switching-venting relief valves or unloading valves with accumulators to effectively pump no more oil than needed (see Figure 5.9). Variable displacement pumps are now very common and these incorporate controls to reduce pump flow to standby levels using pressure, flow, or horsepower limits (see Figure 5.10). Even with these advanced systems, some oil cooling is usually required as standby power consumption can be 5–10% of the pump power.

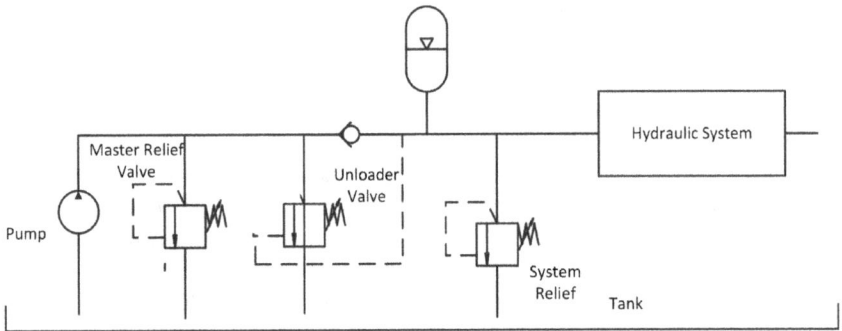

FIGURE 5.9   Basic hydraulic circuit with unloader system.

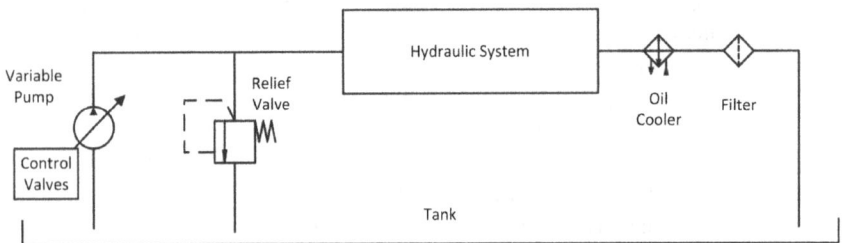

FIGURE 5.10   Basic hydraulic circuit with variable pump.

TABLE 5.1
Kinematic Viscosity of Hydraulic Oil Atmospheric
Pressure [7]

Temperature (C)	HM 32 (mm²/s)	HM 46 (mm²/s)	HM 68 (mm²/s)
20.5	80.1	146.9	217.4
40	30.3	48.5	71.5
100	5.2	6.9	8.8

5.3.2 HYDRAULIC FLUIDS PROPERTIES

The primary concern of hydraulic system design is the dependence of oil viscosity on temperature. As the oil provides both lubrication and power transfer, oil temperature must be controlled and oil viscosity must be chosen to suit the conditions. The dependency of oil viscosity on temperature is shown in Table 5.1 and a chart showing typical oil viscosity ranges is shown in Figure 5.11. High temperature and low viscosity are not the only extreme to be managed. Low temperatures and very high viscosity levels can also damage pumps and other components due to cavitation. The grade number of the oil is usually nominally the kinematic viscosity at 40°C. The data from [7] and [8] is illustrative.

While oil viscosity is a key concern to designers for all hydraulic circuits, several other parameters are more significant in more advanced hydraulic servo-controlled systems. Typical operating parameters of hydraulic oils are listed in Table 5.2.

Hydraulic Oil Viscosity Chart

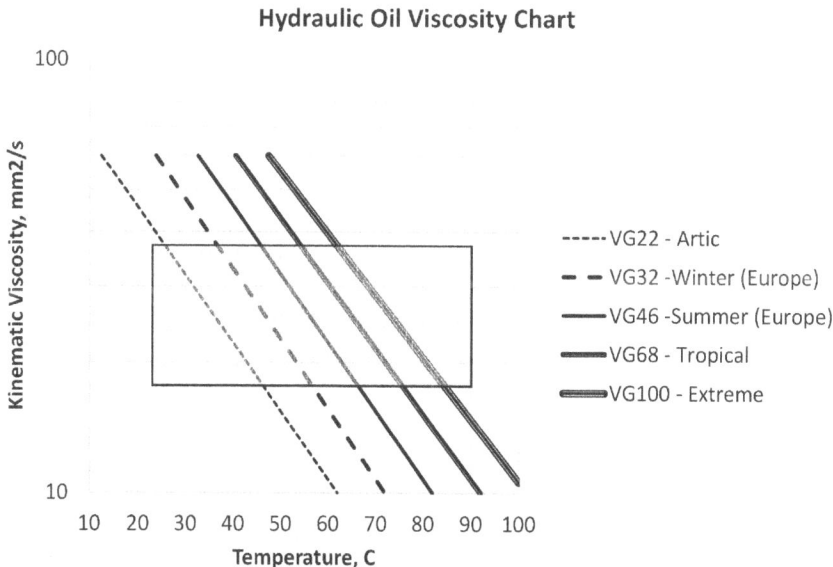

FIGURE 5.11 Typical oil viscosity for common hydraulic oils (see [8]).

TABLE 5.2

Hydraulic Oil Operating Parameters and Coefficients [8]

Parameter	Equation or Notes	Value	Units
Density	m/V	870–890	kg/m³
Absolute viscosity (operation)		0.014–0.032	Pa·s
Kinematic viscosity (operation)	Normal operation	16–36	mm²/s
Kinematic viscosity (maximum)	Maximum for short periods – cold temperatures	1000	mm²/s
Kinematic viscosity (minimum)	Minimum for short periods, marginal lubrication	10	mm²/s
Sonic speed in oil		1320	m/s
Bulk modulus B	$P = B (\triangle V/V)$; P in GPa, V in m³	2.0–3.3	GPa
Compressibility b		$3–8 \times 10^{-5}$	[1/bar]
Viscosity index VI	$VI = (v_{o-} - v_u)/(v_{o-} - v_{100})$	~100	
Thermal expansion coefficient, C	$\triangle V = C V \triangle T$; V in m³, T in Celsius	0.0007 / 0.7×10^{-9}	m³ per C / m³ per C
Air dissolvability in oil	$V_L = C_b V_{oil}$. P; P in bar, V_L air volume in m³, V_{oil} oil volume in m³,	0.09	[1/bar]
Bunsen coefficient, C_b			

5.3.3 Managing Hydraulic Fluids

Hydraulic fluids can be either based on mineral or synthetic fluids for applications where fire is a risk. In either case the importance of the cleanliness and viscosity of the working fluid cannot be overemphasized. As components are built to fine tolerances to prevent leakage, contamination of the oil with wear particles or other contaminants accelerates wear and degradation of the system. Every hydraulic system must have a well-designed filtration system. The simplest method is to filter the return oil because this means that the filter design only needs to cope with low pressure. Generally, such filters are made large enough to filter all returning oil. There are variations on this approach. In some cases, a dedicated pump filter circuit might be deployed to ensure a clean reservoir of oil at all times and has the advantage that it can be run for a period before the main system is operated, ensuring only clean oil is available to the pump. Another variation is the use of in-line pressure filters to protect more sensitive equipment such as servo valves. In line pressure filters require very robust construction and are generally only used where flows are small. A large flow application would require a large filter and hence quite large wall thickness for the cylindrical filter container, noting that hoop stress $\sigma = Pd/(2t)$ where P is pressure, d is diameter, and t is wall thickness. Oil filtering technologies work on the basis of a filter media catching particles as the oil passes through the media. Typically, ratings are given in microns indicating the size of a particle that will pass through the media. Generally, it is desired to keep particles in hydraulic oil smaller than 10 μm. How filtering is specified and performs is another discussion. Some filters are given nominal ratings. A nominal rating is an arbitrary size of particle that should be retained in the filter, but it does not say how many such particles will be retained per pass. A better

TABLE 5.3
Typical Hydraulic Permitted Viscosity Operational Ranges

Condition	Kinematic Viscosity	Comment
Start up	1000 mm²/s	Short periods only
Operating	16–100 mm²/s	Possible range
Optimum	16–36 mm²/s	Highest efficiency and functionality
Over heating	≤16 mm²/s	Marginal Lubrication
Extreme	≤10 mm²	Short periods only

rating is what is known as an absolute rating. An absolute rating is the largest opening in the filter media. Again, it is a manufacturer's rating and there is no standard way of testing the performance. The most accepted rating of filters is known as the beta rating as per [9]. The beta rating compares the number of particles above a certain size (the target size) upstream of the filter divided by the particle count downstream.

$$\beta_x = \frac{(Number\ particles\ upstream > x\ microns)}{(Number\ particles\ downstream > x\ microns)}$$

Filters are therefore specified as $\beta x < R$, e.g., $\beta 10 < 75$, meaning only one 10 μm particle out of 75 passes. The efficiency of the filter can also be thought of as:

$$\eta = \left(N_{upstream} - N_{downstream}\right)/N_{upstream} * 100 \qquad (5.32)$$

Using the same example, $\eta = (75-1)/75*100 = 98.7\%$

Equally important is the control of viscosity. Hydraulic components will have limits advised by the manufacturer as to what viscosity range is possible without equipment damage. A typical specification is as shown in Table 5.3 from [8].

5.3.4 HYDRAULIC CYLINDERS

Hydraulic cylinders provide translational force and displacement. A basic schematic of a cylinder is shown in Figure 5.12. Cylinders can come in many forms as shown in Figure 5.13.

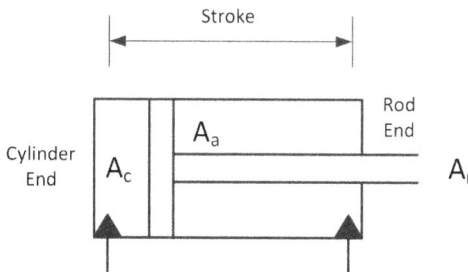

FIGURE 5.12 Double acting hydraulic cylinder.

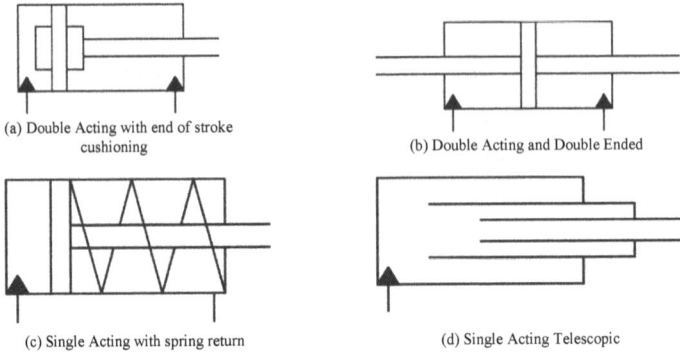

(a) Double Acting with end of stroke cushioning

(b) Double Acting and Double Ended

(c) Single Acting with spring return

(d) Single Acting Telescopic

FIGURE 5.13 Common hydraulic cylinder configurations: (a) double acting with end of stroke cushioning, (b) double acting and double ended, (c) single acting with spring return, (d) single acting telescopic.

Cylinders are described by their pressure rating, cylinder area, A_c, rod area A_r and stroke as shown in Figure 5.12. Speed limits are not usually published, but speeds may need to be a design consideration in extreme applications. The force capability of the cylinder is determined by the pressures applied and the cylinder areas. The force applied by the oil on the *cylinder end*, the left end, in Figure 5.12 is given by $F_c = P_c A_c$. Force applied by the oil on the *rod end*, the right end, in Figure 5.12 is given by $F_r = P_a A_a$, noting that pressure acts on the annulus area giving $A_a = A_c - A_r$. The force delivered by the cylinder is therefore $F = F_c - F_r$. Maximum forces are delivered by the cylinder when pressure is applied to one end with no back pressure at the other end, giving forces of either PA_c or PA_a. Variations on the simple double acting cylinder shown in Figure 5.12, which is the most common, are shown in Figure 5.13. The double ended cylinder Figure 5.13(b) has the feature that the cylinder areas are the same for both directions of travel so will deliver the same maximum force in both directions. Cylinders can also be single acting as shown in Figure 5.13(c) and returned by a spring or returned by the load as is typical of a hoist type application of the telescopic cylinder shown in Figure 5.13(d).

Mathematical modeling of cylinders can be achieved using the parameters indicated in Figure 5.14 and detailed in Table 5.4.

The hydraulic force equation $F = P_c A_c - P_a A_a$ is the general case assuming some back pressure on the exhausting port, as would often be the case when controlled by the servo valve. In simpler on-off controls, oil on the exhausting side of the cylinder will return to tank at low or atmospheric pressure, so cylinder forces are often stated as $F = P_c A_c$ when the cylinder is extending and $F = P_a A_a$ when contracting. The same discussion extends to the oil column stiffness. If the cylinder has pressure on both sides, the two oil columns will act as parallel springs. Oil column stiffness can be derived as follows:

$$\Delta F = \Delta PA \text{ and } \Delta P = B\Delta V/V \text{ where V is the oil volume} \qquad (5.34)$$

$$\Delta F = B\Delta V/VA = B(A\Delta x)A/(Aa) = (BA/a)\Delta x = k\Delta x$$

Hence $$k = (BA/a) \qquad (5.35)$$

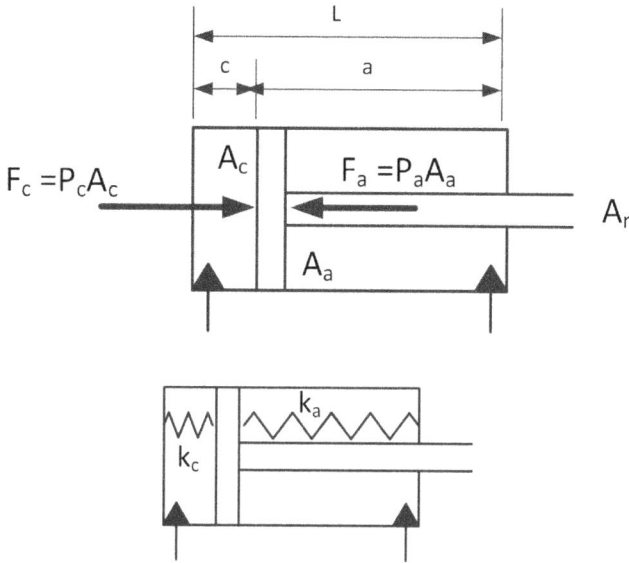

FIGURE 5.14 Hydraulic cylinder parameters.

thus showing the proportionality of the cylinder stiffness to both the cylinder area and the length of the oil column. The position of the piston in the cylinder therefore affects the stiffness and the control response. If the application is single acting or only has pressure on one side of the rod, the stiffness will be governed by the stiffness of just one oil column, i.e., $k_c = B\, A_c/c$ or $k_a = B\, A_a/a$, but the general case is $k_{ac} = k_c + k_a$. It is therefore important to consider cylinder area for both design force and design stiffness requirements. Cylinder length and the position of the piston rod in that length also affects stiffness. Control response of a cylinder will therefore be affected by cylinder and rod diameters, cylinder length and the position of the rod at the time of the control response.

TABLE 5.4
Equations for Mathematically Modeling Hydraulic Cylinders

Input Variables and Data	Output Variables	Equation (5.33)
P – pressure	F – Hydraulic force	$F = P_c A_c - P_d A_a$
Q – Flow		
A_c – Cylinder area	v_{extend} – Rod extension speed	$v_{extend} = Q/A_c$
A_a – Annulus area	$v_{contract}$ – Rod contraction speed	$v_{contract} = Q/A_a$
L – Stroke Length		
B – Oil Bulk Modulus	k – Oil Column stiffness	$k_c = B\, A_c/c$
a – Piston position measured from the rod end		$k_a = B\, A_a/a$
c – Piston position measured from cylinder end		$k_{ac} = k_c + k_a$

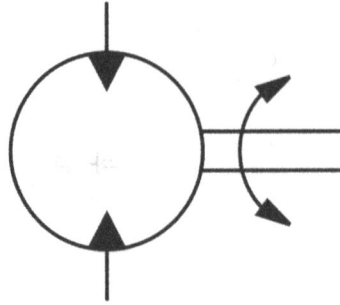

FIGURE 5.15   Schematic of a bi-directional hydraulic motor.

5.3.5   HYDRAULIC MOTORS

Hydraulic motors provide torque and angular displacement. A basic schematic of a motor is shown in Figure 5.15.

The most common form of hydraulic motor is fixed displacement and bi-directional (see Figure 5.15). Like cylinders, motors are described by their pressure rating and size, but they also have a rotational speed limit. Speed limits differ for various designs. Motors are classified by the design type which also tends to set typical pressure ratings. Design types include gears, vanes, and pistons. Gear pumps in small sizes are available up to 200 bar with designs limited by the unbalanced force on the bearings. Larger gear pumps have lower pressure ratings due to the hoop stress on the housing. Similar comment can be made for vane motors. Higher pressure applications are reserved for piston type motors with 350 bar are common in industrial machinery and up to 600 bar in aviation. Modeling equations and parameters are listed in Table 5.5. The theoretical torque capability of fixed displacement motors can be deduced from the power equation:

$$\text{Power} = \Delta P Q = T \omega \qquad (5.36)$$

where ΔP is the pressure difference across the two motor connection ports in Pa

Q is the flow rate in m^3/s
T is the torque in Nm
ω is the angular velocity of the shaft

TABLE 5.5
Equations for Mathematically Modeling Hydraulic Motors

Input Variables & Data	Output Variables	Equation (5.38)
P – pressure	T – Hydraulic torque	$T = \eta_m (P_1 - P_2) V/(2\pi)$
Q – Flow	ω – angular speed	$N = \omega/2 \pi = Q/V$
V – V is the Volume per revolution	N – revolutions per second	
η_m – Mechanical efficiency ratio		

As $Q = VN$ and $\omega = 2\pi N$, where V is the volume of oil used by the motor per revolution and N is the number of revolutions per second,

$$\Delta PVN = T2\pi N; \text{ then } \Delta PV = T2\pi$$

$$T = \Delta PV/(2\pi) \tag{5.37}$$

As might be expected, the theoretical torque delivered is proportional to pressure and motor size. Note that there will be some losses due to friction. An important discussion is the understanding of how wear affects performance. The speed of the motor was defined above as $N = Q/V$. Wear in the motor increases leakage but the volume of the motor is not changed significantly. As the motor wears, the torque capability will remain almost the same, noting $T = \Delta PV/(2\pi)$, but importantly the motor will slow as $N = Q/V$. N will be reduced as the effective flow Q passing through the motor is reduced by the leakage. A worn motor is therefore identifiable by increasing heat on the motor housing.

Variations in motor design and complexity are considerable. Piston motors can be constructed with variable displacement allowing torque and speed control. Torque and speed control is achieved by varying the swashplate angle to get different amounts of cylinder stroke. A diagram of a variable displacement motor is shown in Figure 5.16.

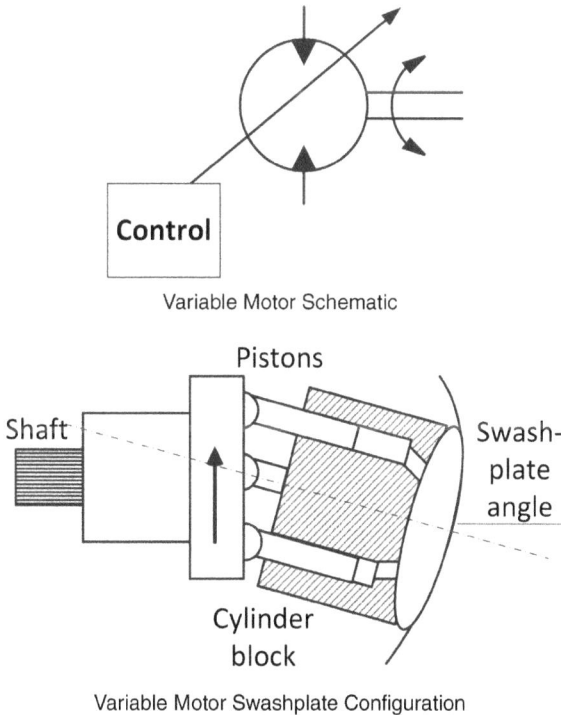

Variable Motor Schematic

Variable Motor Swashplate Configuration

FIGURE 5.16 Bi-directional variable displacement hydraulic motor.

5.3.6 MODELING CONTROL VALVES

Actuators, whether translational (cylinders) or rotary (motors) need the oil flow to be controlled, that is switched on, switched off, flow regulated, and pressure protected. A vast range of control valves have been developed. For the purpose of this chapter, only the basic introduction to four types of valves will be given.

    a. Directional control valves
    b. Flow restrictors
    c. Servo valves
    d. Proportional valves

The first two types, directional control valves and flow restrictors are common on a vast range of machinery including simple manually controlled machines. Directional control valves, as the name suggests, simply change the direction of oil flow and can be thought of as an on-off control or *bang-bang* control. Control options can be manual, electric, or pilot operated by oil or air. A typical directional control valve is shown in Figure 5.17. The oil connection ports are labeled *P*-pressure (or pump connection), *T*-Tank return connection, *A*-actuator connection, and *B*-actuator connection. In the case shown, control is via solenoid valves S_1 and S_2. Energizing S_1 can be seen as moving the valve flow paths to the right, thus connecting $P \rightarrow A$ and $B \rightarrow T$. Conversely, energizing S_2 connects $P \rightarrow B$ and $A \rightarrow T$. If nether S_1 and S_2 are energized all paths are blocked.

    A few of the many variations of the directional control valve are shown in Figure 5.18 and control variations in Figure 5.19 and flow path variations in Figure 5.20.

    Simple directional control often requires additional flow controls to get the desired effects in the circuit. In simple circuits this can be achieved by flow restrictors as shown in Figure 5.21. These valves are manually adjusted and set. It should be remembered that these devices will generate heat according to the power relationship Power $= \triangle P\, Q$, where $\triangle P$ is the pressure difference across the two connection ports in Pa and Q is the flow rate in m³/s. Modeling equations and parameters are listed in Table 5.6.

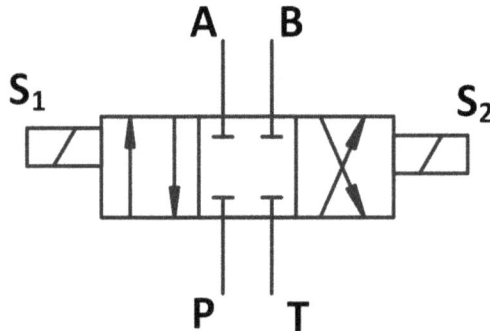

FIGURE 5.17   Typical 3 position directional control valve.

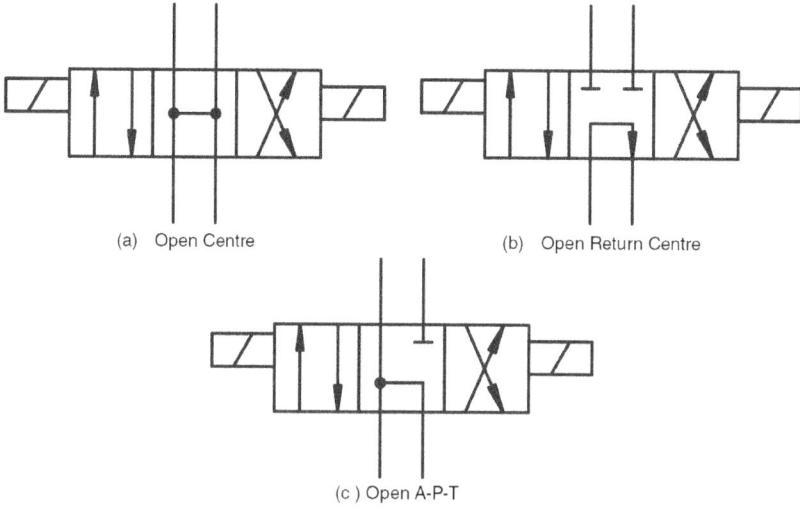

(a) Open Centre

(b) Open Return Centre

(c) Open A-P-T

FIGURE 5.18 Some flow path variations on typical 3 position directional control valve: (a) open centre, (b) open return centre, (c) open A-P-T.

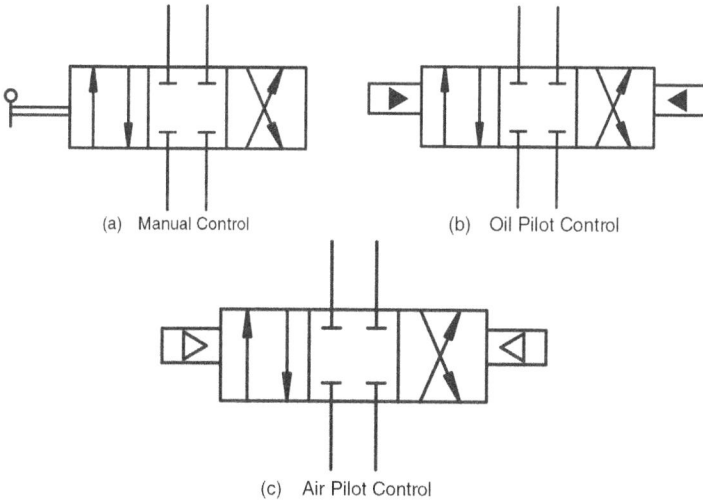

(a) Manual Control

(b) Oil Pilot Control

(c) Air Pilot Control

FIGURE 5.19 Control variations on typical 3 position directional control valve: (a) manual control, (b) oil pilot control, (c) air pilot control.

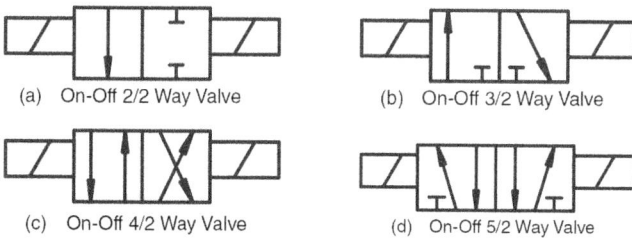

(a) On-Off 2/2 Way Valve

(b) On-Off 3/2 Way Valve

(c) On-Off 4/2 Way Valve

(d) On-Off 5/2 Way Valve

FIGURE 5.20 Further flow path and option variations of directional control valves: (a) on-off 2/2 way valve, (b) on-off 3/2 way valve, (c) on-off 4/2 way valve, (d) on-off 5/2 way valve.

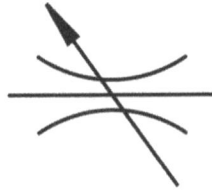

FIGURE 5.21 Flow restrictor.

It will be realized that the valves discussed so far have only crude capabilities in terms of machine control. A special type of directional control valve is known as the servo valve. Servo valves have the property that the flow allowed through the valve is proportional to the signal supplied to the valve. Servo valves can have useful frequency response up to 400 Hz. Common industrial applications and traditional nozzle flapper valve systems can retain full response up to 100 Hz with usable response between 100 and 300 Hz. Servo valves were traditionally constructed with an on-board solenoid (termed a servo motor) and a mechanical feedback system to give closed-loop control of the valve spool. An example of this type of valve construction is shown in Figure 5.22. The valve spool differs from directional controls in that the center always has some leakage. The spool is said to be *underlapped* if flow is allowed all the time, whereas an *overlapped* spool would have a dead zone as the spool is adjusted from one flow direction to the other. As it is desired to minimize all non-linearities, servo valves generally have minimal or no lapping. In some applications, a tiny dither is added to the control signal so that the spool is constantly moving and changing the oil flow direction in the zero signal state.

Such high frequency specifications are not always needed and, in the past three decades, proportional valves have been developed for lower frequency applications. Proportional valves have a much simpler construction and utilize a position transducer to sense where the valve spool as shown in Figure 5.23. Useful frequency response is approximately and order of magnitude less than servo valves, (i.e., 10–20 Hz). Proportional valves can also have *overlapped* spools.

Mathematical modeling of control valves for a control simulation as might be done in Simulink or similar software is quite simple for the simple on-off directional controls. Servo and proportional valves however introduce significant complexity. Mathematical models are listed in Table 5.7.

TABLE 5.6
Equations for Mathematically Modeling Direction Control Valves and Flow Restrictors

Input Variables & Data	Output Variables	Equation (5.39)
P – pressure	Q – Flow	$Q = \sqrt{q(P_1 - P_2)}$
Q – Flow		
q – Resistance coefficient		

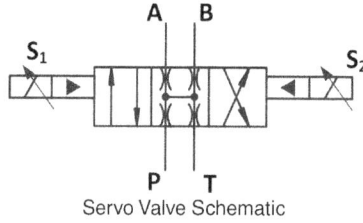

Servo Valve Schematic

Typical Servo-Motor Valve

FIGURE 5.22 Electro-hydraulic servo valve.

5.3.7 CLOSED-LOOP CIRCUITS

Section 5.3 starts with a basic and general hydraulic circuit, Figure 5.8. It was, in fact, what is called an open circuit. In pump-motor traction systems, a different approach, known as a closed-loop circuit, is sometimes used. Such systems generally have a bi-direction flow piston pump, achieved by swashplate control. The bi-directional flow provides a forward/reverse function for the traction system. The motors can either be fixed or variable. Closed-loop systems differ from open loop designs in that most of the oil flow recirculates in the traction system rather than to and from the tank (see Figure 5.8). An example of a closed-loop traction system is shown in Figure 5.24. Referring to Figure 5.24, the main pump P provides flow to the traction system in either of two directions. The closed circuit needs to have oil added to the system when changes of flow are required, and to account for leakage. This added oil is provided by a much smaller charging pump (CP). The charging pump runs at a low pressure as oil need only be added to the low pressure side of the main loop. The two check valves allow flow into whichever loop is at low pressure.

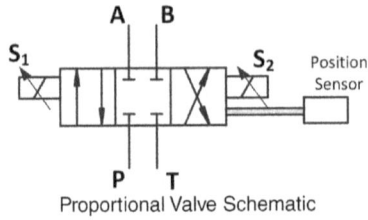

Proportional Valve Schematic

Typical Proportional Valve

FIGURE 5.23   Proportional valve.

TABLE 5.7

Equations for Mathematically Modeling Hydraulic Servo and Proportional Valves

Input Variables & Data	Output Variables	Equation (5.40)
P – pressure	Q – Flow	$Q = \dfrac{S}{100}\sqrt{q(P_1 - P_2)}$
$\triangle P$ – Pressure Difference		
$\triangle P_N$ – Rated pressure Difference		
Q – Flow		$Q = Q_N\sqrt{\dfrac{\Delta P}{\Delta P_N}}$
Q_N – Rated Flow[a]		
q – Resistance coefficient (or function representing flow characteristic)		
S – Signal percentage		
[a]Rated flow is the low across the valve at the rated pressure difference and at 100% signal		

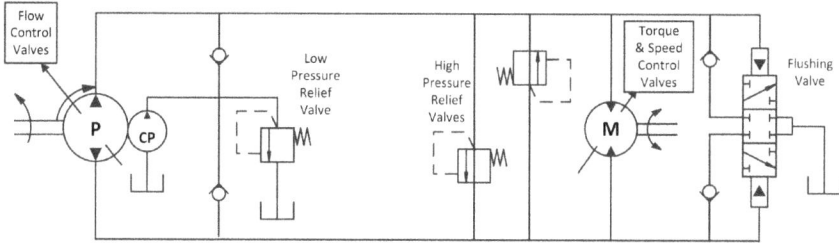

FIGURE 5.24 Closed-loop traction circuit.

The charging pump also provides for another function of circuit cooling. Oil is continually added to the main closed loop by the charge pump and removed from the loop by the flushing valve. Traction is applied by the hydraulic motor M and, if this motor is variable displacement, it can have torque and speed controls fitted. The main circuit is protected from over-pressure by the high pressure cross line relief valves which also serve as a brake on the motor when flow from the main pump P is reduced.

5.3.8 DYNAMIC PERFORMANCE MODELING OF ACTUATOR SYSTEMS

Simple modeling of a servo or proportional valve control can be achieved in control simulation software such as Simulink. A simple servo valve cylinder circuit is shown in Figure 5.25.

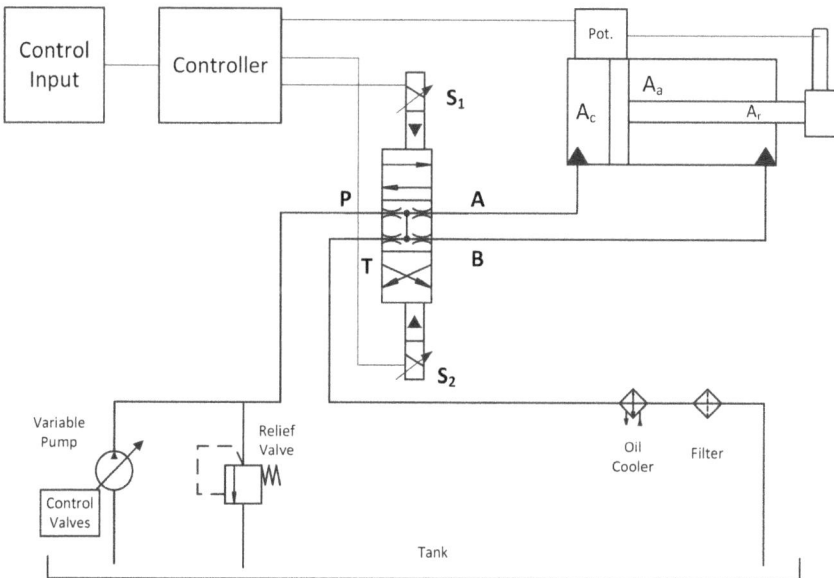

FIGURE 5.25 Simple servo valve cylinder circuit (displacement control).

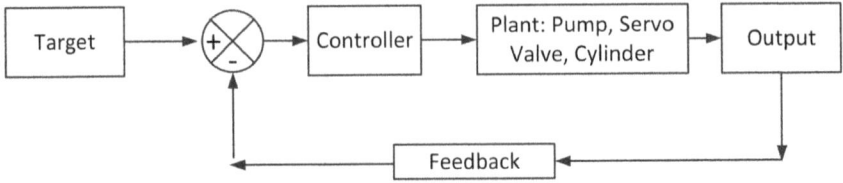

FIGURE 5.26 Simple servo valve cylinder circuit.

It will be seen that the controller requires some type of feedback system to enable setting of the required voltages on the servo valve. The feedback can be either force or distance measurements, but it is important to realize that, in either case, it is actually cylinder velocity that is controlled by the servo valve. Either position or force is achieved by the servo continually adjusting flows to follow the target signal. The control representation is the familiar closed loop as in Figure 5.26.

Assuming all components are linear and the system is not limited by pressure or flow constraints, a simple Simulink model is shown in Figure 5.27 to introduce basic control concepts.

The step response of the model is shown in Figure 5.28, noting that, for simplicity and clarity, only a proportional controller stage is used.

A force control version of the servo control is shown in Figure 5.29, with the associated simple model in Figure 5.30. Note that, although the feedback used for control is via a force transducer, the control of the cylinder is still by means of flow

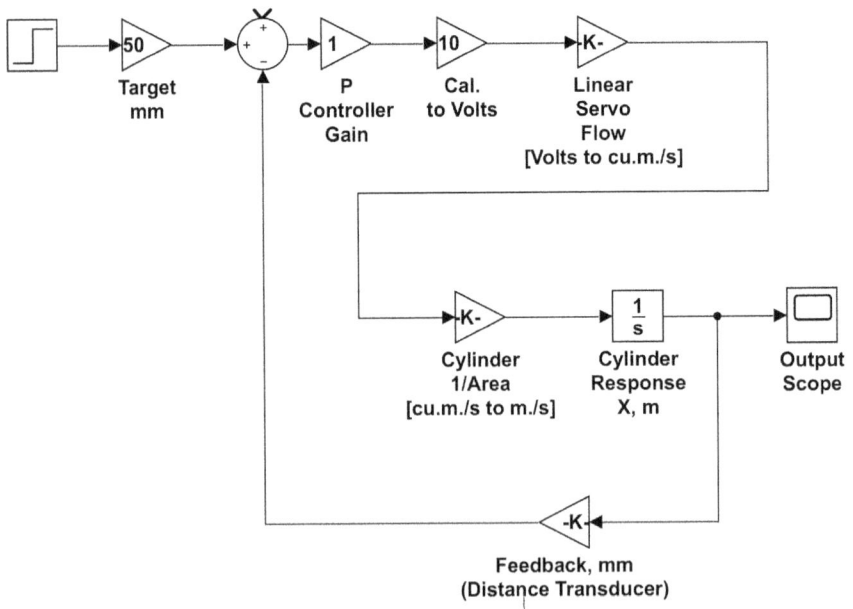

FIGURE 5.27 Simple Simulink model servo valve cylinder circuit (displacement control).

Step Response

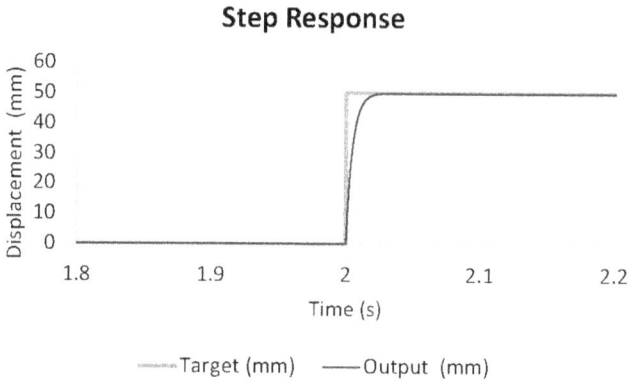

FIGURE 5.28   Step response of the position control servo cylinder circuit.

regulation. The flow control results in displacement control of the cylinder that gives rise to forces due to the stiffness of the structure and the compressibility of the oil. Effectively, force control gives a very much finer displacement control. It should be noted that, in force control, the cylinder would be uncontrolled if the load were suddenly taken away. It is a safety issue that an unconstrained force control cylinder, with a force target, will rapidly move to full stroke at start up. Even if zero were the target force, instrument noise and/or small electrical offsets in the system would result in the cylinder moving to full stroke and staying there. As this behavior is problematic to system set up and safety, control systems often have displacement control to allow set up controls and flow limits to prevent runaway feed rates.

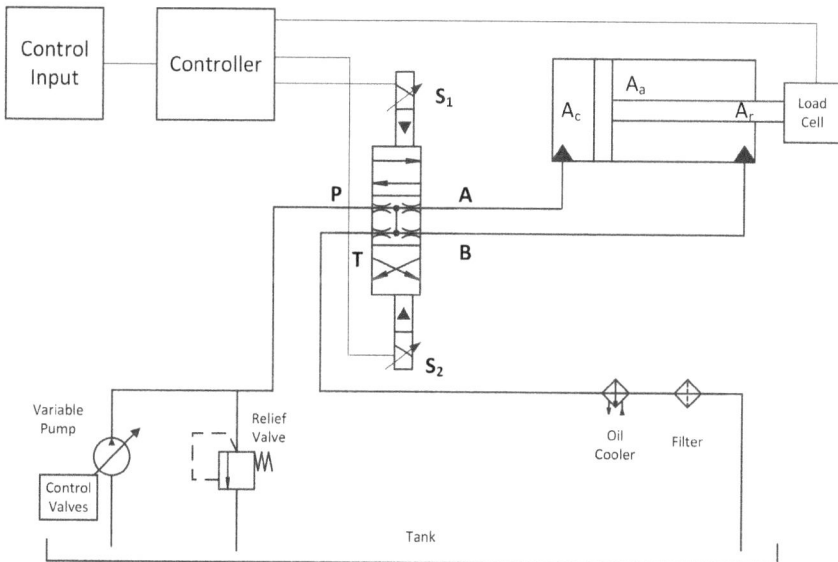

FIGURE 5.29   Simple servo valve cylinder circuit (displacement control).

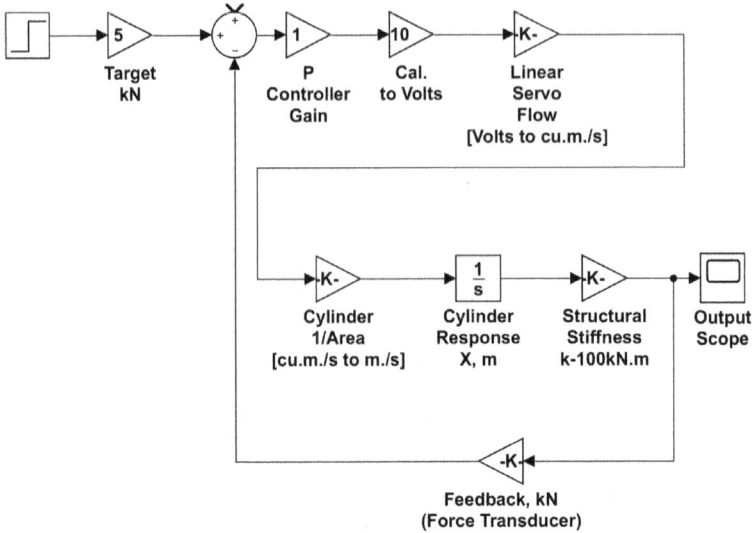

FIGURE 5.30 Simple Simulink model servo valve cylinder circuit (force control).

Of course, the above models are simplified. In practice most components will have limits and some non-linearities. Applications of double ended cylinders are relatively rare, so usually two different cylinder areas need to be modeled. Flow across the servo valve is proportional to the pressure drop across it and systems have a maximum pressure. These features can be added to the modeling. To consider the effects, the system needs a load so that pressure is raised by the load resistance (see Figure 5.31).

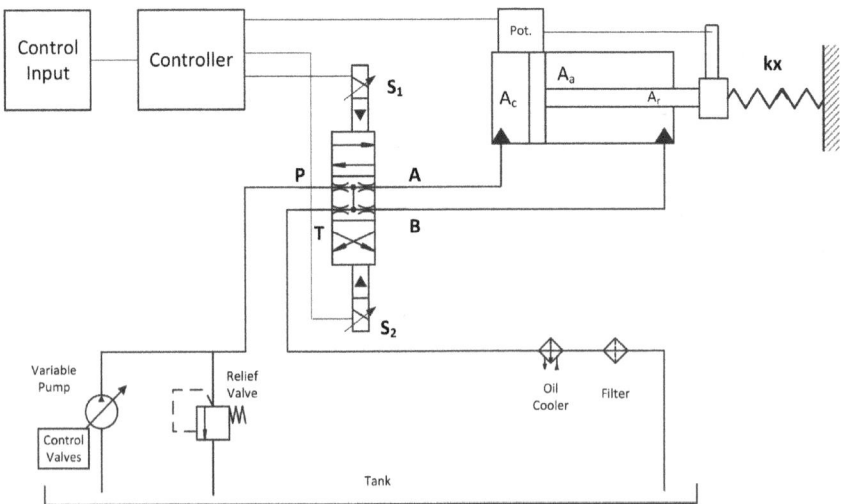

FIGURE 5.31 Servo valve cylinder circuit with spring load (displacement control).

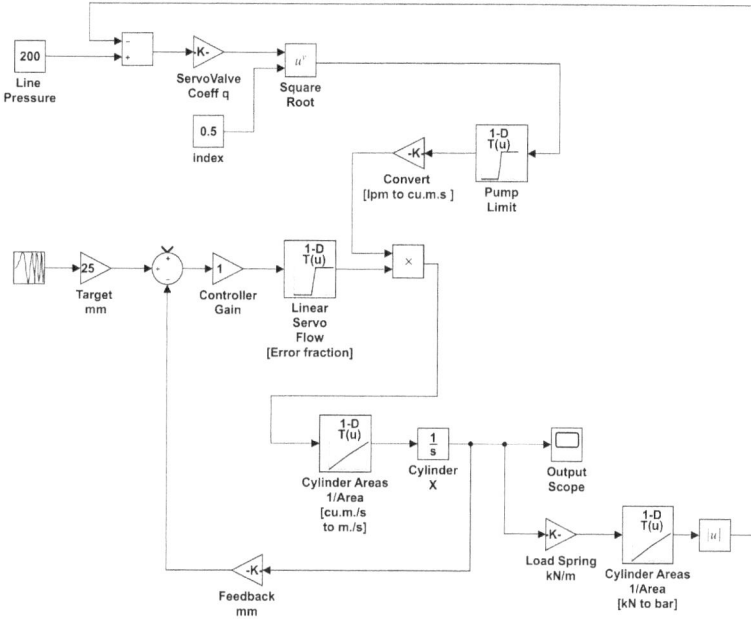

FIGURE 5.32 Simulink model servo valve cylinder circuit with spring load (displacement control).

An example is modeled in Simulink for a 100 lpm pump with a 200 bar pressure limit (see Figure 5.32). The servo valve is chosen as having 75 lpm flow at 100% signal at 70 bar pressure drop. The cylinder diameter is selected as 100 mm and the rod size as 25 mm. The cylinder is modeled with different areas for the cylinder end and rod end of the cylinder. The servo valve delivery is now modified by the pressure difference available. Despite the complexity, the model still does not capture the effects of back pressure on the cylinder, but it allows illustration of the effects of pressure and flow limits as given in Figure 5.33. It will be noted that as the operational frequency increases the servo valve flow is limited by available oil flow after $t \approx 13$ s. Only minor attenuation can be seen in Figure 5.33. Longer plots in Figure 5.34 show how the attenuation develops.

5.3.9 APPLICATIONS

Railway applications of hydraulics can be found in railway suspensions and draft gears. For purposes of mechatronic discussion in this chapter, only suspension controls are considered as several examples of mechatronic-hydraulic control can be cited. The reader is firstly referred to the excellent book chapters of Goodall and Mei [10, 11] for an overall treatment of vehicle dynamics and control issues associated with active suspensions. Discussion in this section will be limited to hydraulic design and control. Goodall and Mei [11] note that substantial applications of active suspensions are found in tilting trains [1, 12–14]. Specific hydraulic applications are noted for lateral secondary suspensions as presented in [15]. The authors

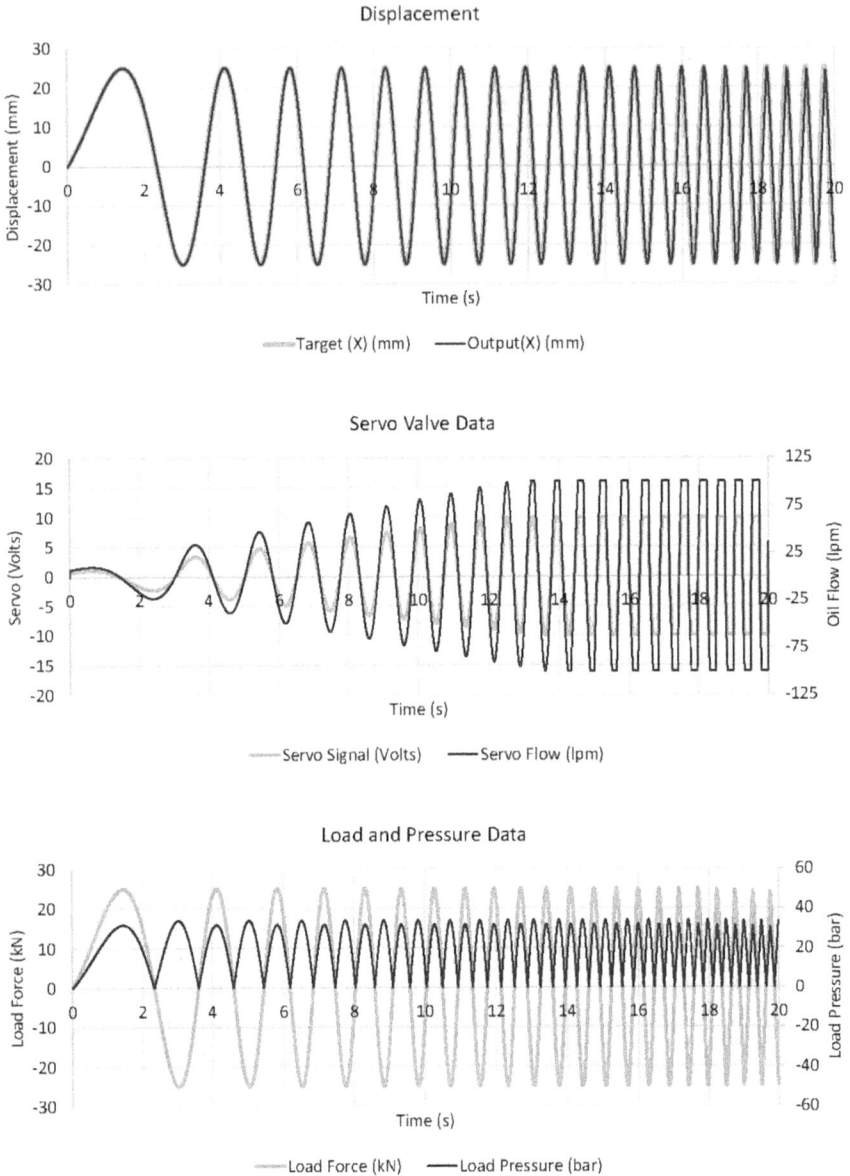

FIGURE 5.33 Frequency response (0.1–2 Hz) – Simulink model servo valve cylinder circuit with spring load.

noted a substantial improvement in ride quality with a 50% reduction in vibrations compared to the passive suspension. The first commercial use of electro-hydraulic systems in railway suspensions was developed by Sumitomo for the E2-1000 and E3 Shinkansen vehicles. The Sumitomo development was also aimed at controlling lateral vibration as noted in [16]. Goodall and Mei [11] also cite the latest development

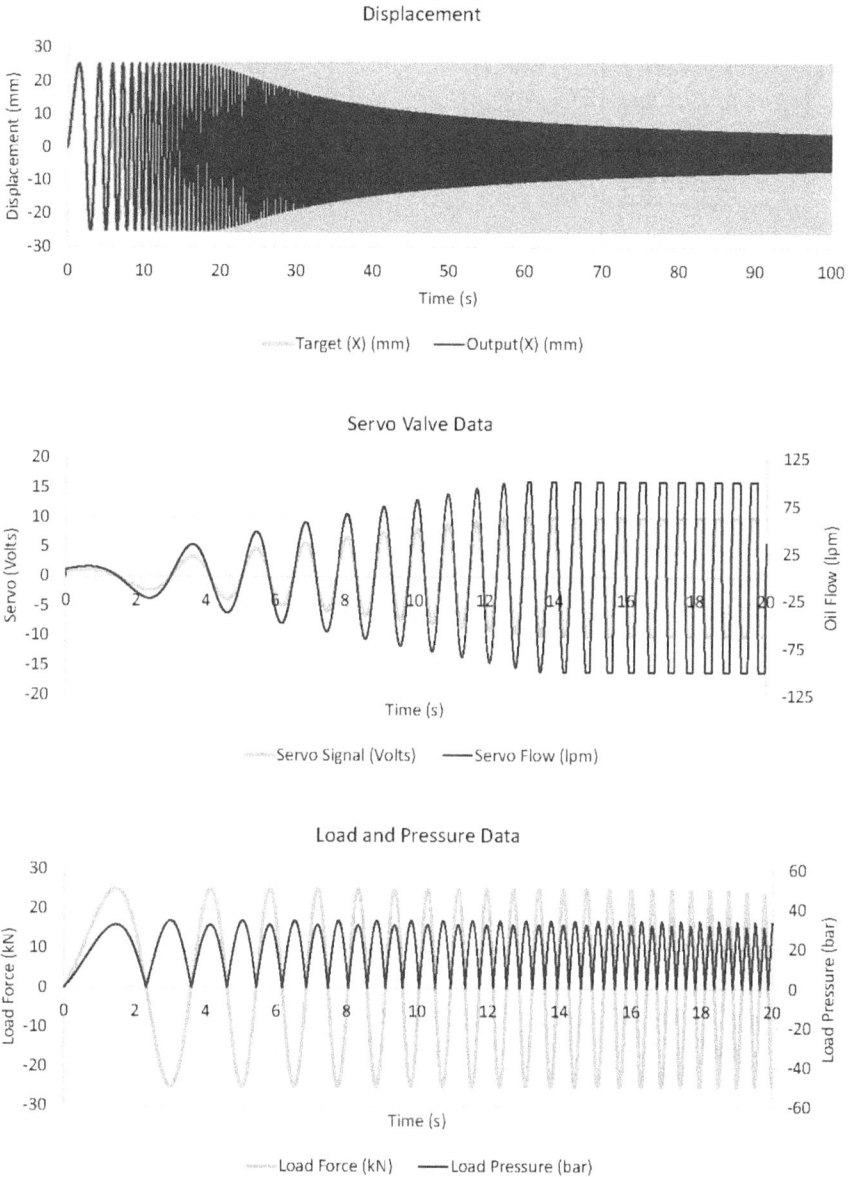

FIGURE 5.34 Frequency response (0.1–10 Hz) – Simulink model servo valve circuit with spring load.

as the Swedish Green Train Project, [17, 18]. The Green Train is a project involving Bombardier Transportation and KTH Royal Institute of Technology that aims at ride improvement in both lateral and vertical directions.

Suspension controls can be broadly classified as semi-active or fully-active. Semi-active controls are defined as where the existing connection element (e.g.,

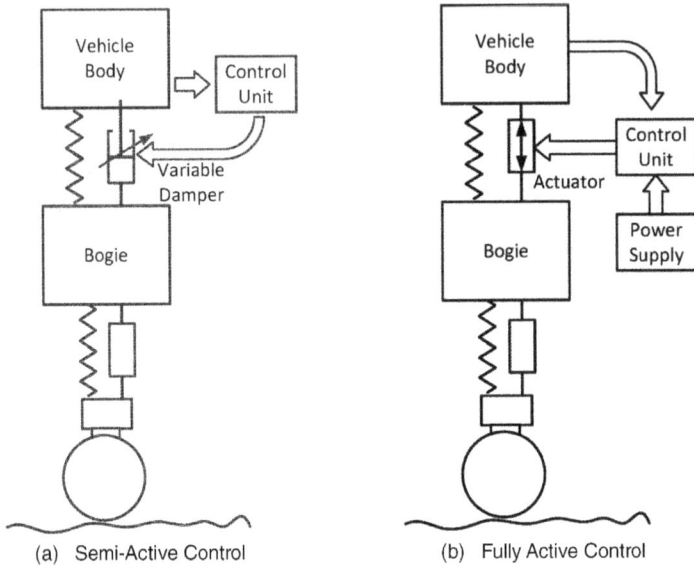

(a) Semi-Active Control   (b) Fully Active Control

FIGURE 5.35   Schematics of semi and fully active control: (a) semi-active control, (b) fully active control. (Concepts reproduced from [10, 11].)

a damper) has its characteristics altered by means of an external control as shown in Figure 5.35(a). For example, a linear damper might become a variable damper. Fully-active controls are defined as where the component is replaced by an actuator (see Figure 5.35(b)). The fully-active control therefore allows a wider range of input force characteristics and can achieve the greatest benefits [10, 11]. The authors also note the improved capability of the fully-active control in that energy can be added or subtracted to the suspension system to effect better performance, while semi-active control can only regulate energy that is present in the suspension.

Fully active control can also be implemented in three different ways as noted by [10, 11] and reproduced in Figure 5.36. Goodall and Mei [11] note that, in practical terms, there are advantages to including passive components. For example, in the option in Figure 5.36(a) the spring can provide most of the supporting force allowing the actuator to be smaller. The option in Figure 5.36(b) assists with controlling high frequency components.

Focusing on the possible hydraulic circuits, these semi-active and fully active concepts are re-drawn with hydraulic details. A possible mechatronic-hydraulic system for semi active control is shown in Figure 5.37(a) and a fully active version in Figure 5.37(b). These diagrams are functionally the same as those presented in [19] but have more detailed presentation of the hydraulic controls.

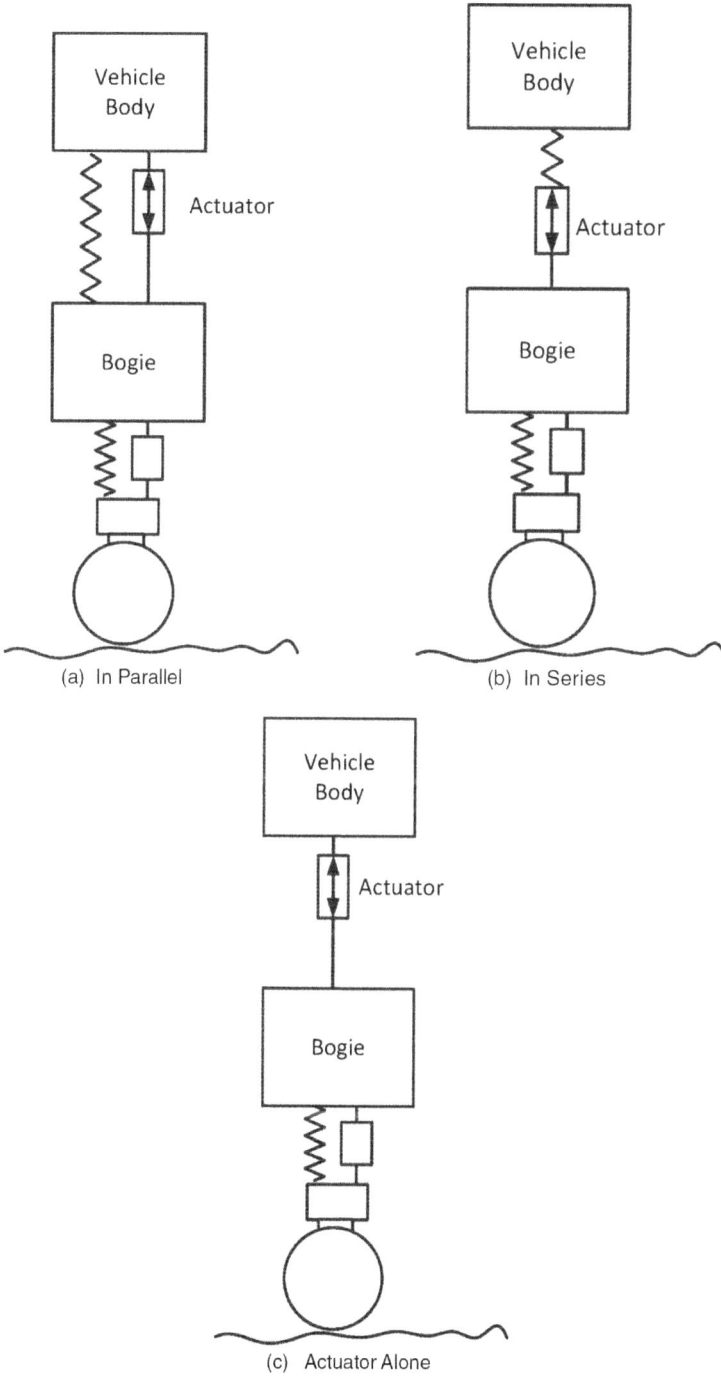

FIGURE 5.36 Three possible configurations fully active suspension control: (a) in parallel, (b) in series, (c) actuator alone. (Concepts reproduced from [10, 11].)

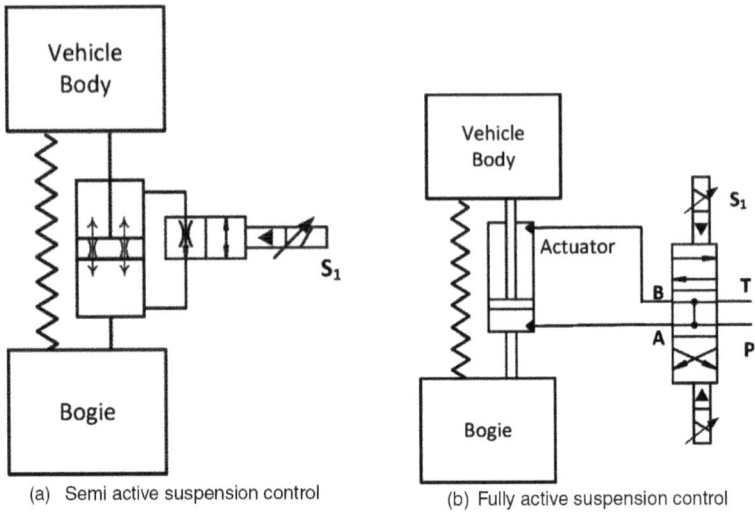

(a) Semi active suspension control   (b) Fully active suspension control

FIGURE 5.37   Possible electro-hydraulic system for (a) semi-active and (b) fully-active suspension control.

5.3.10 OVERALL SUMMARY

Electro-hydraulic systems offer compact high force actuators that have found some applications in improving railway suspensions. Servo control technology coupled with advanced control is capable of suitable frequency responses and can produce adequate forces. It is noted by Goodall [14] that the actuators require force control for suspension applications. Force control is achievable with attention to stiffnesses, precision, and detailed design. The presentation of the hydraulic components gives background to the complex issues behind hydraulic design showing that a whole system approach is needed to ensure the electro-hydraulic system can function across the range of frequencies required.

5.4 PNEUMATIC ACTUATORS

5.4.1 PNEUMATIC POWER SYSTEM BASICS

Pneumatics is the engineering discipline of transferring energy via pressurized air. Many of the considerations for fluid hydraulic systems also apply to pneumatic systems, but there are important differences due to the very different nature of the working fluid, air versus oil.

Operating pressures typically range up to 7 bar but pressures up to 200 bar are possible (as is done in special applications such as aviation). Generally speaking, though, air pressures in industrial applications are two orders of magnitude lower than fluid hydraulic systems. Air supply is obtained from compressors of various types, both positive displacement (piston, rotary vane, rotary lobe) and rotodynamic (centrifugal, axial flow, etc.). Like hydraulic systems the issue of pressure control is of utmost importance for both the equipment and the safety of personnel, but the equipment is quite different. An advantage of pneumatic systems is that the pressure

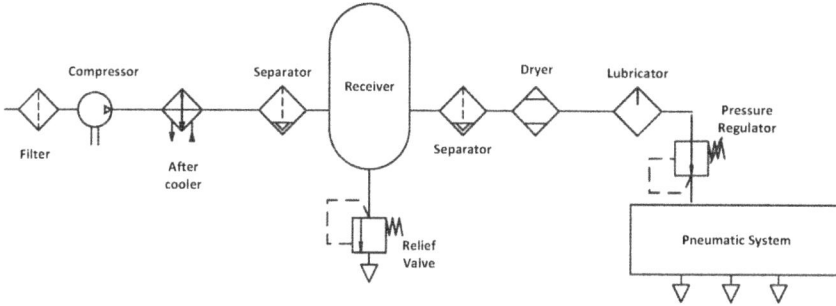

FIGURE 5.38   Basic components of a pneumatic circuit.

controls can vent or unload air flow simply to atmosphere. Apart from the convenience in less plumbing being required, this also means that the heat generated and stored as energy in the air is also released to atmosphere. Pneumatic systems do not need cooling systems to cool the exhausted air. Conversely, several problems are introduced by the air itself. Firstly, air is not clean so some type of intake filtering is needed before compression. Secondly, the air is not dry; it will usually contain some water vapor as expressed as humidity. Thirdly, air is not a suitable lubricant for industrial components so a small amount of lubricant must be added into the air stream. Yet a fourth issue is that it may be necessary to cool the air after compression. A very simple air system could be envisaged as shown in Figure 5.38.

Like hydraulic equipment, pneumatic equipment functions are also based on the transfer of force, being proportional to pressure and area, and on the pressure drop that occurs when the air flows through a restriction. Force or torque can be increased by increasing either the pressure or size of the actuator. An important difference is that the pressures are much lower than for industrial hydraulics and so the forces are likewise much lower or, alternatively, the actuators much larger. Unlike oil, air is compressible and not very stiff. Pneumatic actuators are therefore well suited to force and torque control applications while position control is more difficult to achieve. The movement achieved in an air actuator will depend on the mass flow to/or from the actuator and the state of pressure and temperature of the air in the actuator at the time (i.e., not just dependent on the air mass flow). Furthermore, the response will also depend on the heat transfer during the process. The governing gas laws are:

$$PV = mRT \tag{5.41}$$

where P is absolute pressure in Pa, V is volume in m^3, m is mass in kg, R is the gas constant for air, i.e., 287 J/(kg·K), and T is absolute temperature in degrees Kelvin.

Processes involving gases can be isothermal (constant temperature), adiabatic (no heat transfer), or polytropic (changes in temperature and having some heat transfer).

The set of relationships can be derived from Equations (5.41) and (5.42) and are as given in Table 5.8.

$$PV^n = \text{Constant} \tag{5.42}$$

where $n = 1$ is isothermal, $n = \gamma = 1.4$ is adiabatic and n = other values is polytropic

TABLE 5.8

Governing Gas Laws

Isothermal	Adiabatic	Polytropic
Equation (5.43)	Equation (5.44)	Equation (5.45)
$n = 1$	$n = \gamma = 1.4$	$n = 1.6$
$PV = mRT$	$PV = mRT$	$PV = mRT$
$PV = \text{Constant}$	$PV^\gamma = \text{Constant}$	$PV^n = \text{Constant}$
$P_1 V_1 = P_2 V_2$	$P_1 V_1^\gamma = P_2 V_2^\gamma$	$P_1 V_1^n = P_2 V_2^n$
$T_1 = T_2$	$T_2 / T_1 = (V_1/V_2)^{\gamma-1}$	$T_2/T_1 = (V_1/V_2)^{n-1}$

The control of pneumatic systems and the pressures generated, like the case of hydraulics, is fundamentally about controlling flow, but is more complicated due to the changing density of the working fluid (i.e., air) with both temperature and pressure. It is customary to express air flow rates in terms of volume flow rate at a specified air temperature and pressure, although a less ambiguous parameter would be simply the mass flow rate. A common air flow parameter is liters per minute FAD or m³/s FAD in SI units where FAD is designated as *Free Air Delivery* and is taken as at a standard atmospheric condition such as $P = 101.3$ kPa and $T = 20°C$. The 101.3 kPa/20°C condition gives the well-known value of 1.2 kg/m³ for air density.

Example:

The flow requirement is 100 liters per minute (lpm) at a delivery pressure of 700 kPa.
   Volume flow = 100 lpm = 100/(60 min per sec*1000 liters per m³) m³/s = 0.00167 m³/s
   Assuming isothermal conditions and taking FAD conditions for P_1 and V_1

$$P_1 V_1/t = P_2 V_2/t \tag{5.46}$$

Example:

101.3 * V_1/t = 801.3* 0.00167

$V_1/t = 801.3/101.3 * 0.00167 = 0.0133 \, \text{m}^3/\text{s} = 791 \, \text{liters/min (FAD)}$

Pressure control in pneumatics is provided by relief valves which release air flow when a certain pressure is reached and by pressure regulators that control the pressure supplied to a circuit. Pressure regulators can also be thought of as pressure reducing valves.

5.4.2 AIR PROPERTIES

Air properties are given in Table 5.9.

TABLE 5.9
Air Properties, Coefficients, and Parameters [20]

Parameter	Equation or Notes	Value	Units
Gas Constant, R	$PV = mRT$[a]	287	J/(kg·K)
Specific Heat (constant pressure), C_p	C_p	1005	J/(kg·K)
Specific Heat (constant volume), C_v	C_v	718	J/(kg·K)
Isentropic Index	$\gamma = C_p/C_v$	1.4	
Density at 101.3 kPa (0 kPa gauge) at 20°C	$\rho = P/RT$[a]	1.205	kg/m³
Density at 801.3 kPa (700 kPa gauge) at 20° C	$\rho = P/RT$[a]	9.529	kg/m³

[a] P is absolute pressure in Pa, V is volume in m³, m is mass in kg, and T is temperature in degrees Kelvin = Celsius + 273, ρ is in kg/m³.

5.4.3 PNEUMATIC CYLINDERS

Pneumatic cylinders provide translational force and displacement. A basic schematic of a cylinder is shown in Figure 5.39. Cylinders can come in many forms as shown in Figure 5.40.

Cylinders are described by their pressure rating, cylinder area, A_c, rod area A_r, and stroke as shown in Figure 5.39. Speed limits are not usually published, but speeds may need to be considered in extreme applications. The force capability of the cylinder is determined by the pressures applied and the areas. The force applied by the oil on the *cylinder end*, the left end in Figure 5.38 is given by $F_c = P_c A_c$. Force applied by the oil on the *rod end*. the right end in Figure 5.38 is given by $F_r = P_a A_a$, noting that pressure acts on the annulus area $A_a = A_c - A_r$. The force delivered by the cylinder is therefore $F = F_c - F_r$. Maximum forces are delivered by the cylinder when pressure is applied to one end with no back pressure at the other end, giving forces of either PA_c or PA_a.

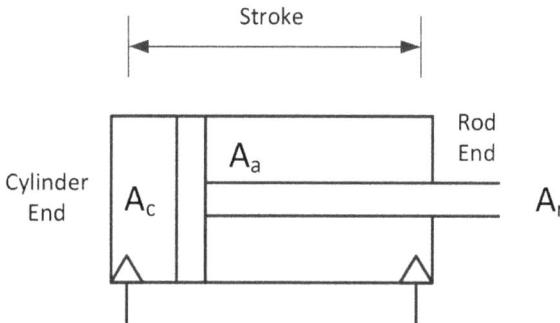

FIGURE 5.39 Double acting pneumatic cylinder.

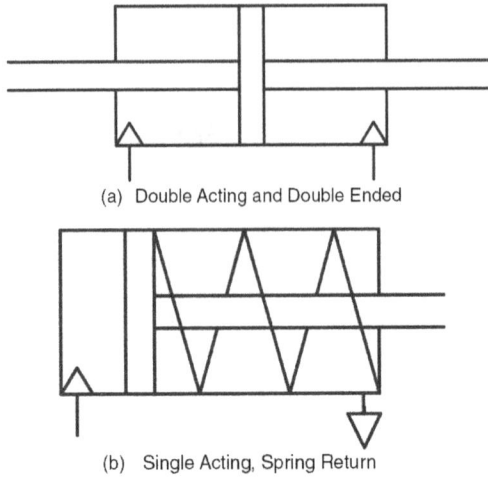

(a) Double Acting and Double Ended

(b) Single Acting, Spring Return

FIGURE 5.40 Other pneumatic cylinder configurations: (a) double acting and double ended, (b) single acting, spring return.

Mathematical modeling of cylinders can be achieved using the parameters in Figure 5.41 and detailed in Table 5.10.

The pneumatic force equation $F = P_cA_c - P_aA_a$ is the general case assuming some back pressure on the exhausting port, as could be the case when controlled by a directional control valve. In simpler controls, air on the exhausting side of the cylinder will be released to return to atmosphere, so cylinder forces are often stated as $F = P_cA_c$ when the cylinder is extending and $F = P_aA_a$ when contracting and with pressures noted as gauge pressures rather than absolute.

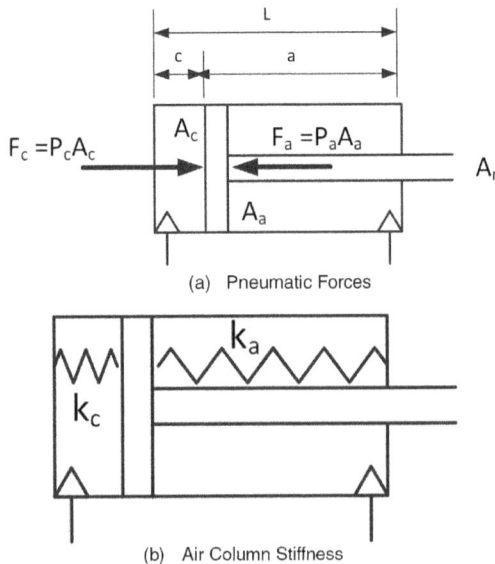

(a) Pneumatic Forces

(b) Air Column Stiffness

FIGURE 5.41 Pneumatic cylinder parameters: (a) pneumatic forces, (b) air column stiffness.

TABLE 5.10
Equations for Mathematically Modeling Pneumatic Cylinders

Input Variables and Data	Output Variables	Equation (5.47)[a]
P – pressure	F – Hydraulic force	$F = P_c A_c - P_a A_a$
m – air mass inside a cylinder volume [b]		
$\dot{m}$ – air mass flow	v_{extend} – Rod extension speed	$v_{extend} = \dot{m}/\rho A_c$
ρ – air density ($\rho = P/RT$) inside a cylinder volume [b]	$v_{contract}$ – Rod contraction speed	$v_{contract} = \dot{m}/\rho A_a$
Q – Flow		
A_c – Cylinder area		$k_c = m\,R\,T/c^2 = P_c A_c/c$
A_a – Annulus area	k – Air column stiffness	$k_a = m\,R\,T/a^2 = P_a A_a/a$
L – Stroke Length		
a – Piston position measured from the rod end		$k_{ac} = k_c + k_a$
c – Piston position measured from the cylinder end		

[a] P is absolute pressure in Pa, V is volume in m³, m is mass in kg, and T is temperature in degrees Kelvin = Celsius + 273.

[b] refers to volume inside the cylinder in the cylinder end, or the volume inside the cylinder in the rod end.

The same discussion extends to the air column stiffness. If the cylinder has pressure on both sides, the two air columns will act as parallel springs. The air column stiffness can be derived as follows:

$F = P\,A$ and for a gas, $PV = mRT$

$F = (mRT/V)\,A$ where V is the air volume and A is the cylinder area.

$F = (mRT/A\,x)\,A = mRT/x$ where x is the length of the cylinder air volume

Differentiating

$$dF/dx = (mRT/A\ x)A = -mRT/x^2$$

$$\Delta F = k\,\Delta x = -mRT/x^2\ \Delta x; \quad k = -mRT/x^2 \tag{5.48}$$

Recalling that $PV = mRT$ and $PAx = mRT$

$$\Delta F = k\,\Delta x = -PA/x\ \Delta x; \quad k = -PA/x \tag{5.49}$$

thus showing the proportionality of the cylinder stiffness to three variables: cylinder pressure, cylinder area, and the length of the air column. Both the position of the piston in the cylinder and the pressure therefore affect the stiffness and the control response. The relationships are illustrated in Figure 5.42.

If the application is single acting (i.e., has pressure applied on only one side of the piston), the stiffness will be governed by the stiffness of just one air column,

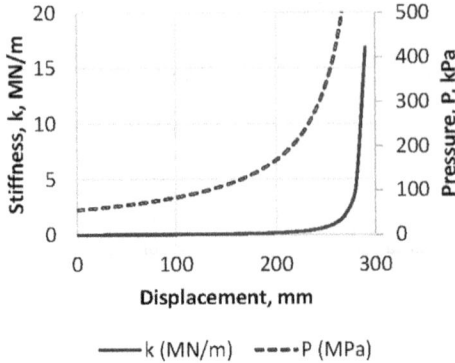

FIGURE 5.42 Pneumatic cylinder stiffness (example: Diameter 35 mm, stroke 300 mm): (a) single acting cylinder stiffness and pressure, (b) double acting cylinder stiffness.

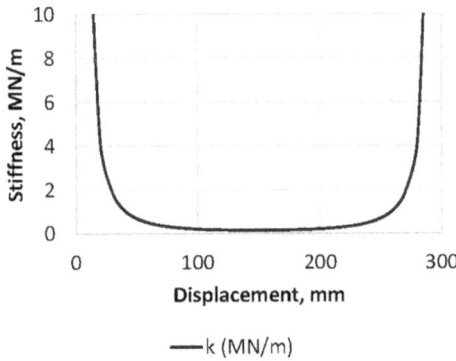

i.e., $k_c = m R T/c^2 = P_c A_c/c$ or $k_a = m R T/a^2 = P_a A_d/a$ but the general case is $k_{ac} = k_c + k_a$. It is therefore important to consider cylinder area for both design force and design stiffness requirements. Cylinder length and the position of the piston in that length also affects stiffness. Control response of a cylinder will therefore be affected by cylinder and rod diameters, cylinder length and the position of the rod, and of course air pressure at the time of the control response. Note that the pressure will be subject to changes in temperature due to work done and heat transfer. While it is common to assume isothermal behavior because it greatly simplifies calculations, fast processes will approximate more closely to adiabatic (no heat transfer) or polytropic processes as detailed in Table 5.8.

5.4.4 AIR MOTORS

Air motors provide torque and angular displacement. A basic schematic of a motor is shown in Figure 5.43.

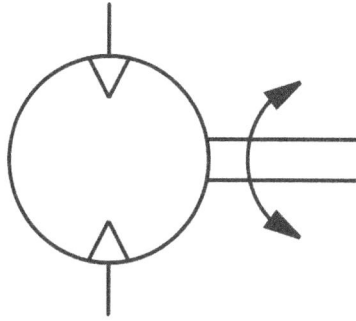

FIGURE 5.43 Bi-directional pneumatic motor.

The maximum theoretical torque capability of fixed displacement motor can be deduced from the power equation:

$$\text{Power} = P_1 Q_1 - P_2 Q_2 = T\omega \qquad (5.50)$$

where P_1, P_2 are inlet and exhaust pressures at the two motor connection ports in Pa.

Q_1, Q_2 are the volumetric flow rates in m³/s in the motor at the two motor connection ports. Importantly, this parameter must consider air density noting $Q = \dot{m} / \rho$ and $\rho = P/RT$ from Table 5.10

T is the torque in Nm

ω is the angular velocity of the shaft in rad/s $\omega = 2\pi N$ where N is revolutions per second.

As might be expected, the theoretical torque delivered is proportional to pressure and motor size. Note also that speed of the motor will vary with load because changes in load will be reflected in changes in pressures. Increases in pressure within the motor will increase air density and therefore reduce the volumetric flow rate and shaft speed. Note that there also will be some losses due to friction and leakage. An important discussion is the understanding of how wear affects performance in positive displacement motors (i.e., vane of lobe types). The speed of the motor in revolutions per second can be defined as $N = Q/V$ where V is the volume per revolution. Wear in the motor increases leakage but the volume of the motor is not changed significantly. As the motor wears, the torque capability will remain almost the same, but noting importantly that the motor will slow because the actual volumetric flow rate Q passing through the motor is reduced by the leakage. A worn motor is therefore identifiable by increasing heat on the motor housing and slowing speed.

Mathematical modeling of air motors can be achieved using the parameters detailed in Table 5.11, noting that due to the changes in volumetric efficiency performance curves are needed to ensure a basic understanding and correct modeling of the volumetric and mechanical efficiencies. In Table 5.11 this complexity is simplified to the single parameter η_{mv}.

TABLE 5.11
Equations for Mathematically Modeling Air Motors

Input Variables & Data	Output Variables	Equation (5.51)
P – pressure	T – Pneumatic torque	$T = \eta_{mv}\,(P_1Q_1 - P_2Q_2)/(2\pi N)$
m – air mass inside a cylinder volume [b]	ω – angular speed	
$\dot{m}$ – air mass flow	N – revolutions per second	
ρ – air density ($\rho = P/RT$) inside a cylinder volume [b]		$\rho = P/RT$
Q – Flow		
V – V is the Volume per revolution		
η_{mv} – efficiency ratio (mechanical and volumetric)		

5.4.5 CONTROL VALVES

Actuators, whether translational (cylinders) or rotary (motors), need the air flow to be controlled, i.e., switched on, switched off, flow regulated and pressure protected. A vast range of control valves have been developed. For the purpose of this chapter, only the basic introduction to three types of valves will be given.

   a. Directional control valves
   b. Flow restrictors
   c. Proportional valves

The first two types, Directional control valves and flow restrictors are common on a vast range of machinery including simple manually controlled machines. Directional control valves, as the name suggests, simply change the direction of air flow and can be thought of as an on-off control or *bang-bang* control, control options can be manual, electric, or pilot operated by air. A typical directional control valve is shown in Figure 5.44. The air connection ports are labeled P – pressure (airline or compressor

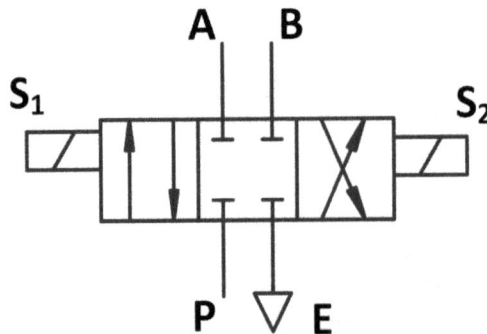

FIGURE 5.44 Typical 3 position directional control valve.

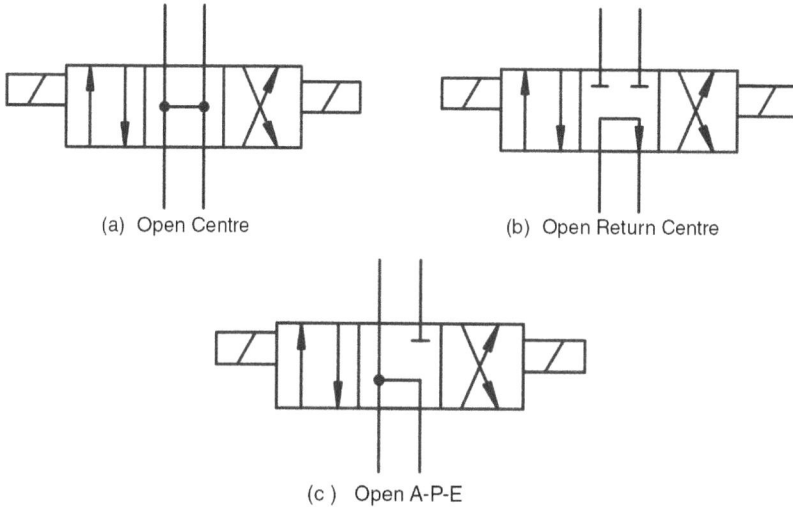

(a) Open Centre   (b) Open Return Centre

(c) Open A-P-E

FIGURE 5.45   Some flow path variations on typical 3 position directional control valve: (a) open centre, (b) open return centre, (c) open A-P-E.

connection), E – Exhaust to atmosphere, A – actuator connection, and B – actuator connection. In this case shown control is via solenoid valves S_1 and S_2. Energizing S_1 can be seen as moving the valve flow paths to the right, thus connecting $P \rightarrow A$ and $B \rightarrow T$. Conversely, energizing S_2 connects $P \rightarrow B$ and $A \rightarrow T$. If neither S_1 nor S_2 is energized all paths are blocked.

A few of the many variations of the directional control valve are shown in Figure 5.45 and control variations in Figure 5.46 and flow path variations in Figure 5.47.

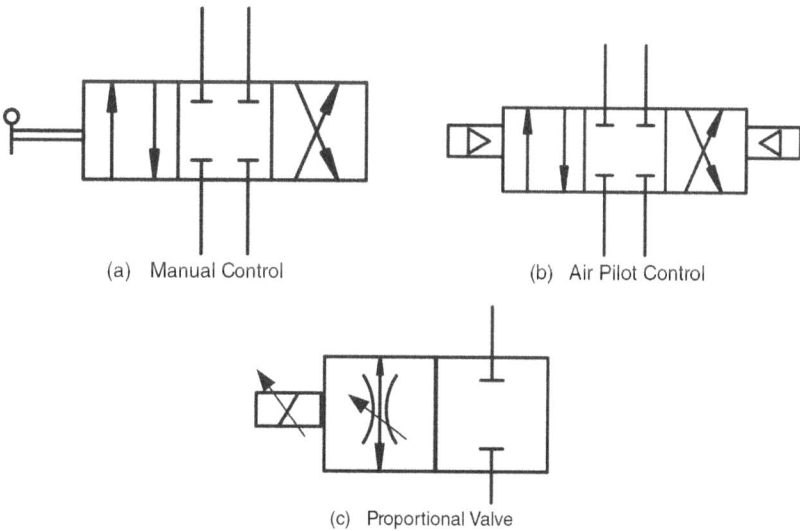

(a) Manual Control   (b) Air Pilot Control

(c) Proportional Valve

FIGURE 5.46   Control variations on typical 3 position directional control valve: (a) manual control, (b) air pilot control, (c) proportional valve.

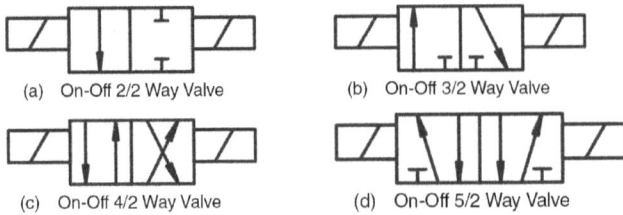

(a) On-Off 2/2 Way Valve   (b) On-Off 3/2 Way Valve

(c) On-Off 4/2 Way Valve   (d) On-Off 5/2 Way Valve

FIGURE 5.47 Further flow path and option variations of directional control valves: (a) on-off 2/2 way valve, (b) on-off 3/2 way valve, (c) on-off 4/2 way valve, (d) on-off 5/2 way valve.

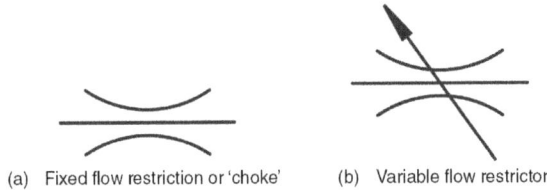

(a) Fixed flow restriction or 'choke'   (b) Variable flow restrictor

FIGURE 5.48 Flow restrictors: (a) fixed flow restriction or "choke," (b) variable flow restrictor.

5.4.6 RESTRICTIONS AND CHOKES

Simple directional control often requires additional flow controls to get the desired effects in the circuit. In simple circuits this can be achieved by flow restrictors as shown in Figure 5.48. These valves are either of fixed size (see Figure 5.48(a)) or are manually adjustable as shown in Figure 5.48(b). It should be remembered that these devices will generate heat according to the power relationship Power = $\triangle PQ$, where $\triangle P$ is the pressure difference across the two connection ports in Pa and Q is the flow rate in m³/s. In pneumatic systems, fixed restrictions are often called *chokes*. To fully understand the phenomena of *choking*, the reader should consult a fluid mechanics reference on compressible flow. A useful introductory chapter in [21] is recommended. Choking is not just associated with restrictions or orifices. It should be realized that plumbing and pipes are also circuit restrictions. Flow in an air pipe or in a duct with parallel sides becomes what is said to be *choked* if the pipe or duct is supplied with sufficient back pressure and the duct is sufficiently long [21].

For low Mach numbers, simple fluid equations that assume incompressible flow as shown in Table 5.12 can be used with relatively small errors. The Mach number

TABLE 5.12

Equations for Mathematically Modeling Direction Control Valves and Flow Restrictors (Incompressible and Low Subsonic Applications)

Input Variables & Data	Output Variables	Equation (5.52)
P – pressure	Q – Flow	$Q = \sqrt{q(P_1 - P_2)}$
Q – Flow		
q – Resistance coefficient		

TABLE 5.13

Compressible Flow Regimes [21]

Flow Regime	Mach Number
Incompressible	$M \ll 1.0$
Subsonic	$M < 1.0$
Transonic	$M \approx 1.0$
Supersonic	$M > 1.0$
Hypersonic	$M \gg 1.0$

is defined as the ratio of the speed to speed of sound in the material—gas, liquid, or solid. [21]. The speed of sound in an ideal gas is given by:

$$c = \sqrt{\gamma \frac{P}{\rho}} = \sqrt{\gamma RT} \qquad (5.53)$$

Mach number is then:

$$M = \frac{V}{C} \qquad (5.54)$$

Flow regimes are defined in relation to Mach number [21] and the terminology is shown in Table 5.13.

Of particular interest to the use of air in control circuits and actuators is the flow of air in ducts/pipes and through orifices. In both cases, conditions can easily be envisaged where *choked* flow can occur. Choked flow can be simply thought of as the condition where the flow is supersonic as it leaves a pipe, duct, or orifice. It can be shown that compressible flow in a pipe cannot exceed sonic. If there is sufficient back pressure and the pipe is sufficiently long, the air will accelerate along the pipe and reach a steady flow state such that the speed at the pipe exit is sonic. This condition is said to be *choked* flow. If the back pressure on the pipe is increased further or the pipe is made longer, the conditions at the end of the pipe remain choked. If the back pressure is reduced, at some point the exit velocity of the air from the pipe will become subsonic. This feature can also be exploited for control purposes by the use of very small diameter drillings that provide choking conditions and therefore simplify air flow control and timing. Of interest is the relationship when $M = 1.0$ [21]

$$\frac{\dot{m} \sqrt{RT_o}}{P_o A} = \text{Constant} \qquad (5.55)$$

where T_o is the stagnation temperature and P_o is the stagnation pressure of the conditions upstream from the duct. For an ideal gas and isentropic assumptions $n = \gamma$.

$$P_o = P \left(1 + \frac{\gamma - 1}{2} M^2 \right) \text{ and } \frac{P_o}{P} = \left(\frac{T_o}{T} \right)^{\left(\frac{\gamma}{\gamma - 1} \right)} \qquad (5.56)$$

TABLE 5.14

Value of the Choked Mass Flow Function in Equation (5.55) at $M = 1.0$ [21]

Ratio of Specific Heats	Constant
$n = 1.3$	0.6673
$\gamma = 1.4$	0.6847
$n = 1.67$	0.7267

The values of the Constant therefore depend only on the ratio of specific heats γ or n. Values for (5.55) are presented in Table 5.14 and a much more detailed presentation is published in [21].

Conditions for choking of air flow for a simplified case of a back pressure in a duct exiting abruptly to a low back pressure are:

$$\frac{P_{exit}}{P_o} < \left(\frac{2}{\gamma+1}\right)^{\left(\frac{\gamma}{\gamma-1}\right)} = 0.528, \tag{5.57}$$

$$\text{or } P_o > 1.89\, P_{exit}$$

so choked flow can occur for conditions where the stagnation pressure upstream of the exhaust duct exceeds exit back pressure by a factor of 1.89. Obviously, the above is a significant simplification of the underlying theory, equations, and behavior, but is illustrative of the concept of choked flow and why the behavior is so useful in pneumatic controls. It means that as long as the pressure ratio across the choke is higher than given by Equation (5.57), the mass flow will be proportional to just upstream conditions and the duct area.

5.4.7 APPLICATIONS

Railway applications using pneumatic actuators can be found in railway braking and suspensions.

5.4.7.1 Railway Air Braking

Since the mid 1800's railways have used pneumatic principles to achieve train brake control without electrical systems. These systems exploit the behavior of chokes which has allowed cylinder fill and exhaust rates to be finely tuned by careful selection of the diameters of orifices and ducts in the valving system. Using the mass flow Equation (5.55), railway engineers have been able to produce very sophisticated brake controls. A major problem with the fully pneumatic train brake system was the need to convey both air supply and control signals in the air. As explained, the velocity in a pipe cannot exceed sonic velocity which occurs at the pipe exhaust. Brake engineers exploited this feature, firstly by making the brake application trigger on exhausting air. The problem remained, however, that sonic velocity is not achieved

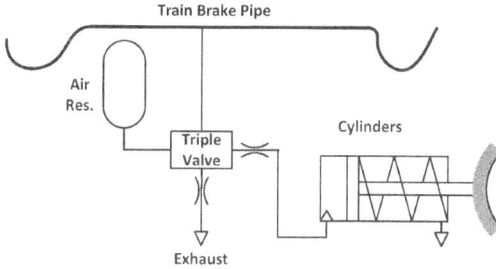

FIGURE 5.49 Basic pneumatically controlled wagon brake circuit.

along the pipe, being slower the further away from the air exhaust. This problem was addressed in two ways. One was to use the pressure drop at each vehicle to trigger the opening of further exhaust ports, effectively keeping the distance between the exhaust and the braking system short. A second strategy was to use more sensitive valves to sense a pressure wave running through the pipe. Using one or both of these approaches, railway brakes have been able to achieve signal propagation speeds of up to ~300 m/s. A very basic schematic of a fully pneumatically controlled brake system is shown in Figure 5.49 with control logic in Figure 5.50. It is emphasized that this is the simplest of systems. The triple valve, first patented in 1873 [22] evolved considerably and differently in different continents. These valves are now very complicated. Improvements have included relay systems, compensation for empty and loaded wagons, boosted pressures for emergency applications, accelerated release systems, and accelerated pipe exhausting systems. The systems so produced allowed operation on longer trains, however the propagation speed limitation in the train bake

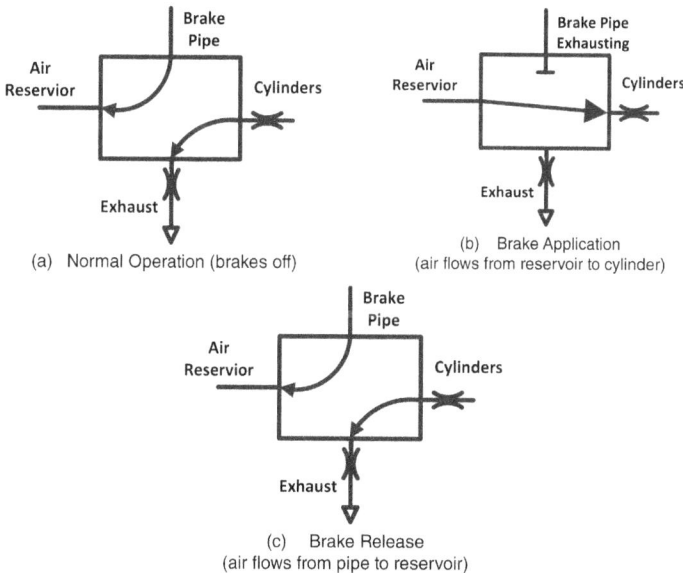

FIGURE 5.50 Basic pneumatically controlled wagon brake circuit: (a) normal operation (brakes off), (b) brake application (air flows from reservoir to cylinder), (c) brake release (air flows from pipe to reservoir).

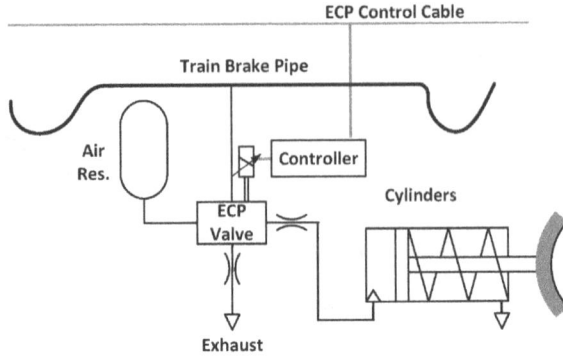

FIGURE 5.51   Basic electronically pneumatically controlled wagon brake circuit.

pipe due to choking cannot be solved as propagation can never be at a greater speed than sonic velocity, i.e., 350 m/s. This means that very long trains, even with the most sophisticated triple valve braking systems, still have very long delays in braking application; for example, a 2.4 km long train will have a delay of at least 6.9 s. This constitutes a substantial amount of unbraked momentum for several seconds, causing inter-wagon impacts and large coupling forces. Recently there has been wide take-up of electronically controlled brakes as shown in Figure 5.51. Again, a greatly simplified diagram is given. Of particular note is that the brake pipe is not used for control and can be charging the onboard reservoir all the time (see Figure 5.52(b)).

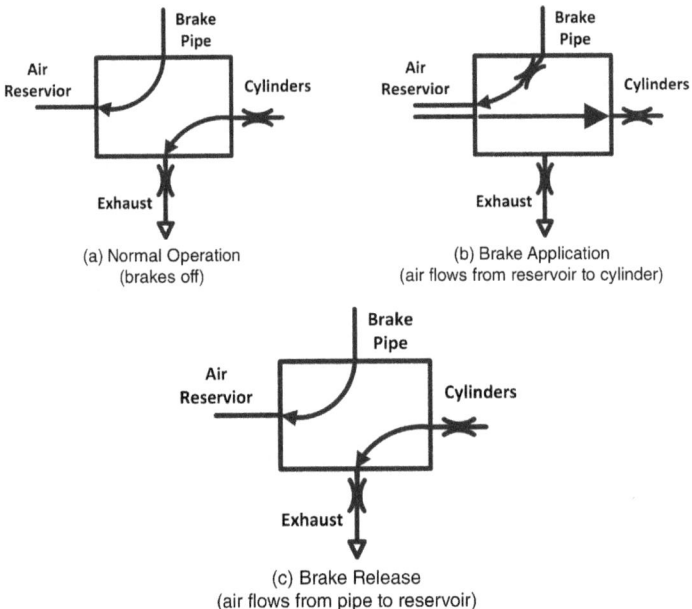

FIGURE 5.52   Basic electronically controlled wagon brake circuit : (a) normal operation (brakes off), (b) brake application (air flows from reservoir to cylinder), (c) brake release (air flows from pipe to reservoir).

5.4.7.2 Railway Air Suspensions

Air suspensions are commonly found in passenger rollingstock. As shown in Figure 5.53, the typical arrangement consists of a bellows which acts as a spring and an air reservoir which maintain pressure.

As discussed in the sections on cylinders, the stiffness of an air column is dependent on pressure as illustrated in Figure 5.54.

The stiffness can therefore be made more stable by adding an air reservoir; the increased volume means that a larger range of displacement can be utilized before the suspension gets too stiff (see Figure 5.54). Damping can be added by a restrictor or choke added to the air pipe connecting the bellows with the air reservoir as shown in Figure 5.53(a). The suspension can be made semi-active by adding some control to this restriction with a proportional valve as shown in Figure 5.53(b). An obvious advantage of air suspensions is thus illustrated, as suspension stiffness can be adjusted by simply adjusting pressure. The semi-active control can be used to change both the pressure and the amount of damping by regulating flow to and from a surge reservoir (see Figure 5.53(b)).

For purposes of modeling and controlling vehicle dynamics, equivalent mechanical models taking in the characteristics of the pneumatic system are needed. The suspension characteristic of the air spring or bellows is a complex combination of air

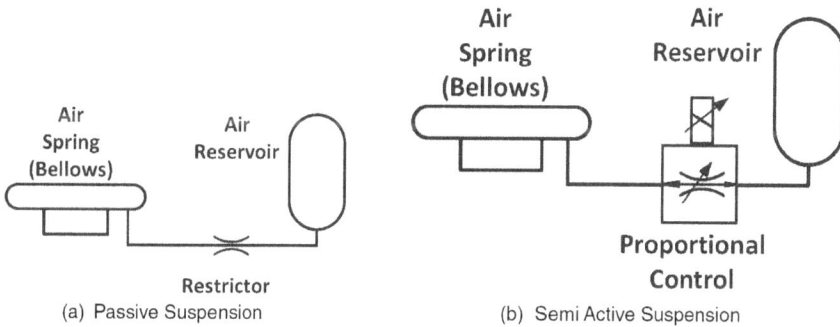

FIGURE 5.53 Schematics of (a) passive and (b) semi-active railway air suspensions.

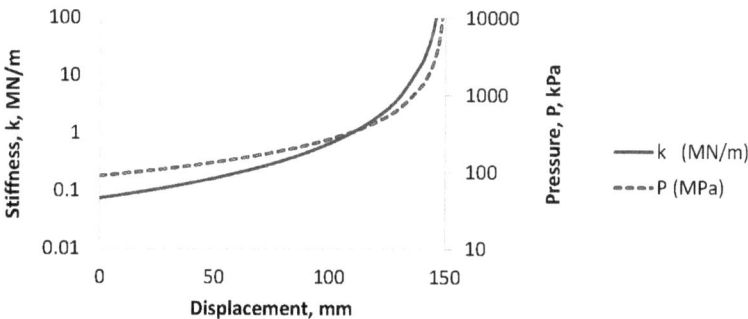

FIGURE 5.54 Pneumatic cylinder stiffness (example: Diameter 400 mm, stroke 150 mm).

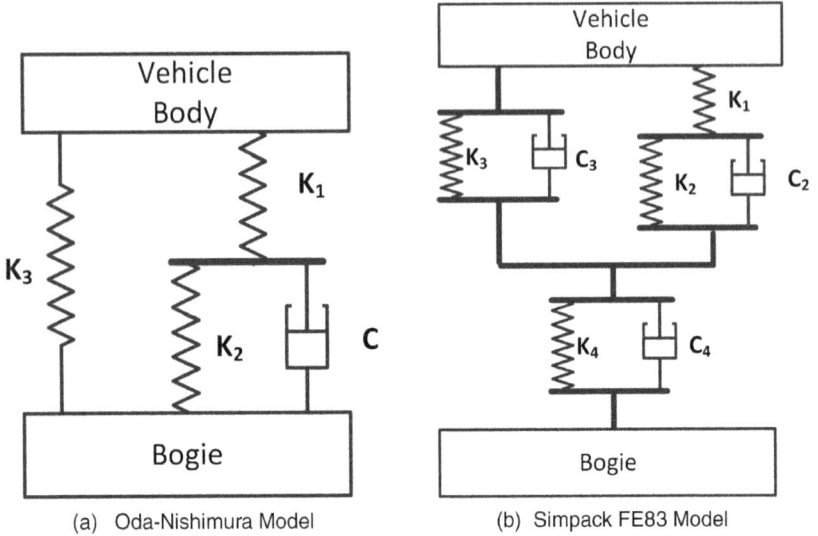

(a) Oda-Nishimura Model   (b) Simpack FE83 Model

FIGURE 5.55 Mechanical models of air spring suspensions: (a) Oda-Nishimura Model, (b) Simpack FE83 Model (see [24, 25]).

and polymer stiffness/damping, but the stiffness of the bellows can be approximated as indicated by Docquier et al. [23]:

$$K = \gamma \frac{A_e^2 p_b}{V_b} \tag{5.58}$$

where γ is the isentropic coefficient, p_b the mean pressure in the bellows in Pa, V_b the volume in the bellows in m³, and A_e the effective area in m².

There are various mathematical models in the literature as summarized in Figures 5.55 and 5.56.

As explained in [24], the simplest model that could be used would consist of just one spring and viscous damper, but this would only represent quasi-static conditions and the choice of parameters would be difficult. The system is quite non-linear. In the Oda-Nishimura Model shown in Figure 5.55(a), K_1 represents the stiffness of the bellows which is placed in series with spring K_2 and damper C which model the compressibility of the air in the bellows and surge reservoir and the damping provided by the restriction in the surge pipe. The additional stiffness K_3 was identified as necessary due to changes in the effective area of the bellows changing with height variation due to changes in loading. The authors identify a similar model provided in Simpack. The additional K_4, C_4 component is an emergency spring for operation with the pneumatic system flat so that the vehicle can be removed from service along the track after suspension failure. The authors of [24] point out that the Oda-Nishimura and Simpack models do not model the internal suspension resonances due to the movement of the air mass in the surge pipe. There is consensus that models such as those shown in Figure 5.56(a–c) are required to properly model and study pneumatic

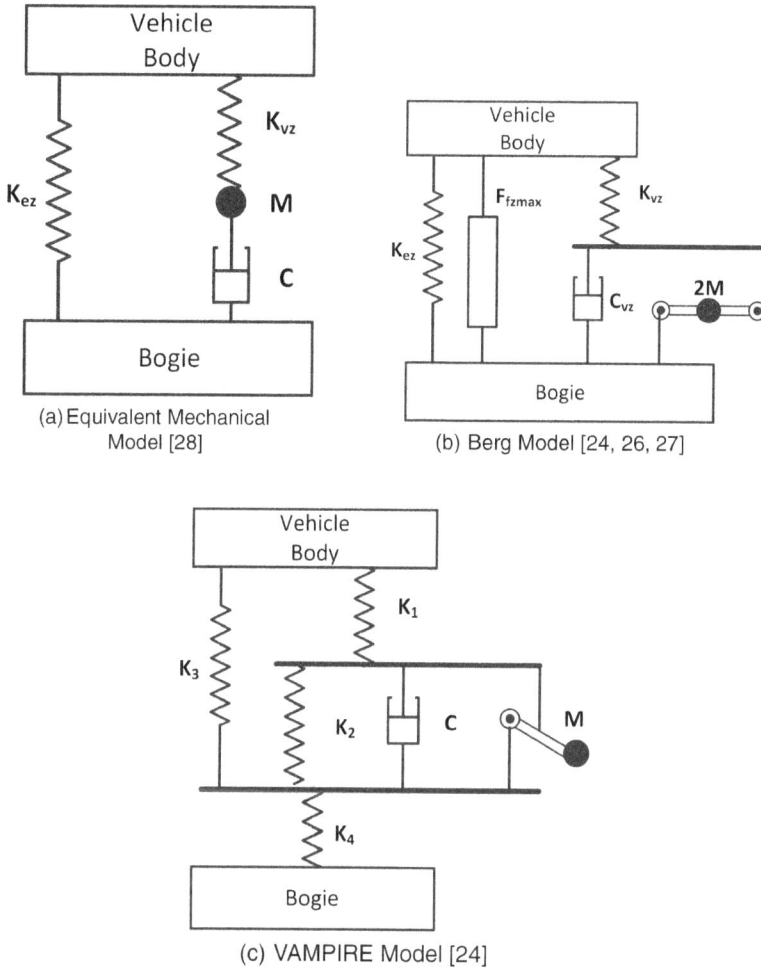

FIGURE 5.56 Mechanical models of air spring suspensions that include modeling of the air surge pipe: (a) Equivalent Mechanical Model (see [28]), (b) Berg Model (see [24, 26, 27]), (c) VAMPIRE Model (see [24]).

suspension behavior. These models derive mainly from the work of Presthus [25] and of Berg [26, 27], and such modeling also appears in VAMPIRE as noted in [25]. The Berg Model [26, 27] and the *Equivalent Mechanical Model* [28] utilize a different nomenclature where K_{ez} is the elastic stiffness, frictional components are modeled as F_f and viscous components are modeled by k_{vz}, $C_{z\beta}$, and the air surge mass, M or $2M$.

Active control of air spring suspensions combining both the mechanical stiffness implications and the gas dynamics is explored in [29]. The control was applied to reservoir pressure as shown in Figure 5.57 which differs from the two possibilities introduced in Figure 5.53. An improvement (reduction) of 17% in the RMS value of car body accelerations was reported for the simulated vehicle at speed of 300 km./h.

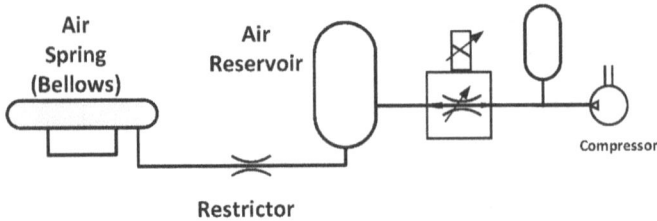

FIGURE 5.57 Semi-active railway air suspensions as proposed in [29].

5.4.8 OVERALL SUMMARY

Fully pneumatic and electro-pneumatic systems offer force actuators and suspension components that have found some applications in railway brakes and in improving railway suspensions.

The application of fully controlled air braking, although now an increasingly obsolete technology, illustrates the possibility of pneumatics in the provision of a fully pneumatic control and actuator system.

In suspensions, only semi-active control was observed in the literature. It would seem that variable flow control technology for pneumatics is less precise due to the changing properties of air with temperature, so adjustments and semi-active controls are better suited to quasi-static adjustments for varying operating conditions such as loading, track quality, temperature, etc. It is noted by Goodall and Mei [11] that the actuators require force control for suspension applications. Force control is easily achievable in pneumatic suspension systems due to the compressibility of air making them less responsive to small displacements and vibration inputs. The presentation of the pneumatic components gives background to the complex issues behind pneumatic design, showing that a whole system approach is needed to ensure the electro-pneumatic system can function in the operating conditions required.

REFERENCES

1. S. Bruni, R. M. Goodall, T. X. Mei, H. Tsunashima, Control and monitoring for railway vehicle dynamics, Vehicle System Dynamics, 45(7–8), 743–779, 2007.
2. C. W. de Silva, Mechatronics: An Integrated Approach, CRC Press, Boca Raton, FL, 2004, ISBN 9780849312748.
3. P. Vas, Electrical Machines and Drives: A Space-Vector Theory Approach, Oxford Science Publications, Oxford, UK, 1993, ISBN 9780198593782.
4. G. Diana, S. Bruni, F. Cheli, F. Resta, Active control of the running behaviour of a railway vehicle: Stability and curving performances, Vehicle System Dynamics, 37(Sup 1), 157–170, 2003.
5. F. Braghin, S. Bruni, F. Resta, Active yaw damper for the improvement of railway vehicle stability and curving performances: Simulations and experimental results, Vehicle System Dynamics, 44(11), 857–869, 2006.
6. A. Pacchioni, R. M. Goodall, S. Bruni, Active suspension for a two-axle railway vehicle, Vehicle System Dynamics, 48(Sup 1), 105–120, 2010.

7. D. Knežević, V. Savić, Mathematical modeling of changing of dynamic viscosity, as a function of temperature and pressure, of mineral oils for hydraulic systems, Facta Universitatis: Mechanical Engineering, 4(1), 27–34, 2006.

8. Rexroth, Hydraulic fluids, RE 90220206.82, Lohr am Main, Germany, 1982.

9. ISO 16889:2008:Hydraulic fluid power – Filters – Multi-pass method for evaluating filtration performance of a filter element

10. R. M. Goodall, T. X. Mei, Active Suspensions, Chapter 11, Handbook of Railway Vehicle Dynamics, 327–358, CRC Press, Boca Raton, FL, 2006, ISBN-13: 978-0-8493-3321-7.

11. R. M. Goodall, T. X. Mei, Active Suspensions, Chapter 11, Handbook of Railway Vehicle Dynamics, 2nd Edition, 579–613, CRC Press, Boca Raton, FL, 2020, ISBN-13: 978-1-1386-0258-4

12. J. K. Hedrick, D. N. Wormley, Active Suspensions for Ground Transport Vehicles – A State of the Art Review, In: B. Paul, K. Ullman, H. Richardson (Eds), Mechanics of Transportation Suspension Systems, ASME, AMD, 15, 21–40, 1975.

13. R. M. Goodall, W. Kortüm, Active controls in ground transportation – A review of the state-of-the-art and future potential, Vehicle System Dynamics, 12(4–5), 225–257, 1983.

14. R. M. Goodall, Active railway suspensions: Implementation status and technological trends, Vehicle System Dynamics, 28(2–3), 87–117, 1997.

15. R. M. Goodall, R. A. Williams, A. Lawton, P. R. Harborough, Railway Vehicle Active Suspensions in Theory and Practice, In: A. H. Wickens (Ed), Proceedings 7th IAVSD Symposium Held at Cambridge University, UK, 7-11 September 1981, Swets and Zeitlinger, Lisse, The Netherlands, 301–316, 1982.

16. M. Tahara, K. Watanabe, T. Endo, O. Goto, S. Negoro, S. Koizumi, Practical Use of an Active Suspension System for Railway Vehicles, Proceedings International Symposium on Speed-up and Service Technology for Railway and Maglev Systems, STECH'03, 19-22 August 2003, Japan Society of Mechanical Engineers, A503, Tokyo, 225–228, 2003.

17. A. Orvnäs, S. Stichel, R. Persson, On-track tests with active lateral secondary suspension: A measure to improve ride comfort, ZEV Rail Glasers Annalen, 132(11–12), 469–477, 2008.

18. A. Qazizadeh, R. Persson, S. Stichel, On-track tests of active vertical suspension on a passenger train, Vehicle System Dynamics, 53(6), 798–811, 2015.

19. R. Goodall, G. Freudenthaler, R. Dixon, Hydraulic actuation technology for full-and semi-active railway suspensions, Vehicle System Dynamics, 52(12), 1642–1657, 2014.

20. Y. A. Cengel, R. H. Turner, Fundamentals of Thermal-Fluid Sciences, McGraw Hill, New York, NY, 2005.

21. P. Gerhart, R. Gross, J. Hochstein, Fundamentals of Fluid Mechanics, 2nd Edition, Addison-Wesley Publishing, New York, NY, 1992.

22. G. Westinghouse, USA Patent No. 144006, Improvement in Steam and Air Brakes, 1873.

23. N. Docquier, P. Fisette, H. Jeanmart, Model-based evaluation of railway pneumatic suspensions, Vehicle System Dynamics, 46(Sup 1), 481–493, 2008.

24. S. Bruni, J. Vinolas, M. Berg, O. Polach, S. Stichel, Modelling of suspension components in a rail vehicle dynamics context, Vehicle System Dynamics, 49(7), 1021–1072, 2011.

25. M. Presthus, Derivation of Air Spring Model Parameters for Train Simulation, Master's Thesis, Luleå University of Technology, Sweden, 2002.

26. M. Berg, An Air Spring Model for Dynamics Analysis of Rail Vehicles, TRITA-FKT Report 1999:32, Division of Railway Technology, Department of Vehicle Engineering, Royal Institute of Technology, Stockholm, Sweden, 1999.

27. M. Berg, A three-dimensional air spring model with friction and orifice damping, in: Proceedings of the 16th IAVSD symposium on dynamics of vehicles on roads and tracks, Pretoria, South Africa, 1999, Journal of Vehicle System Dynamics, 33(Sup 1), 528–539, 2000.
28. N. Docquier, P. Fisette, H. Jeanmart, Multiphysic modelling of railway vehicles equipped with pneumatic suspensions, Vehicle System Dynamics, 45(6), 505–524, 2007.
29. S. Alfi, S. Bruni, E. D. Gaialleonardo, A. Facchinetti, Active control of air spring secondary suspension for improving ride comfort in presence of random track irregularity, Journal of Mechanical Systems for Transport and Logistics, 3(1), 143–153, 2010.

6 Sensors

6.1 INTRODUCTION

Sensors are an important part of any mechatronic system, as they allow the measurement of physical quantities representative of the system's state that can be used as the feedback variables in a closed-loop control implementation (see Chapter 4) or can be used to monitor the system and to identify the occurrence of faults. Sensors translate the magnitude of a physical parameter of interest such as force, position, speed, or acceleration to an electric signal (a tension or a current) that can be managed by the mechatronic system's control unit. Depending on the specific measuring principle on which the sensor is designed, this involves different laws of physics, so that a sensor can be seen as a dynamic system having its own frequency response function (FRF) and pass-band. Furthermore, the measuring principle and the design of the sensor affects the range and the accuracy of the sensor. Therefore, the selection of appropriate sensors is critical to the performance of mechatronic systems and sensor properties need to be considered in the design of closed-loop control systems. In a railway vehicle, the capability of the sensor to withstand harsh environmental conditions is also extremely important and drives the selection of the sensors for a given application. In this regard, particularly important are the degree of protection from contaminants such as dust and water, the range of temperature in which the sensor can work properly, and the magnitude of mechanical shocks and vibrations the sensor can withstand without compromising its integrity.

In this chapter, we will confine ourselves to the measurement of quantities typically used in mechatronic applications for railway vehicles: linear and angular position, linear and angular speed, acceleration, pressure, force, and torque. The measurement of other quantities such as temperature, stress and strain, and sound is relevant to vehicle monitoring applications and also to measuring set-ups used in the testing for qualification/certification of railway vehicles but are not covered here for the sake of brevity.

The aim of this chapter is not to provide a comprehensive description of all types of sensors in use in practical applications, which would require a book on its own, but rather to introduce the working principles of sensors most relevant to the scope of this book, and to point out the aspects affecting the correct functioning of the sensors in the context of a mechatronic application for railway vehicles. Due to space limitations, general issues related to the measurement process like the evaluation of uncertainty and calibration of sensors are not covered here and the interested reader is referred to specialized books such as [1]. The principles of data acquisition and data processing techniques, relevant to the management of measured signals in a mechatronic system, are covered in Chapter 10 of this book.

DOI: 10.1201/9781003028994-6

6.2 DISPLACEMENT SENSORS

Displacement sensors are particularly relevant to mechanical systems featuring active and semi-active controlled components, as they are used to measure the output of the controlled system in the case of a feedback control loop using position as the feedback variable. In many cases, displacement sensors are directly fitted inside the actuator: in this way, the connection of the actuator and sensor to the same control unit is facilitated and the sensor is effectively protected from environmental factors such as dust, water, and solar radiation.

Several transduction principles can be used to design displacement sensors. In this section, we will confine ourselves to the ones that are most used in mechatronic systems and, specifically, in mechatronic railway vehicles, while more comprehensive coverage of other sensors and transduction principles can be found in specialized books.

6.2.1 RESISTIVE SENSORS

Resistive displacement sensors (potentiometers) consist of an electric resistance (resistor) with a movable contact (wiper). A known voltage e_{in} is applied to the resistor. The wiper is connected to a movable end of the sensor, often a rod, which slides with respect to the housing of the sensor, and contacts the body whose displacement is the measurand, as shown in Figure 6.1. A movement of the rod caused by a change of the measurand results in the sliding s of the wiper over the resistor, thus changing the output voltage e_{out} being read between one end of the resistor and the wiper.

The above-described arrangement is for a linear displacement transducer, but rotary potentiometers are also available. In this latter case, the resistor is laid along an arc of the circumference and the wiper performs a rotation so that the angular position of the wiper relative to the resistor is transduced.

Ideally, the electric resistance in the circuit including the wiper is linearly varying with the position of the wiper and can be measured from the difference of potential between one end of the resistor and the wiper. However, it must be noted that the voltmeter measuring the output voltage e_{out} creates a second electric circuit in parallel to the portion of resistance to be measured. Hence, out of the total current i_1

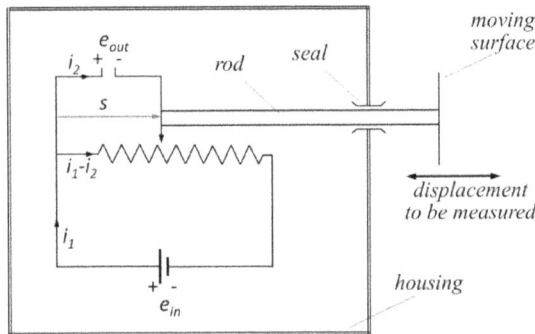

FIGURE 6.1   Schematics of a resistive linear sensor (rod-type).

circulating in the circuit of the resistor, a portion i_2 is diverted to the circuit of the potentiometer, so that only the remaining current $i_1 - i_2$ flows through the portion of the resistance to be measured and affects the output voltage e_{out}. This means the ratio of e_{out} to e_{in} is affected by the position of the wiper, leading to a non-linear relationship between the relative motion of the two ends of the sensor and the output voltage e_{out}. This effect can be kept under tolerable limits by using a sufficiently high internal resistance of the voltmeter.

Wire-wound resistors are often used to increase the electric resistance of the sensor. According to this construction, the electric wire realizing the resistor is wound in a helicoidal arrangement around a core which is either cylindrical for a linear sensor or toroidal for a rotational sensor. With this construction, the change of resistance is not continuous with the position of the wiper but is instead varying in steps when the wiper moves from one winding to another one. This means the sensitivity of the sensor is affected by the distance between the windings. More details about sensitivity and resolution issues in resistive displacement sensors can be found in [1].

Resistive displacement sensors are simple, relatively inexpensive and may be designed to have quite large measuring ranges, up to tens of centimeters. On the downside, resistive sensors are subject to friction and wear due to the solid contact between the resistor and the wiper and have relatively low protection to ingress of contaminants as the sensor is designed to allow the relative movement of the rod relative to the housing.

6.2.2 Capacitive Sensors

Capacitive displacement sensors are based on the change of capacitance of a parallel-plate capacitor with the air gap between the plates. In general, the sensor provides one plate of the variable capacitor, while the other plate is provided by the surface for which the displacement measure is taken, provided this is made of a conductive material. Different methods can be used to transform the variation of capacitance into a readable electric signal: in general, the variable capacitor is driven by an electronic circuit to produce an output either as a sequence of pulses whose width is affected by the distance between the plates of the capacitor or a high-frequency amplitude-modulated harmonic signal with amplitude depending on the variable capacitance of the capacitive transducer. More details on capacitive displacement sensors and their conditioning systems can be found in [2].

Capacitive displacement sensors provide high resolution and a wide bandwidth in the kHz range. They are contactless sensors and therefore are insensitive to friction and wear. They are of small size and work in a wide range of temperatures. However, they may suffer linearity issues, are quite expensive due to the complex electronics required, and their measuring range is relatively small, in the order of few millimeters.

6.2.3 Linear Variable Differential Transformers

The linear variable differential transformer (LVDT) works according to the principles of an electric transformer. The main components are a primary coil, at least

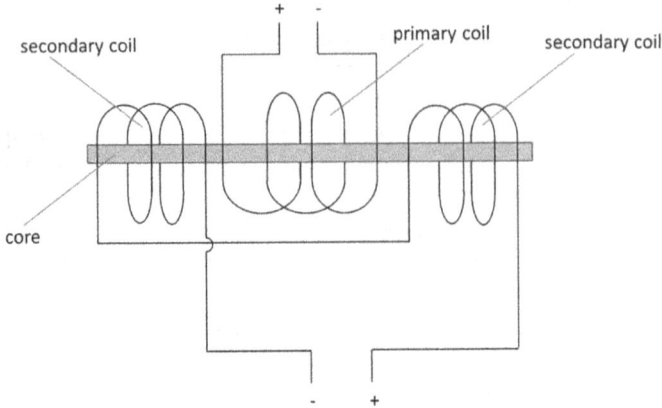

FIGURE 6.2 Schematics of a linear variable differential transformer (LVDT).

two secondary coils and a magnetically permeable core (see Figure 6.2). The core is connected to a rod that slides with respect to the housing of the sensor, so that the displacement to be measured is replicated by the displacement of the core with respect to the coils. An input AC voltage e_{in} is applied to the primary coil by a driving electronic circuit and an AC voltage with the same frequency is induced in each one of the secondary coils. The amplitude of the AC voltage induced in each secondary coil depends on the mutual inductance between the primary and secondary coils, which in turn depends on the position of the core. The two secondary coils are usually connected in series opposition, as shown in Figure 6.2. Therefore, when the position of the core is such that the same mutual inductance is realized between the primary coil and the two secondary coils, the voltages induced in the two secondary coils have the same amplitude and phase and the output voltage e_{out} is zero. This position of the core is called the null. A displacement of the core from the null in either direction causes a different mutual inductance between the primary coil and the two secondary coils, so the net output voltage is non-zero with its amplitude being proportional to the displacement of the core over the linear range of the sensor, while the direction of the displacement of the core from the null is indicated by the phase of the output voltage.

Rotary LVDT sensors, known as rotary variable differential transformers (RVDT) are also available. These sensors are based on the same transduction principle described above but provide the measure of rotary displacements. In this case, the core is connected to a shaft and has a shape that provides variable mutual inductance between the primary and secondary coils of the transformer. More details on this special construction can be found in [2].

In order to produce a DC output signal from the sensor, the output voltage e_{out} is fed into a demodulation circuit such as a diode bridge and then low-pass filtered to remove the fluctuations of the signal in output from the demodulation circuit. The cutoff frequency of the low-pass filter should be low enough to allow the effective rejection of spurious harmonic components in the rectified wave in the output from the demodulation circuit and, at the same time, should be high enough to preserve as much as possible the high-frequency harmonic components of the signal being measured. Therefore, the frequency of the input voltage e_{in} should be at least ten times the desired pass-band of the sensor. There is however a trade-off between the pass-band and other performance parameters of the sensor, as a high frequency of the input signal leads to increased losses due to eddy currents and hence to reduced sensitivity and possibly to temperature drifts. Typical values of the excitation frequency of this type of sensor are in the range of 2–5 kHz [2], but lower and higher frequencies are possible. More information regarding the design and manufacturing features, demodulation and signal conditioning for LVDTs can be found in [2].

LVDT displacement sensors have good accuracy, very high resolution, a wide bandwidth, and their measuring range spans from tenths to hundreds of millimeters, depending on the design of the sensor. LVDTs do not include sliding contacts, apart from the one between the rod and the seals protecting the sensor from the ingress of contaminants, hence they are weakly subject to wear. These sensors are however more expensive compared to resistive or capacitive ones, due to the greater complexity of the electronics and conditioning system and may be subject to linearity issues. Variants of LVDT sensors are variable-inductance and variable-reluctance pickups, for which the reader is referred to [1].

6.3 ENCODERS

Encoders are digital displacement transducers for both linear and rotary displacements, although the rotary configuration is more frequently used. Encoders convert the linear or rotary displacement to be measured into a sequence of pulses which can be treated as a digital word. Encoders are relatively inexpensive sensors compared to other types of position sensors and are natively digital transducers, so their interface to digital controllers is straightforward. Therefore, rotary encoders are very frequently used as internal sensors to provide angular position and/or angular speed signals to be used as feedback signals in position and velocity control loops for electric motors. In these cases, a rotary encoder is installed directly on the shaft of the electric motor, resulting in effective protection of the sensor from the environment.

Optical encoders are by far the most common type of encoder, although other construction types exist, such as magnetic encoders. Optical encoders use couples of light-emitting diode (LED) photoemitters and phototransistor light receivers. Light emitters and receivers are separated by a solid layer made of a material which is transparent to light, arranged in the form of a bar for a linear encoder and of a disk in the case of a rotary encoder. A pattern of opaque sectors is applied on the surface of the solid layer, so that the light emitted by the LEDs may or may not reach the corresponding light receiver, depending on the position of this body. This construction is exemplified in Figure 6.3 for the more common case of a rotary encoder: the solid layer is realized by a disk whose surface is divided into tracks, each track consisting of an alternation of light-transparent and opaque sectors. The disk is driven by a shaft that rotates according to the rotary motion to be measured by the encoder. While the disk rotates, the opaque sectors interrupt the path from the light emitter to the light receiver in each track, producing a set of digital pulses from which the angular speed or the angular position of the shaft can be reconstructed within the resolution allowed by the number of sectors.

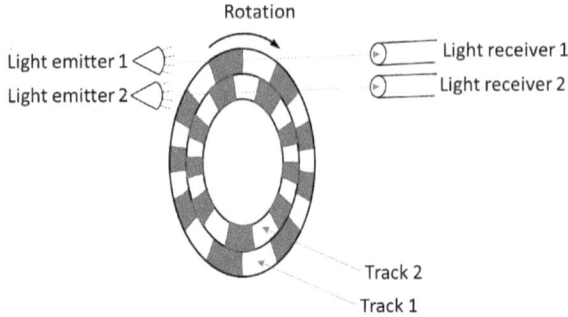

FIGURE 6.3   Scheme of a rotary optical encoder.

There are three types of encoders: optical tachometers, incremental encoders, and absolute encoders (see Figure 6.4). An optical tachometer only requires one emitter-receiver couple and one track in the encoder disk. The track is divided into alternating opaque and transparent sectors having the same angular width, so that the difference in time between two pulses allows to define the angular speed of the shaft. With this construction, the sensor does not provide information about the direction of motion, as the time between pulses will be the same for both directions. Therefore, this type of encoder is suitable only for applications in which reversing of rotation is not possible.

Incremental encoders use two or three tracks. Tracks 1 and 2 are arranged in the same way as the single track of an optical tachometer, but the sectors are shifted by one fourth of the period, so that there is a 90° phase difference between the two pulse sequences. The sign of the 90° phase depends on the direction of rotation of the shaft: in this way, the incremental encoder can be used to measure the rotation and angular speed of the shaft, including those cases where reversing of motion takes place. The resolution of an incremental encoder is determined by the number of sectors in Tracks 1 and 2. For rotary encoders, the resolution is defined in terms of pulses per revolution (PPR).

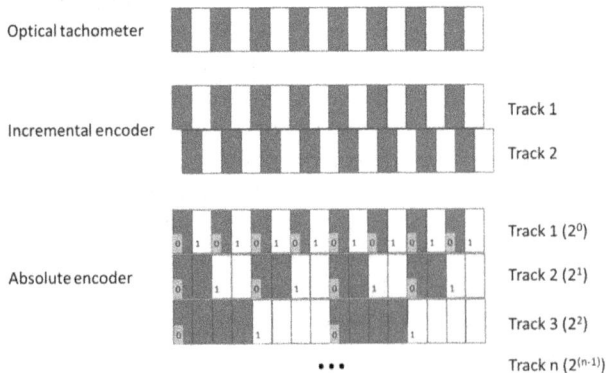

FIGURE 6.4   Schematic representation of tracks and sectors in encoders.

Optionally, a third track may be included in an incremental encoder to provide a reference for the absolute angular position of the shaft. Note that an incremental encoder does not provide as output the absolute position of the shaft, but rather the increment in the rotation of the shaft since the start of pulse counting. The 1 × revolution signal provided by the third track can be used to initialize pulse counting at a known angular position of the shaft, thus allowing to determine the absolute angular position of the shaft. Still, the absolute angular position provided by the incremental encoder through this technique will be affected by errors in the pulse count due to the effect of noise, and the absolute position will be lost in the case of loss of power to the encoder.

Absolute encoders use multiple tracks and emitter-receiver couples to translate the absolute angular position of the encoder to multiple binary values, one for each track, forming a digital word. The absolute angular position of the shaft is defined by the digital word, within a resolution depending on the number of bits in the word, which is ultimately the number of tracks in the encoder. The advantage of absolute encoders versus incremental encoders is that they natively provide the absolute angular position of the mechanical system or shaft to which they are connected and are therefore less sensitive to noise. However, incremental encoders are less expensive and usually provide a finer resolution compared to absolute encoders.

6.4   SPEED SENSORS

Angular speed sensors are frequently used in railway vehicles, particularly for wheel slide protection. The angular speed of a shaft can be measured using a toothed wheel called the *tone wheel* and a pickup generating one pulse at the passage of each tooth in the tone wheel. The angular speed of the tone wheel Ω is then obtained from the distance Δt between two pulses, given as follows for the number of teeth n in the tone wheel:

$$\Omega = \frac{2\pi}{n\Delta t}$$

The device sensing teeth passage in the tone wheel is often a variable-reluctance proximity pickup, sensing the change of magnetic reluctance caused by the change in the gap separating the pickup and the rotating wheel. The measuring principle of a tone wheel is the same as that of an optical tachometer, but a tone wheel is more suitable for use in an environment subject to contaminants which might impair the operation of an optical device.

A special type of angular speed sensor are gyroscopes. In some tilting trains, gyroscopes are used to measure the yaw or roll angular speed of the bogie frame and these signals are used to define a feed-forward component of the tilt command, enabling the achievement of faster actuation.

Modern gyroscopes are most frequently designed and built in the form of a Coriolis vibratory gyroscope; the measuring principle of this sensor is based on the Coriolis effect (see Figure 6.5). A driving force is applied to a proof mass along a direction called the *drive direction* and this will produce an oscillation of the mass along the drive direction. If an angular velocity component is applied to the mass along a direction perpendicular to the drive direction, a Coriolis force is generated

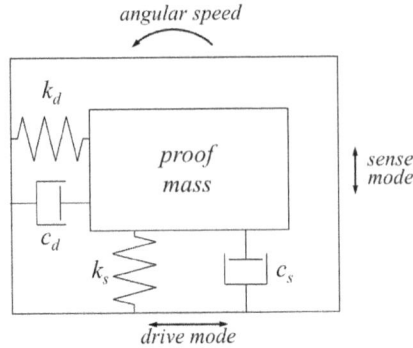

FIGURE 6.5 Principle of a Coriolis vibratory gyroscope.

on the mass along the *sense direction* perpendicular to both the drive direction and the axis of rotation. For a harmonically varying driving force and constant angular velocity component, the motion of the mass along the sense direction is also harmonic with amplitude proportional to the magnitude of the angular velocity. The motion of the proof mass along the sense direction can be measured by measuring the variable capacitance of a capacitor formed between the fixed frame of the gyroscope and the proof mass. This design of a gyroscope is particularly suitable for microelectromechanical systems (MEMS); more details on MEMS vibratory gyroscopes can be found in [3, 4].

Linear speed sensors are less frequently used in mechatronic systems compared to linear position and acceleration sensors. A commonly used technology is based on the principles of electromagnetic induction. The main components of an inductive linear velocity sensor are a coil and a permanent magnet where these two parts are arranged so that the coil slides with respect to the permanent magnet and an electronic circuit reads the voltage generated in the coil by the rate of change of the magnetic flux linked with the circuit of the coil which is proportional to the speed of the coil relative to the magnet. An example of use of inductive sensors in mechatronic railway vehicles is the active yaw damper concept described in [5]. In this application, an electromechanical actuator is used to mimic a viscous damper and an inductive linear velocity sensor is used to define the reference force for the actuator as proportional to the relative yaw speed of the bogie and car body.

6.5 ACCELEROMETERS

Accelerometers are the most widely used sensors in mechatronic railway vehicles. They are used to sense the lateral acceleration of the car body in active secondary lateral suspensions, define the desired amount of car body tilt in trains using an active tilting system, estimate the curvature of the track in the horizontal plane in active steering systems, and provide the feedback signal for active stabilization systems; see Chapter 11 for more details on these applications. Furthermore, acceleration measurements are often used for monitoring the condition of various components in the vehicle including wheels [6], bearings [7], suspension components [8],

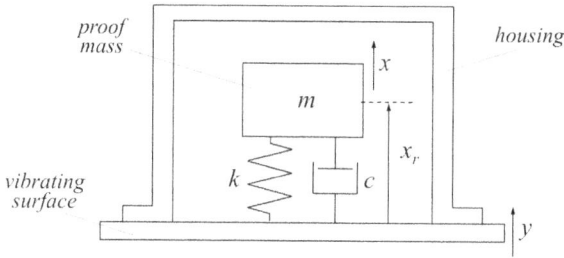

FIGURE 6.6 Measuring principle of a mono-axial deflection-type accelerometer.

and integrity of axles [9]. Additionally, accelerometers can be used to monitor the quality of track geometry from on-board measurements [10, 11]. Typical measuring ranges for accelerometers in use in railway vehicles are from ±1 g for the measure of car body acceleration to ±200 g which can be appropriate to measure the vertical acceleration of the vehicle's un-sprung masses, typically at the axle box.

Several types of accelerometers are available on the market. In the majority of the cases, the sensing principle is based on measuring the displacement of a rigid mass connected by a stiff spring to the body at the position where the acceleration is to be measured, as shown in Figure 6.6. The arrangement shown in the figure allows measurement of the scalar component of the acceleration of the body along the direction of movement of the mass, called the *sensing axis* of the accelerometer. In order to measure multiple components of acceleration, the same measuring principle can be used combining two or three sensors with orthogonal sensing axes; in this case the accelerometer is called bi-axial or tri-axial, depending on the number of measured components of acceleration.

According to the notation introduced in Figure 6.6, the motion of mass m in the accelerometer is governed by the second-order differential equation:

$$m\ddot{x} + c\dot{x} + kx = c\dot{y} + ky \tag{6.1}$$

with x the absolute displacement of the mass and y the absolute displacement of the base. Introducing coordinate x_r representing the displacement of the mass relative to the base:

$$x_r = x - y \tag{6.2}$$

Equation (6.1) can be rewritten as:

$$m\ddot{x}_r + c\dot{x}_r + kx_r = -m\ddot{y} \tag{6.3}$$

Introducing the following notation:

$$\omega = \sqrt{\frac{k}{m}}; \; \xi = \frac{c}{2m\omega} \tag{6.4}$$

with ω the natural frequency and ξ the non-dimensional damping ratio of the mass-spring system, Equation (6.3) can be rearranged in the form:

$$\ddot{x}_r + 2\xi\omega\dot{x}_r + \omega^2 x_r = -\ddot{y} \tag{6.5}$$

Assuming the base undergoes a harmonic motion, which can be expressed in complex-valued form as:

$$y = Ye^{j\Omega t}; \ddot{y} = \left(j\Omega\right)^2 Ye^{j\Omega t} \tag{6.6}$$

with Ω the circular frequency of the harmonic motion and j the imaginary unit, the following expression is obtained for the complex-valued FRF between the acceleration of the base $\ddot{y}$, and the displacement x_r of the mass relative to the base:

$$\frac{x_r}{\ddot{y}} = -\frac{1}{\omega^2}\frac{1}{-\left(\dfrac{\Omega}{\omega}\right)^2 + j2\xi\left(\dfrac{\Omega}{\omega}\right) + 1} \tag{6.7}$$

In the quasi-static frequency range of the oscillator, i.e., for $\Omega \ll \omega$, the FRF described by Equation (6.7) is well approximated by a constant real value, i.e., a linear relationship is established between the acceleration of the base $\ddot{y}$, which is the measurand, and the displacement x_r of the mass relative to the base which can be measured according to the measuring principles described in the sub-sections below.

6.5.1 PIEZOELECTRIC ACCELEROMETERS

Most deflection-based accelerometers use the piezoelectric effect to measure the displacement of the oscillating mass in Figure 6.6. Piezoelectric materials are natural or synthetic crystals. The property of piezoelectric materials is that, when subject to deformation along a specific direction, they produce a polarization of electric charge across their volume. Piezoelectric materials also show the dual effect, i.e., when a polarization of electric charge is applied on its surface, the crystal undergoes a change of length; this effect can be used to produce small lightweight actuators with a wide pass-band, enabling multiple input control of motion in flexible systems. In this context, however, we are interested in piezoelectric materials as displacement transducers.

An exemplary schematic representation of a piezoelectric accelerometer is shown in Figure 6.7 where the piezoelectric crystal is enclosed between the base of the accelerometer and the oscillating mass. A spring is used to keep the mass in contact with the crystal and to provide pre-compression of the crystal. Other arrangements of the components shown in the figure are possible, and a description of different piezoelectric accelerometer designs can be found in specialized books.

The base and housing of the accelerometer follow the movement of the vibrating surface whose acceleration is to be measured. The fixing of the accelerometer to

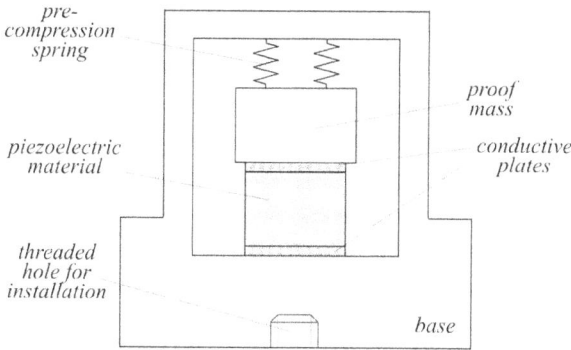

FIGURE 6.7   Scheme of a piezoelectric accelerometer.

the vibrating surface is generally obtained by a grub screw to prevent detaching of the two surfaces even under high levels of acceleration. Alternatively, gluing can be used. The piezoelectric crystal and the precompression spring act as springs in parallel, providing the stiffness k corresponding to the scheme of the accelerometer shown in Figure 6.6. The damping provided by the piezoelectric material and precompression spring is usually very low, therefore the housing is sometimes filled with oil to increase the damping associated with the motion of the proof mass.

Conductive metal plates are fixed to the upper and lower faces of the piezoelectric crystal. Since piezoelectric materials have a high electric resistance, the crystal and the two metal plates form a planar capacitor. When the base of the accelerometer moves, the piezoelectric crystal generates a polarization of charge in the two metal plates proportional to the motion of the mass relative to the base which is, in turn, proportional to the absolute acceleration of the base provided the frequency falls in the oscillator's quasi-static range. The two plates are connected to a charge amplifier which converts the charge of the capacitor to a voltage in a desired range, often ±10 V or ±5 V. In recent piezoelectric accelerometers, the sensor incorporates built-in microelectronics performing the charge amplifier function. This construction simplifies the accelerometer's conditioning system and reduces the sensitivity to noise from the cables connecting the sensor to the conditioning system which means longer cables can be used without introducing excessive corruption of the measured signal.

Piezoelectric accelerometers cannot measure constant or slowly varying acceleration because of the discharge effect in the capacitor through the voltage-measuring circuit. Therefore, the usable frequency range of the sensor will be from a lower limit depending on the time constant of capacitor discharge and an upper limit which is a fraction of the natural frequency ω of the oscillator defined by Equation (6.4).

Piezoelectric accelerometers are built for a wide variety of measuring ranges, require a simple conditioning system and are easily designed as rugged sensors which makes them particularly suitable for applications in railway vehicles involving the measure of motion in the un-sprung masses or bogies. Their main drawback is that they cannot measure constant or slowly varying accelerations.

6.5.2 CAPACITIVE ACCELEROMETERS

Capacitive accelerometers work under the same principle as piezoelectric ones, but the displacement pickup measuring the relative displacement of the proof mass relative to the base exploits the change of capacitance in a parallel plate capacitor with the two plates attached to the base and to the moving mass. The advantage of using this principle of transduction is that a constant acceleration signal can be correctly measured and, more generally, there is no lower boundary to the bandwidth of the accelerometer. However, the conditioning system required is more complex compared to piezoelectric accelerometers.

In recent years, MEMS capacitive accelerometers have been progressively introduced and are now extensively used for a wide range of applications including automotive, but also for railway systems. MEMS capacitive sensors are obtained from a printed microcircuit board. The base of the accelerometer is a portion of the board's substrate, and a micro-machined proof mass is elastically suspended from the base by flexible legs which may be designed with folds to increase their flexibility. The proof mass forms two capacitors with the fixed part of the sensor, so that the change of the capacitance is proportional to the acceleration component to be sensed. The advantages of a MEMS sensor with respect to other constructions consist of a lower cost per unit, lower mass, and lower size of the sensor. More details on MEMS capacitive accelerometers can be found in [3].

6.6 PRESSURE SENSORS

Pressure sensors are frequently used in conventional railway vehicles to measure pressure in passive systems such as the pneumatic braking system or to monitor air pressure in pneumatic suspensions. They are also used in mechatronic railway vehicles when pneumatic or hydraulic actuators are used. Several types of pressure sensors are available, but those used in railway applications are generally based on measuring the deflection caused by pressure to an elastic membrane. The membrane's deflection is often measured using piezoresistive strain gauges, i.e., transducers made of semiconductor materials that change their electrical resistance when subjected to strain. Alternatively, metal strain gauges or a displacement transducer can be used to measure the deflection of the membrane.

Pressure pickups are compact and rugged sensors, so they can be conveniently used in a railway vehicle. Their frequency response is good enough for capturing transients in pneumatic and hydraulic circuits relevant to mechatronic applications in railway vehicles, e.g., pneumatic and hydraulic cylinders used for active and semi-active vehicle suspensions.

6.7 MEASUREMENT OF FORCE AND TORQUE
IN MECHATRONIC RAILWAY VEHICLES

Some active control implementations require the application of a controlled force or torque to the plant. Examples relevant to mechatronic railway vehicles are a hold-off type active lateral suspension or the use of independently driven wheels to apply differential driving/braking torques on the different wheels of the vehicle, producing a guiding action, see Chapter 11 for examples of both. In these applications, the force applied by the actuator shall be measured to be appropriately controlled.

Force sensors in use in mechatronic systems, also known as *load cells*, are generally based on measuring some effect of the deformation caused by the measurand force to an elastic body incorporated in the sensor. This effect can be either a linear displacement or, more frequently, a combination of strain components. In the first case, a displacement transducer is used to measure the elastic displacement; although different types of transducers can be used, the most frequent choice is to use a piezo-electric transducer as they meet the requirements of the load cell in terms of stiffness, linearity, pass-band, and ruggedness. Alternatively, load cells are frequently designed using strain gauges to measure strains which are then combined through one or more resistive bridges to eventually provide a voltage signal proportional to the measurand force. Load cells based on strain gauge bridges also have very good linearity and pass-band, but may be sensitive to temperature effects and may also show some cross-axis sensitivity, i.e., the transduced signal will be affected by force components orthogonal to the measuring direction.

Torque measurements can be performed using multiple force sensors, combining the measured forces based on their lines of action to obtain the desired torque measurement. However, torques in shafts are more simply measured using strain gauges bonded on the surface of the shaft along directions inclined by 45° with respect to the axis of rotation. The measure of torques in shafts poses the additional challenge of transmitting the signals measured from the rotating body to a non-rotating receiver: this is accomplished using wireless data transmission, based either on analog or digital technology.

In general, force and torque transducers are expensive sensors, require periodic calibration to ensure good accuracy, and are not suited for operation in a harsh environment. Therefore, in many cases, forces or torques are derived from an indirect measure. For instance, in a pneumatic or hydraulic actuator like a hold-off device or a semi-active damper, pressure pickups can be used to measure the pressure in the two chambers of the actuator and the force F applied by the actuator can be obtained as:

$$F = p_1 A_1 - p_2 A_2 \tag{6.8}$$

with p_1 and p_2 the measured pressures and A_1 and A_2 the area of the piston's surface in the two chambers.

For electric motors or electro-mechanical actuators, the torque generated by the motor can be derived from the measure of the motor current. In the case of a linear mechanical actuator formed by an electric motor and a ball screw mechanical transmission, the linear force F applied by the actuator can be related to the mechanical torque T produced by the motor based on the generalized gear ratio that relates the rotation of the shaft to the linear elongation of the actuator:

$$F = T \frac{2\pi}{s} \tag{6.9}$$

with s the linear displacement of the actuator corresponding to one turn of the motor's shaft.

REFERENCES

1. E. Doeblin, Measurement Systems: Application and Design, Fifth Edition, McGraw-Hill Science, New York, NY, 2003, ISBN: 978-0072922011.
2. D. S. Nyce, Position Sensors: Theory and Application, John Wiley & Sons Ltd, Hoboken, NJ, 2016, Print ISBN 9781119069164, eBook ISBN 9781119069362.
3. A. Corigliano, R. Ardito, C. Comi, A. Frangi, A. Ghisi, S. Mariani, Mechanics of Microsystems, John Wiley & Sons Ltd, Hoboken, NJ, 2018, Print ISBN:9781119053835, Online ISBN:9781119053828.
4. G. Langfelder, A. Tocchio, MEMS Integrating Motion and Displacement Sensors, Chapter 13, In: S. Nihtianov and A. Luque (Eds), Smart Sensors and MEMS Intelligent Devices and Microsystems for Industrial Applications, Woodhead Publishing Limited, Cambridge, UK, 2014, 366–401, Print ISBN 9780081020555, eBook ISBN 9780081020562.
5. F. Braghin, S. Bruni, F. Resta, Active yaw damper for the improvement of railway vehicle stability and curving performances: Simulations and experimental results, Vehicle System Dynamics, 44(11), 857–869, 2006.
6. E. Bernal, M. Spiryagin, C. Cole, Onboard condition monitoring sensors, systems and techniques for freight railway vehicles: A review, IEEE Sensors Journal, 19(1), 4–24, 2019.
7. P. Pennacchi, S. Chatterton, A. Vania, L. Xu, Diagnostics of bearings in rolling stocks: Results of long lasting tests for a regional train locomotive, Mechanisms and Machine, 61, 321–335, 2019.
8. X. Y. Liu, S. Alfi, S. Bruni, An efficient recursive least square-based condition monitoring approach for a rail vehicle suspension system, Vehicle System Dynamics, 54(6), 814–830, 2016, DOI: 10.1080/00423114.2016.1164869.
9. M. Hassan, S. Bruni, Experimental and numerical investigation of the possibilities for the structural health monitoring of railway axles based on acceleration measurements, Structural Health Monitoring, 18(3), 902–919, 2019, DOI: 10.1177/1475921718786427.
10. C. Li, S. Luo, C. Cole, M. Spiryagin, An overview: Modern techniques for railway vehicle on-board health monitoring systems, Vehicle System Dynamics, 55(7), 1045–1070, 2017, DOI: 10.1080/00423114.2017.1296963.
11. P. Weston, C. Roberts, G. Yeo, E. Stewart, Perspectives on railway track geometry condition monitoring from in-service railway vehicles, Vehicle System Dynamics, 53(7), 1063–1091, 2015, DOI: 10.1080/00423114.2015.1034730.

7 Modeling of Complex Systems

This chapter is focused on the complex system modeling techniques and their implementation in a computer environment and software packages designated for rail vehicle multi-body and control system studies. No unique technique exists and each task requires an individual approach to build a complex model. It includes software development and interfacing between different software packages and codes. However, it is always based on some existing principles that are described in the next section.

7.1 BASIC PRINCIPLE OF COMPLEX SYSTEM DESIGN

At the present time, when a rail vehicle design needs to be modeled as a mechatronic complex system, it should include a detailed analysis of designs for running gear, traction power and auxiliary equipment, primary and secondary suspension systems, pantograph and monitoring systems, the braking system, driver control and advisory system, communication system and interfaces, software, and control algorithms. That means that all chains for existing processes should be considered and all limitations or simplifications should be explained and justified.

In terms of rail vehicle mechatronics, railway vehicle dynamics is one of key elements that is based on the mechanical engineering discipline and it is increasingly starting to include sensors, electronics, and computer processing for the design and modeling of complex systems [1]. In addition, there are very tight connections with other electrical, hydraulic, and pneumatic processes that necessitate considering the overall system as a multi-disciplinary modeling task. For example, in the case of electrical and mechanical processes, it needs consideration as an electromechanical complex system. These processes might be included in the mechatronic design of complex running gear systems with the following components [2]:

- Traction power equipment and traction drives, including the diesel-engine, alternators, inverters, traction motors, etc. The main task in the modeling is focused on the transmission of electrical energy to mechanical energy which can be represented by traction or dynamic braking efforts applied to the wheels of a railway vehicle.
- Active or semi-active suspension systems that can be implemented as electromechanical, pneumatic, or hydraulic elements introduced in the design. This generally allows providing a better dynamic performance and ride comfort, e.g., control of displacements of car bodies during a rail vehicle's run on the curved railway track.

DOI: 10.1201/9781003028994-7

- Active damping elements that improve rail vehicle behavior in operational train configurations by means of improvements of ride control, in-train force distribution, etc.
- Elements of tilting train systems acting on a car body to increase the operational speed on regular rail tracks. A similar scaled design of actuator elements can be also used on pantograph systems.
- Air springs and their valve control elements.
- Valve and actuator control elements of brake systems.
- Elements of other control systems that prevent rail vehicle derailments and collisions.

The design list can be further extended based on specific requirements for a complex rail vehicle mechatronic system and this can have an effect on basic design principles. The design is always based on the modeling concept that can be simply represented as shown in Figure 7.1. However, the concept might change when the integrated rail vehicle design concept is developed in use with the implementation of specialized software products. An example of an integrated design concept is shown in Figure 7.2.

The common software packages involved in the integrated design and virtual prototyping processes can be divided into the following groups:

1. Computer-aided design software packages:
   a. Pro/Engineer: http://www.ptc.com;
   b. Siemens NX: https://www.plm.automation.siemens.com/global/en/products/nx/
   c. DS CATIA: https://www.3ds.com/products-services/catia/
   d. DS SOLIDWORKS: https://www.solidworks.com/
   e. AUTODESK (AutoCad and Inventor): https://www.autodesk.com/products

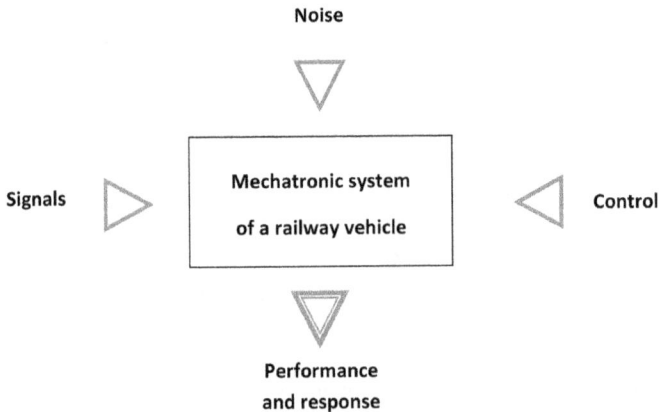

FIGURE 7.1 Simplified design concept of a mechatronic system.

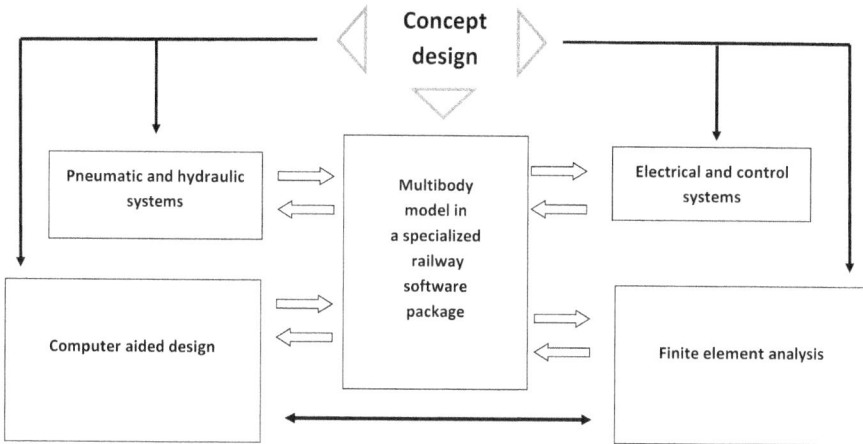

FIGURE 7.2 Example of an integrated concept in a design of complex mechatronic system for a rail vehicle.

2. Railway vehicle specialized commercial multi-body software (MBS) packages:
   a. Simpack: https://www.3ds.com/products-services/simulia/products/Simpack/
   b. VI-Rail: http://www.vi-grade.com
   c. VAMPIRE: http://www.vampire-dynamics.com
   d. Gensys: http://www.gensys.se
   e. NUCARS: http://www.aar.com/nucars/
   f. Universal Mechanism: http://www.universalmechanism.com
3. Industrial application software packages for a vast spectrum of tasks:
   a. Simulia ABAQUS: https://www.3ds.com/products-services/simulia/products/abaqus/
   b. Ansys: http://www.ansys.com
   c. MSC Software Products: http://www.mscsoftware.com
4. Software packages for electric and electronic control systems, actuator design, and data processing:
   a. MathWorks (Matlab®/Simulink): http://www.mathworks.com
   b. National Instruments (LabVIEW): http://www.ni.com
   c. GNU Octave: https://www.gnu.org/software/octave/

Considered from the point of view of the development of mechatronic systems, the major interest is focused on the usage of groups 2 and 4. The literature published in this area [3, 4] show that Matlab/Simulink software in combination with a MBS package can provide all required needs for the development of complex mechatronic systems at the concept design and testing stages. The subsequent further development might involve the usage of other software packages, including in-house software products, when the detailed analysis of a whole mechatronic system is needed.

7.2 INTRODUCTION OF CO-SIMULATION

Due to the high cost of physical testing and the increasing acceptance of computer simulation modeling, the design and validation has become more and more reliant on computer simulations to advance the state of the art. The development of new rail vehicles and the investigation of their mechatronic systems requires the application of an advanced modeling methodology because a rail vehicle is a complex system which includes not only mechanical, but also electrical, hydraulic, and other subsystems.

In a real vehicle, all systems should *communicate* between themselves. However, in the simulation world it is quite difficult to find a software product which can work in multi-disciplinary areas and cover all the areas of interest. Therefore, it is necessary to have different software products for each discipline and to allow them to talk to each other. Some ideas on how this can be organized have been published by Körtum and Vaculín [5]. The simplified scheme for the multi-disciplinary simulation is shown in Figure 7.3. Moreover, not only communication needs to be achieved,

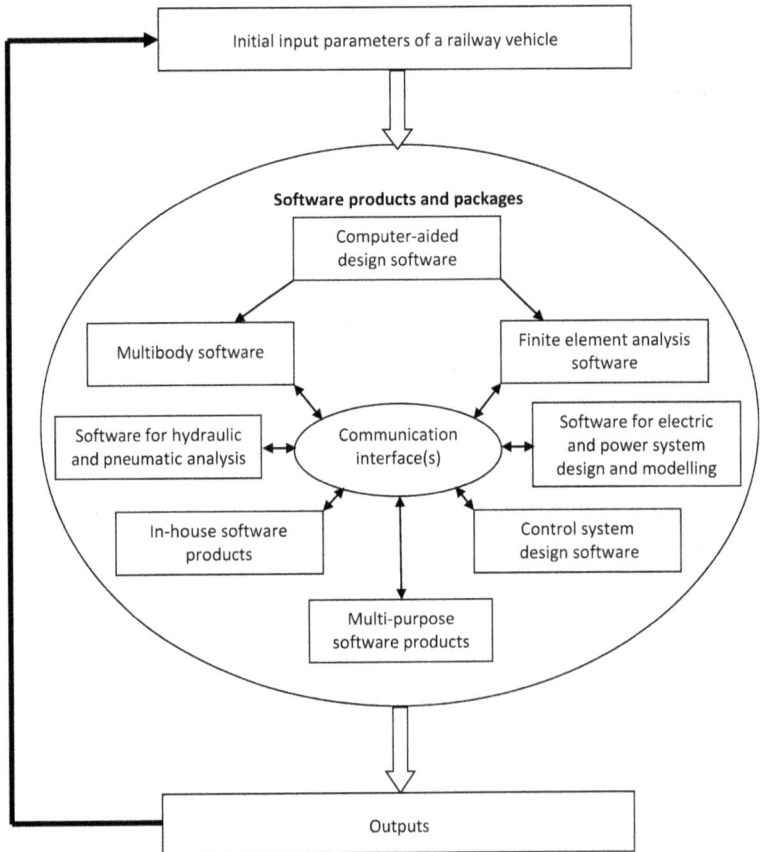

FIGURE 7.3 The application of different software products for computer simulation in the rail vehicle design process.

but different software packages should also be synchronized in the time-domain in order to provide accurate results. There are few cases where time independent parameters are in use. All these issues are included in the advanced simulation methodology which can be covered by the co-simulation process.

Here is a reasonable question: what does co-simulation mean? Co-simulation has been defined in several papers and documents. Arnold et al. [6, 7] define a co-simulation approach as one in which "the subsystems are handled or integrated by different integration methods or solvers and each of these methods or solvers can be or is tailored to the corresponding subsystems." In addition, it should be equipped with a data exchange algorithm which has discrete synchronization points with a sampling time of 0.001 s for vehicle system dynamics studies. Another definition can be found in [8], which states that "co-simulation is used to solve a coupled system by simulating each part with its own couplable simulation tool." A further comprehensive definition is presented in [9] as "co-simulation exploits the modular structure of coupled problems in all stages of the simulation process beginning with the separate model setup and pre-processing for the individual subsystems in different simulation tools."

7.3   CO-SIMULATION TECHNIQUES

The data exchange processes are usually based on three types of communication techniques:

- Integrated memory shared communication between software products
- Network data exchange
- Exporting code from one package to another

All these techniques have their own advantages and disadvantages; the decision about which is best to use is usually based on the initial requirements and existing hardware or software limitations.

The integrated memory shared communication technique is a good approach in the following cases:

- When in-house product is in use and an engineer or a researcher has knowledge of programming and access to source codes for software products. This situation is uncommon for commercial products because they are usually supplied as a black box software package without any proper access provided to users;
- When both commercial software packages are produced by one company, in which case the company can use this approach;
- When a commercial software product allows users to implement a user specific subroutine that can be compiled for the usage inside of the code. However, this technique is limited in terms of programming capabilities and strict implementation requirements set for such subroutines.

Network data exchange is a good approach when commercial or in-house software has the possibility to communicate with each other through the established network connection. For general applications, two network protocols are in use: Transmission Control Protocol/Internet Protocol (TCP/IP) and User Datagram Protocol (UDP). The latter requires the development of some additional data verification algorithms because this protocol is designed for low-latency and loss-tolerating connections between software packages. For both protocols, the usage of the client-server architecture should be implemented where a server is defined as a resource or service and a client is a service requester.

However, now consider a complex system with multi-client agents for multiple subsystems implemented in other software packages where one main software package, as a server, provides computational services for all these multi-clients. In such an architecture, the data exchange between each multi-client and their servers makes the co-simulation process too complex in terms of the modeling system design and therefore slows down the computational speed significantly. In such cases, it is better to use an exporting code technique in order to avoid communication speed and interface processing losses.

The exporting code technique is an approach that allows to both import from and export to models in shared libraries: a compiled *.dll* (Dynamic Link Library for Windows-based platforms) or a. *so* (Shared Object for Linux-based platforms) file. The other variation of this approach is built based on the Functional Mock-up Interface (FMI) [8, 9]. This interface was developed to allow software packages to export a model to the model description script written with Extensible Markup Language (XML) and to a C code or binary code that contains the model's equations and pre-defined parameters. However, it is necessary to mention that the FMI specification specifies not only FMI for Model Exchange (import and export), it also specifies FMI for co-simulation. The difference between these FMI approaches is that, in FMI for co-simulation, each software uses its own solver while, in the other approach, the importing software's solver should be used.

The decision on the appropriate co-simulation technique should be made after considering all software products and operating system environments and no universal approach exists.

7.4 REVIEW OF THE EXISTING MULTI-BODY SOFTWARE PACKAGES AND THEIR CO-SIMULATION FUNCTIONALITIES

It has begun to be a common practice to use Matlab/Simulink software in the design of mechatronic system dynamics studies because this software is a very powerful tool which allows the development of any components based on the existing well-developed libraries. Matlab/Simulink is also very friendly for the development of data exchange connections with multi-body packages [3, 4, 10–20]. The review of possible co-simulation approaches is presented in the next sections.

7.4.1 GENSYS AND MATLAB®/SIMULINK

Gensys MBS uses a script-based interface for the development of multi-body models for rail vehicles or trains. In order to perform a co-simulation though its network

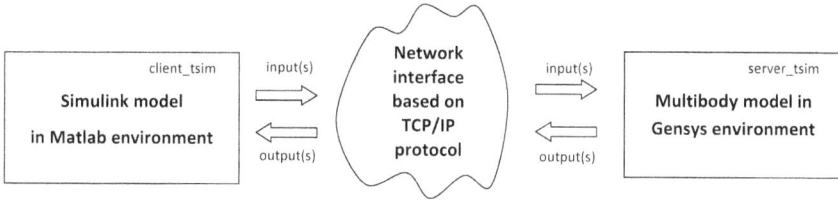

FIGURE 7.4 Architecture of the network data exchange co-simulation process.

data exchange interface, Gensys introduced two input data main commands [21]: *cosim_server* (for a co-simulation using TCP/IP) and *cosim_server_udp* (for a co-simulation using UDP). Since the development of version 10.10 of Gensys software, this package includes a server co-simulation interface which allows the creation of the client interface in the user's programs (in-house software) or in the Simulink environment. In this case, the CALC program used for a time-domain analysis is a server which can be called *server_tsim*. The existing Matlab/Simulink Gensys interface works under TCP/IP protocol version 4. The simple client interfaces for the both main input data commands are available on the Gensys website [21]. In the case of the client for Simulink software, the architecture for a co-simulation process through the TCP/IP interface is shown in Figure 7.4. The co-simulation process for Gensys should be as depicted in Figure 7.5. The Matlab/Simulink environment includes a

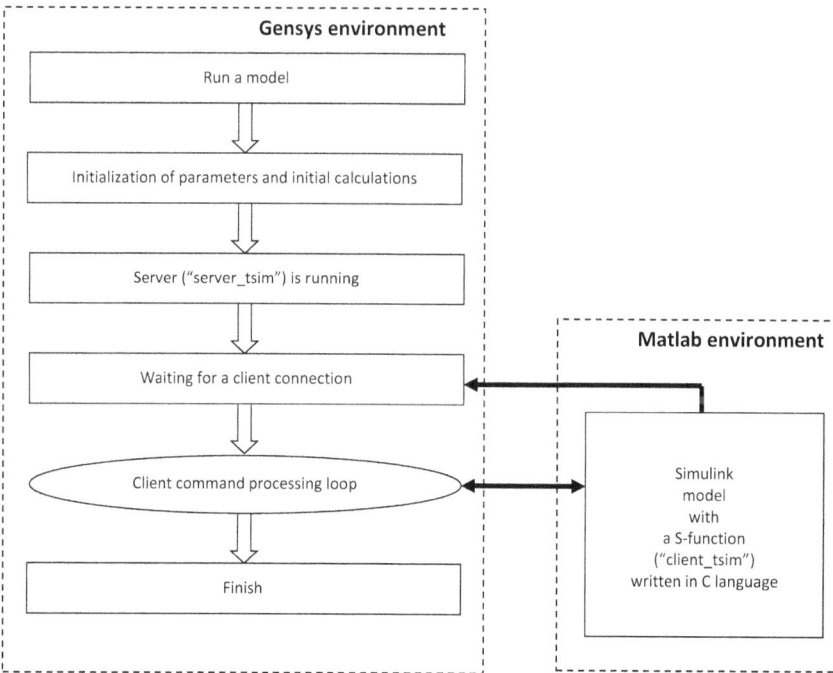

FIGURE 7.5 The flow chart for the co-simulation process in Gensys and Matlab.

Start of the simulation

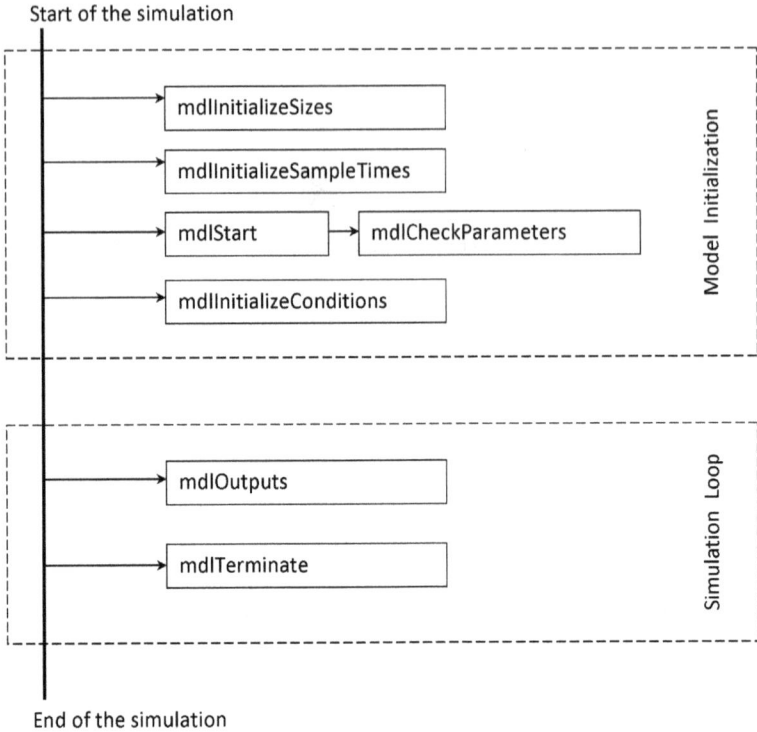

FIGURE 7.6 Structure of model initialization and simulation loop in Simulink. (From [4], with permission.)

very powerful tool for writing program code based on the S-function mechanism. An S-function called *client_tsim* was written in C language [3, 4] and represents a client interface between Gensys and Simulink. After creation, the client_tsim function was compiled with Mex-compiler in order to use it inside Simulink.

For the writing of the tsim_client [3, 4, 17, 21], the following callback functions have been implemented inside the S-function [22] as shown in Figure 7.6:

- mdlStart
- mdlCheckParameters
- mdlInitializeSizes
- mdlInitializeSampleTimes
- mdlInitializeConditions
- mdlOutputs
- mdlTerminate

The overall sequence of implemented functions is presented in Figure 7.6. The figure shows the order in which the Simulink engine invokes the functions. The mdlInitializeSizes and mdlInitializeSampleTimes functions are executed during initialization and at all time steps during the simulation loop. The other functions inside the Model Initialization rectangle are executed only once during the initialization process. The functions from the Simulation Loop rectangle are executed at each time step during the simulation process.

The function mdlStart is needed to define names/values of input and output parameters and store them in arrays for further use by other functions

The mdlCheckParameters is a function which is used for the validation and verification of the function block parameters.

The mdlInitializeSampleTimes function is used in order to specify a value of sample time for a S-function in Simulink.

For the initialization that must be carried out at each time step, the function mdlInitializeConditions is used to initialize the connection between programs. It requires using the parameter's IP address and Port number for the TCP communication process.

The *sizes* information is used by Simulink to determine the S-function block's characteristics (number of inputs, outputs, states, etc.) and these are included in mdlInitializeSizes.

The function mdlOutputs is used to start and stop the co-simulation process and for data exchange between client_tsim and server_tsim at each time step. The flowchart for the data exchange process is shown in Figure 7.7. The client_tsim sends the command *run_tout* to the server_tsim in order to start the calculation process for one time iteration. For receiving output parameters, the following sequence of commands is used:

- Send the command *ask_iadr* <output parameter> in order to find the addresses of the variables which need to be overwritten and the server_tsim replies with the address of the variable (var_address);
- Send the command *put_iadr* <var_address> <value> in order to change/ overwrite a variable. After sending the first argument, the server_tsim sends a request to input the second argument. After sending the second argument (the new value of the variable), the server_tsim replies with a zero, indicating that the new value was successfully inserted in the main memory.

A similar sequence of commands is used for sending input parameters to the server_tsim:

- Send the command *ask_iadr* <input parameter> in order to find the addresses of the variables which need to be read and the server_tsim replies with the address of the variable (var_address);
- Send the command *get_iadr* <var_address> in order to retrieve a value from the server_tsim. The tsim_server replies with the value of the variable.

The function mdlTerminate is a mandatory function. In order to stop the simulation on the server_tsim's side, the command *run_stop* is used to send a stop command inside of this function. As a result, both server_tsim and close the TCP/IP connection between the GENSYS and Simulink software products. The software code of client_tsim used in this approach can be found in [3, 4].

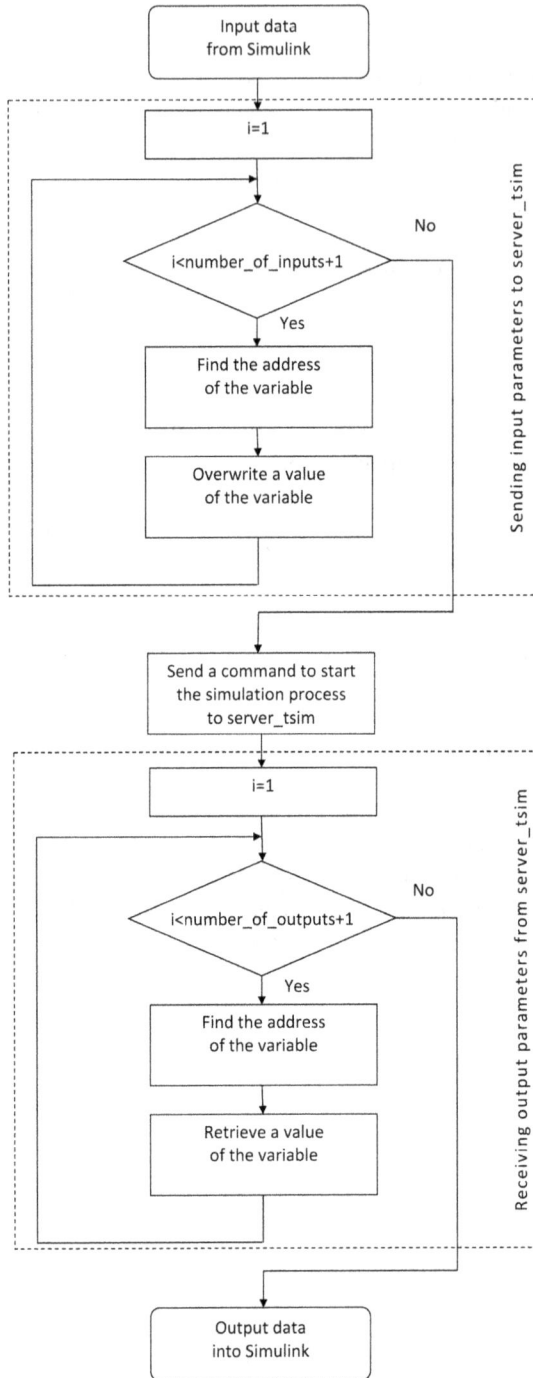

FIGURE 7.7 Algorithm flowchart written inside the mdlOutputs routine. (From [4], with permission.)

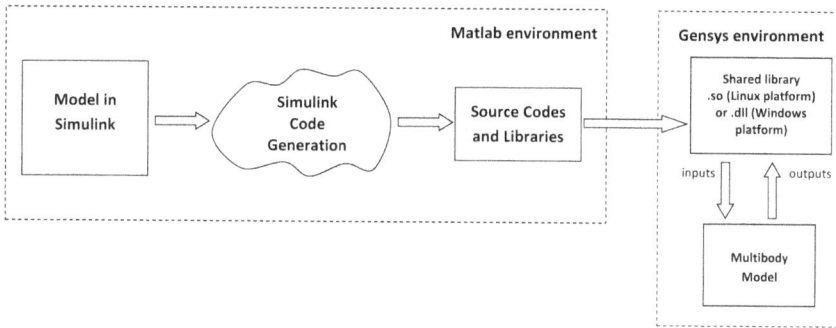

FIGURE 7.8 Architecture of the shared library-based co-simulation process.

The other co-simulation approach that was implemented in Gensys is based on the standard shared library generation approach used in Simulink [19]. For this purpose, the Simulink Code Generation platform is needed, and it allows generation of a shared library for host-based simulation [23, 24]. The concept of this approach is shown in Figure 7.8. Depending on the platform where Matlab is installed, the code generation platforms generate. so (Linux platform) or. dll (Windows platform) files.

In order to work with a Simulink shared library, a special user co-simulation input main data command *simulink_control#* has been introduced in the Gensys MBS package. The following rules have been set for this command:

- The name of the shared library shall be simulink_control#.so and it shall be located at project level, i.e., the level from where the programs in Gensys are launched.
- Only the integrators without a back step computational approach can be used together with simulink_control#.
- The following interface to call this function should be used:

```
func simulink_control# TimeStep_Simulink
noutput output(1) … output(noutput)
ninput input (1) … input(ninput)
```

where #in the library name is any number between 1 and 9;

- *TimeStep_Simulink* is a time step defined in the Simulink model;
- *noutput* represents a number of new variables to be created in Gensys, i.e., they are outputs defined in the Simulink model (the shared library)
- *output*(1.N) represents a newly created or existing variable(s). These variables are outputs of the Simulink model and, at the same time, the same variables are inputs in the Gensys multi-body model.
- *ninput* represents a number of existing input variables to be taken from the Gensys model and then sent to the shared library.
- *input*(1.N) represents an input data variable(s) for the shared library, i.e., inputs defined in the Simulink model.

The co-simulation function *simulink_control#* was written to run in the fixed time mode. For the time step synchronization between the Simulink and Gensys models, the algorithm as shown in Figure 7.9 is used. As shown in the figure, for multi-tasking models developed in Simulink, the names of called functions for one time step vary from *func_simulink_control#_step0()* to *func_simulink_control#_stepN()*.

Examples of applications of such co-simulation approaches for studies of mechatronic systems and rail vehicle performance can be found in [3, 4, 17–19].

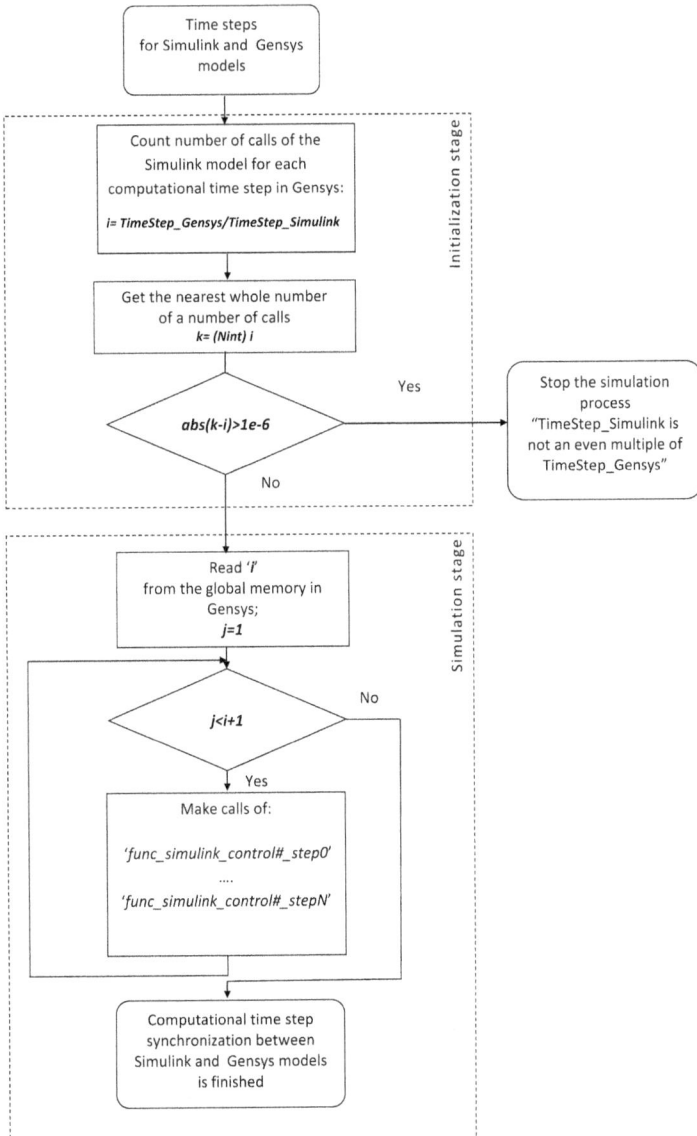

FIGURE 7.9 Algorithm flowchart for time step synchronization.

7.4.2 SIMPACK AND SIMULINK

Regarding the documentation provided by Simpack [25], this product supports six types of communication techniques:

- SIMAT interface (matrix data exchange or network data exchange);
- MatSIM (exporting code from Simulink to Simpack);
- Simpack FMU (Functional Mock-up Unit) Import (adding other models in the Simpack environment);
- Simpack FMU Export (export a Simpack model to the standardized FMU format for co-simulation);
- IPC Co-simulation (coupling Simpack with a virtual simulator produced by third parties);
- Simpack Realtime (enables the use of Simpack models in real-time applications such as Hardware-in-the-Loop and Software-in-the-Loop);
- CodeExport (only supported by the earliest versions of Simpack to export a Simpack model in the code for further use on real-time platforms).

The SIMAT interface allows communication with different software packages during a simulation process. One of these packages is Matlab/Simulink. The data exchange can be realized through a model linearization process and presents a multi-body model as linear state-space matrices. This approach is not recommended for a complex mechatronic system analysis; it is better to use SIMAT's network data exchange. The approach is based on data exchange through the TCP/IP protocol as shown in Figure 7.10. It is necessary to define input and output parameters of the multi-body model inside the Simpack environment. The network co-simulation interface is realized as the S-Function block in Simulink. The network addresses and ports required for the communication should be defined in both systems. This approach is a very good tool for the development of a detailed or full mechatronic system of a railway vehicle and is based on the synchronization method with discrete time points. Examples of applications of such a communication interface can be found in [12, 15, 20, 26].

There is another option contained in SIMAT called *the S-function Export*. This allows the export of a Simpack model for usage as a S-function in the Simulink environment.

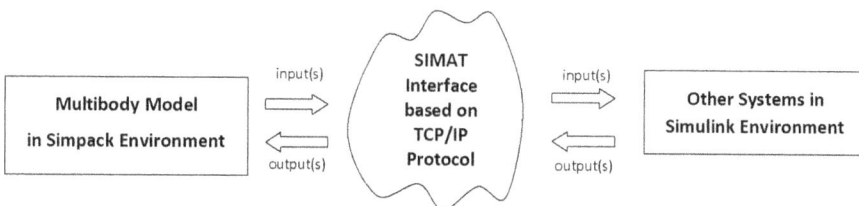

FIGURE 7.10   The co-simulation process between Simpack and Simulink with the network data exchange interface (SIMAT). (From [3], with permission.)

MatSIM [10] allows the creation of a library from a Simulink model by means of the application of Simulink Coder [23]. The special configuration files for such a process are supplied with the Simpack package – Target MatSIM.tlc. The obtained *<model>.matsim* file is then used by Simpack as a plug-in via the specially designed *Control Element 233: MatSIM*.

FMU Import allows importing standardized models that are generated accordingly to FMI versions. The model file in this case can be used via the specially designed *Control Element 238: FMI Import*. It requires the use of Simpack Solver to integrate the continuous states of the FMU.

FMU Export allows exporting a Simpack model for a co-simulation process to the standardized FMU format.

IPC Co-simulation allows coupling Simpack with a virtual simulator or other similar software products produced by third parties and defined by Simpack through TCP/IP-based sockets.

Simpack Realtime enables the use of Simpack models in real-time applications such as Hardware-in-the-Loop and Software-in-the-Loop. This module consists of the solver, animation and logger. The solver can run on single or multiple cores. The logger and animation run on separate cores and both of these units communicate with real-time environments through a UDP interface.

CodeExport (available in the earliest versions of Simpack) allows the export of models from Simpack and generation of a source code in Fortran which can then be converted into C language. The obtained code can be used in Simulink as an S-Function or can find further use in the implementation of software-in-the-loop, hardware-in-the-loop and real-time simulations. This approach has a lot of limitations and restrictions. For example, a great number of force elements are not supported by CodeExport. In the case of simulations for rail vehicles, the existing wheel-rail contact models are not supported when using this approach. However, it is still possible to use this process as shown in Figure 7.11. An example of such an application technique can be found in [27].

To summarize the above, it is possible to say that Simpack is a very powerful tool for multi-body simulations and has a great number of special features for the simulation of complex mechatronic systems for railway vehicles. However, further improvements and the development of enhancements will be very useful for rail vehicle dynamicists.

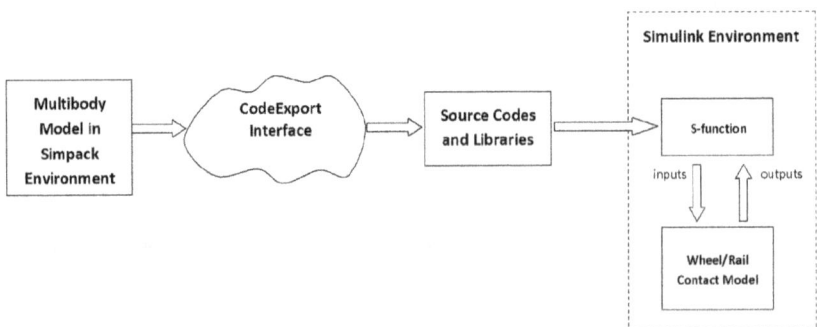

FIGURE 7.11   The implementation of CodeExport in a simulation process. (From [3], with permission.)

7.4.3 VI-RAIL (ADAMS/RAIL) AND SIMULINK

One of the leading and first products in the field of multi-body dynamics is Automatic Dynamic Analysis of Mechanical Systems (MSC.ADAMS). In 1995, the special plug-in called ADAMS/Rail was developed for the study of rail vehicle dynamics. In the middle of 2000, ADAMS/Rail was replaced by the new plug-in called VI-Rail produced by the company VI-grade. VI-Rail is based on the same algorithms, structures and co-simulation principles as its predecessor. The co-simulation between VI-Rail and Simulink is performed by means of the additional plug-in called ADAMS/Controls. This plug-in supports two co-simulation techniques:

- Exporting code from one package to another
- Network data exchange with TCP/IP

The export of code can be done in two ways. The first one is from MSC.ADAMS to Simulink by means of exporting a mechanical model into the Matlab function. The second way is to import a control system model from Simulink to MSC.ADAMS. In this case, the model should be exported by means of Simulink Coder [23] into a file written in C language and then combined with a mechanical model in ADAMS for further simulation processing.

For the network data exchange, ADAMS runs as a server and Matlab/Simulink as a client (see Figure 7.12). In Simulink, the adjustment of simulation parameters can be done through the Function Block Parameters interface developed by MSC. ADAMS.

Before starting simulations, all input and output parameters should be defined in models, depending on what simulation technique is going to be used. In addition, both continuous and discrete modes can be used for the integration of ADAMS and control models. However, for most simulations, it is recommended to use the discrete mode. The continuous mode can be used in some cases when a small-time step is required.

Some examples of the co-simulation application between ADAMS/Rail and Simulink by means of ADAMS/Controls can be found in [28–31].

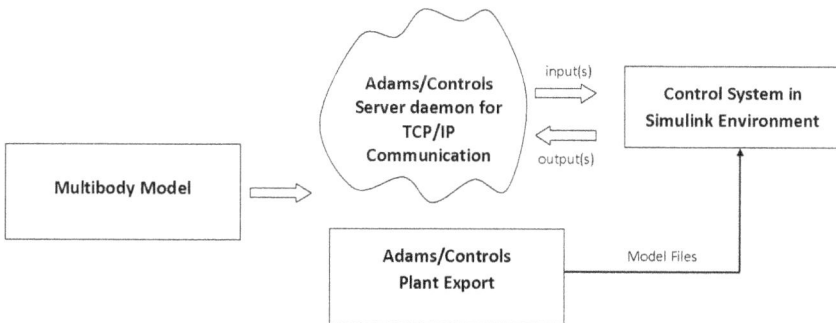

FIGURE 7.12 The co-simulation process between MSC.ADAMS and Simulink using TCP/IP. (From [3], with permission.)

7.4.4 VAMPIRE AND SIMULINK

The communication technique between Vampire and Simulink [22] is performed with a specially developed interface which is implemented as an S function in Simulink and written as an m-file. This function calls a required command from the DLL developed by Vampire (see Figure 7.13). This interface allows users to use a co-simulation process for the development of their own suspension systems but has a limitation in that it allows the Simulink model to use only one Vampire communication function inside of the model. When the Simulink model runs, it calls the Vampire function and Vampire Control analysis is fully controlled by Simulink. The process flow chart is shown in Figure 7.14.

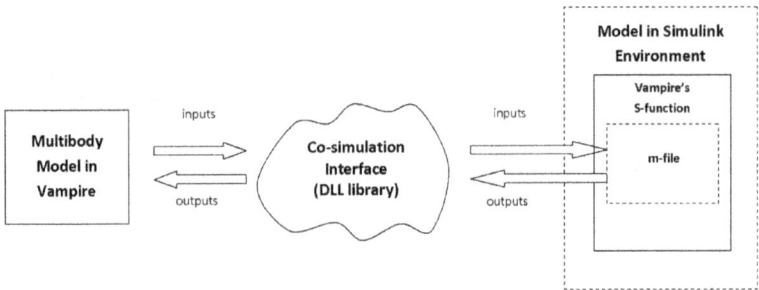

FIGURE 7.13   The co-simulation process between Vampire and Simulink. (From [3], with permission.)

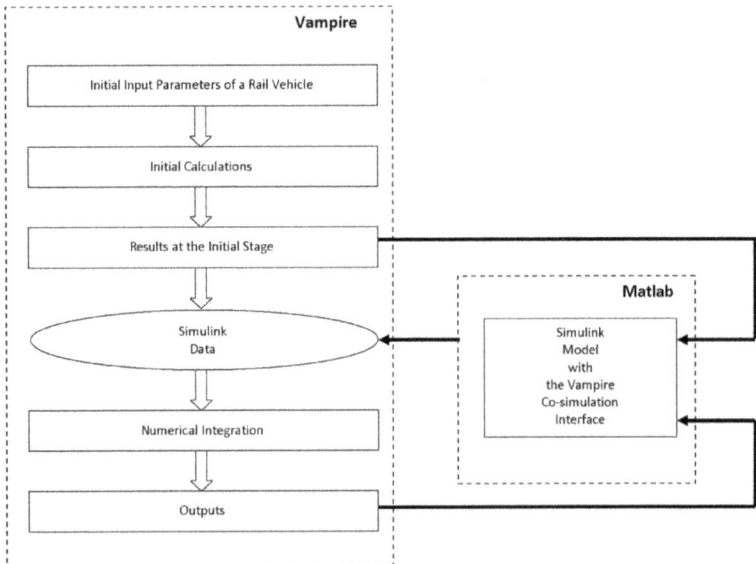

FIGURE 7.14   The flow chart for the co-simulation process between Vampire and Simulink. (From [3], with permission.)

The Vampire package provides several examples of Simulink models for rail vehicle suspension design with a co-simulation process such as a spring element design, a displacement control for a tilting vehicle, etc. [32].

7.4.5 Universal Mechanism and Simulink

Universal Mechanism supports two co-simulation techniques [33, 34]:

- Exporting code from Universal Mechanism to Simulink (UM/CoSimulation)
- Exporting code from Simulink to Universal Mechanism (UM/Control)

The first one generates an m-file and a data file with information about model inputs/ outputs. In Simulink, the S-function is used for the co-simulation process and this function works with the m-file created during the model export. The architecture of such a technique is very similar to Vampire, but the difference is that it allows the use of several multi-body models exported from Universal Mechanism inside the Simulink model. This means that each of the models has its own m-file. During a simulation process, Simulink creates an m-file which contains function calls for three stages: initialization, calculation of output values, and termination.

The second technique, which involves exporting code from Simulink is called Matlab Import in the Universal Mechanism documentation. The Code Generator is used in this case. For each Matlab software version, specific setting files are required. As a result, a DLL file is generated. Subsequently, the library of DLL files obtained for all of the versions should be connected to the Universal Mechanism software as an external library. Some limitations are present for such a technique: only parameters which are parts of force element parameters can be used as inputs for the multi-body model in Universal Mechanism and outputs of the Simulink model, respectively.

Examples of applications of co-simulation processes between Universal Mechanism and Simulink can be found in [13, 14, 16].

7.5 DESIGN OF CO-SIMULATION INTERFACES

Summarizing the review of co-simulation techniques, it is possible to state that the best solutions for co-simulation are based on network data exchange and shared library techniques. For the network communication, the development and operational principles of a TCP/IP network connection for Windows and Linux platforms are well-described in [3, 4]. In this section, the simple example of the development of a dummy approach based on a shared library technique used for the design of the co-simulation interface between the self-developed code (referred to in the following text as *Code*) and the Simulink software product is described.

7.5.1 Design of the Simple Simulink Model and Generation of the Shared Library

The shared library interface version (referred to in the following text as *Library*) presented in this section has been developed for Matlab/Simulink 2018a running in a Linux 64-bit environment. For the development of the Simulink model, it is

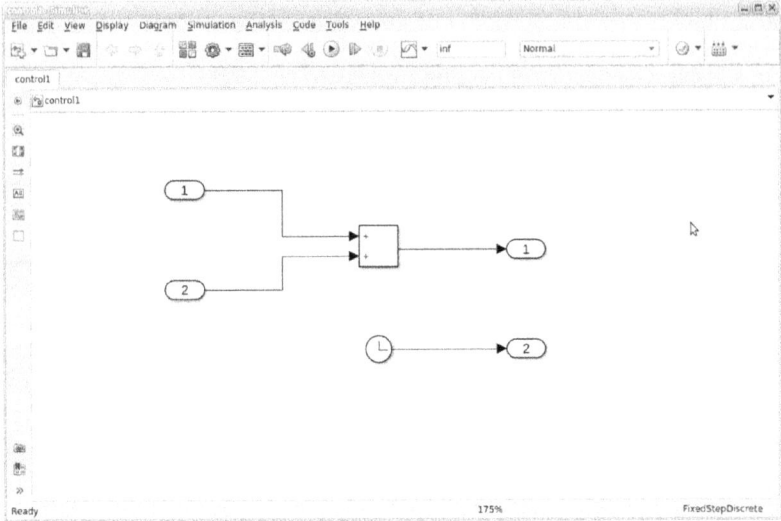

FIGURE 7.15 Simulink model for the Library.

necessary to initially design the input and outputs for the communication between the Code and the Library. Let's assume that we need to design a Simulink model *control1* that is a dummy model with two inputs and two outputs. The two inputs should be represented by two integer numbers that should be summarized in order to obtain a result for the first output. In the case of the second output, it should return a simulation time from the Simulink model. The Simulink model for such requirements can be built as shown in Figure 7.15. As it is possible to see from the figure, we need to define unlimited simulation time in the model in order to avoid any limitation for its call in the Code. Also, for the proper time synchronization, it is necessary to use the discrete (no continuous states) solver with a fixed time step of 1 ms as shown in Figure 7.16.

FIGURE 7.16 The solver configuration parameters in the Simulink model.

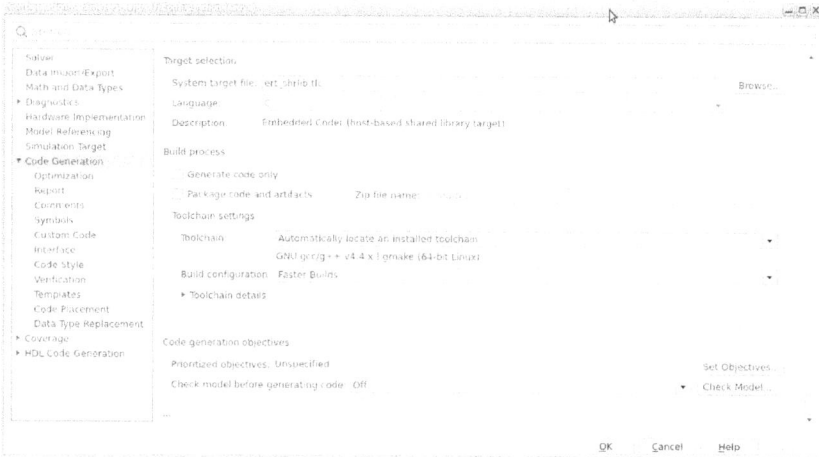

FIGURE 7.17 The code generation settings in the Simulink model.

In order to generate the Library [23, 24] through the System Target File Browser, it is necessary to choose the *ert_shrlib.tlc* file that is designated as an embedded coder for a host-based shared library target for the Linux-based platform. The model should be configured for the code generation through the graphical user interface as shown in Figure 7.17. After that, it is necessary to perform an automatic generation process by means of the following steps in the main Simulink menu (as shown in Figure 7.18): Code → C/C++ Code → Build Model. This call generates a *control1.so* file that is the Library for the designed co-simulation process.

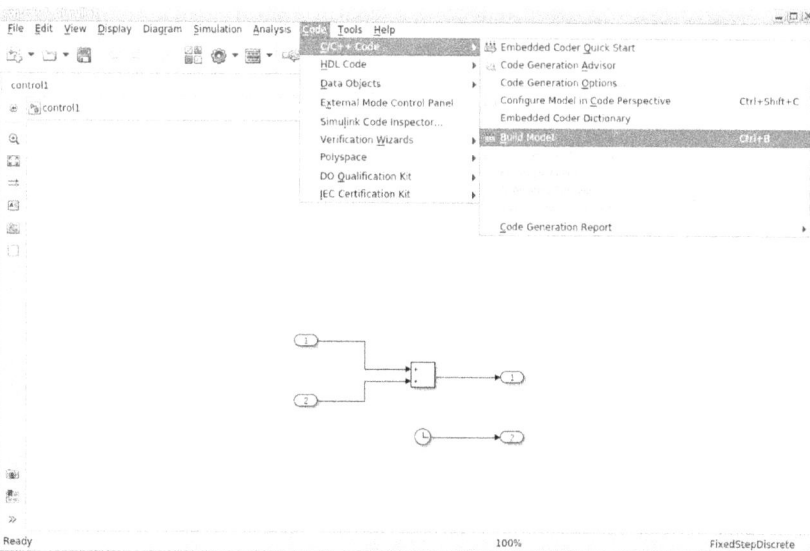

FIGURE 7.18 The code generation process in Simulink.

7.5.2 SHARED LIBRARY INTEGRATION IN THE CODE

The simple dummy code was written in Fortran and C languages. The main program called *example.f90* is listed below:

```fortran
program example
    use: : iso_c_binding
implicit none
    integer(c_int), parameter: : rtld_lazy=1! value extracted
from the C header file
    integer(c_int), parameter: : rtld_now=2! value extracted
from the C header file
    !
    ! interface to linux API
    interface
        function dlopen(filename,mode) bind(c,name="dlopen")
            ! void *dlopen(const char *filename, int mode);
            use iso_c_binding
            implicit none
            type(c_ptr): : dlopen
            character(c_char), intent(in): : filename(*)
            integer(c_int), value: : mode
        end function
        function dlsym(handle,name) bind(c,name="dlsym")
            ! void *dlsym(void *handle, const char *name);
            use iso_c_binding
            implicit none
            type(c_funptr): : dlsym
            type(c_ptr), value: : handle
            character(c_char), intent(in): : name(*)
        end function
        function dlclose(handle) bind(c,name="dlclose")
            ! int dlclose(void *handle);
            use iso_c_binding
            implicit none
            integer(c_int): : dlclose
            type(c_ptr), value: : handle
        end function
    end interface
    ! Define interface of call-back routine.
    abstract interface
        subroutine called_proc (i, i2) bind(c)
            use, intrinsic: : iso_c_binding
            integer(c_int), intent(in): : i
            integer(c_int), intent(out): : i2
        end subroutine called_proc
    end interface
            abstract interface
        subroutine control1_initialize_T() bind(c)
            implicit none
        end subroutine control1_initialize_T
```

```
      end interface
              abstract interface
        subroutine control1_step_T() bind(c)
            implicit none
        end subroutine control1_step_T
    end interface
              abstract interface
        subroutine control1_terminate_T() bind(c)
            implicit none
        end subroutine control1_terminate_T
    end interface
              type, bind(c): : ExtY_control1_T
      real (c_double):: out (100)
    end type ExtY_control1_T
              type, bind(c): : ExtU_control1_T
      real (c_double):: in (100)
    end type ExtU_control1_T
              type (ExtY_control1_T) outdata
              type (ExtU_control1_T) indata
              common/response/outdata
              common/input/indata
              type(c_ptr): : module_handle
              common/ptr/module_handle
              type(c_funptr): : proc_address
              type (ExtY_control1_T), pointer: : data_address
      procedure(control1_initialize_T), bind(c), pointer: :
control1_initialize
              procedure(control1_step_T), bind(c), pointer: :
control1_step
              procedure(control1_terminate_T), bind(c),
pointer: : control1_terminate
              integer::i
              character(len=:), allocatable: : filename
              character(len=:), allocatable: : libpath
              character(len=:), allocatable: : libname
  ! **** Input and outputs parameters defined in Simulink model
              integer::number_of_inputs
              integer::number_of_outputs
              number_of_inputs=2
              number_of_outputs=2
    filename = "control1.so"
              libpath = "./"
              libname=TRIM(ADJUSTL(libpath))//
TRIM(ADJUSTL(filename))
  ! Dummy input parameters
              do i=1,number_of_inputs
    indata%in(i)=i
  !           print *,indata%in(i)
              end do
  ! **** Open library and initialize
    module_handle=dlopen(libname//c_null_char, RTLD_LAZY)
```

```
    if (.not. c_associated(module_handle))then
       print*, 'Unable to load DLL. /*.so'
       stop
    end if
          proc_address = dlsym (module_handle, &
     c_char_'control1_initialize'//c_null_char)
          if (.not. c_associated(proc_address)) &
     stop 'Unable to obtain procedure address'
     call c_f_procpointer(proc_address, control1_initialize)
     call control1_initialize
 ! ***** Send inputs
          call sendlib (number_of_inputs)
 ! **** Run one step
          proc_address = dlsym (module_handle, &
     c_char_'control1_step'//c_null_char)
          if (.not. c_associated(proc_address)) &
     stop 'Unable to obtain procedure address'
          call c_f_procpointer(proc_address, control1_step)
          call control1_step
 ! **** Run more timestep for testing purposes
          call control1_step
          call control1_step
          call control1_step
          call control1_step
          call control1_step
 ! ***** Retrieve outputs
  call recvlib (number_of_outputs)
          do i=1,number_of_outputs
                 print *,outdata%out(i)
          end do
 ! ****** Terminate
          proc_address = dlsym (module_handle, &
     c_char_'control1_terminate'//c_null_char)
          if (.not. c_associated(proc_address)) &
     stop 'Unable to obtain procedure address'
          call c_f_procpointer(proc_address,
control1_terminate)
    !
end program example
```

The next step is to define a code for sending inputs through 'sendlib' into sendlib.c as below:

```
#include <stdio.h>
#include <stdlib.h>
#include <strings.h>
#include <dlfcn.h>
/* External inputs (root in-ports fed by signals with auto
storage) */
typedef struct {
  double In[100];                          /* '100 signals' */
```

```
} ExtU_control1_T;
extern struct {double in[100];} input_;
extern struct {void *libHandle;} ptr_;
void sendlib_ (int* number_of_inputs)
{       int j;
        ExtU_control1_T *inPtr;
    inPtr = (ExtU_control1_T*) dlsym(ptr_.libHandle,
"control1_U");
        if (inPtr == NULL) {
          printf("dlsym failed\n");
          printf("%s\n", dlerror());
          return;
      }
//Store input parameters in the model
    for (j = 0; j < (*number_of_inputs); j++) {
      (*inPtr).In[j]=input_.in[j];
      }
  return;
}
```

In a similar manner, it is now time to define a code for retrieving outputs from the Simulink model, i.e., the Library, through *recvlib* into recvlib.c as below:

```
#include <stdio.h>
#include <stdlib.h>
#include <dlfcn.h>
/* External outputs (root out-ports fed by signals with auto
storage) */
typedef struct {
  double Out[100];                              /* '100 signals' */
} ExtY_control1_T;
extern struct {double out[100];} response_;
extern struct {void *libHandle;} ptr_;
void recvlib_ (int* number_of_outputs)
{       int j;
        ExtY_control1_T *outPtr;
//Retrieve output parameters
   outPtr = (ExtY_control1_T*) dlsym(ptr_.libHandle,
"control1_Y");
        if (outPtr == NULL)
    {
        printf("dlsym failed\n");
        printf("%s\n", dlerror());
        return;
      }
//Store output parameters
    for (j = 0; j < (*number_of_outputs); j++) {
      response_.out[j]=(*outPtr).Out[j];
      }
  return;
}
```

7.5.3 COMPILATION AND EXECUTION OF THE CODE

For the compilation purposes, we need:

- gcc-8 and gfortran-8 compilers or the latest versions of the gcc and gfortran compilers to be installed on the Linux platform.
- The Library file from the Simulink model, *controll.so*, should be copied in the same folder with the source files described in Section 7.5.2.
- We need to write a simple makefile that describes how to compile and link the code which consists of one Fortran source file, two C source files, and one shared library file.

The *Mainfile* is defined at the same folder with other files and contains the following text:

```
SHELL =/bin/sh
##
## Program parameters
## ---------------
code_OBJS= sendlib.o recvlib.o
##
## Target Rules
## ---------------
all:            code
code:           $(code_OBJS) Makefile
                gfortran-8 example.f90 -o example $(code_OBJS) -ldl
                @echo "To execute: "
                @echo "./example"
                @echo " "
%.o:            %.c Makefile
                gcc-8 -o $@ -c $<
```

The compilation of the Code can be done as shown in Figure 7.19.

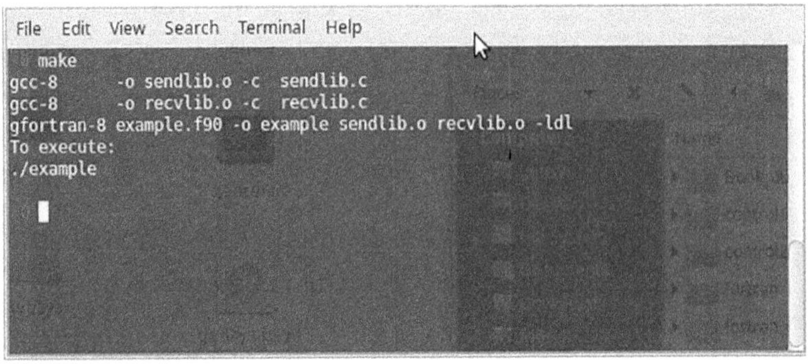

FIGURE 7.19   The compilation process in the Linux terminal window.

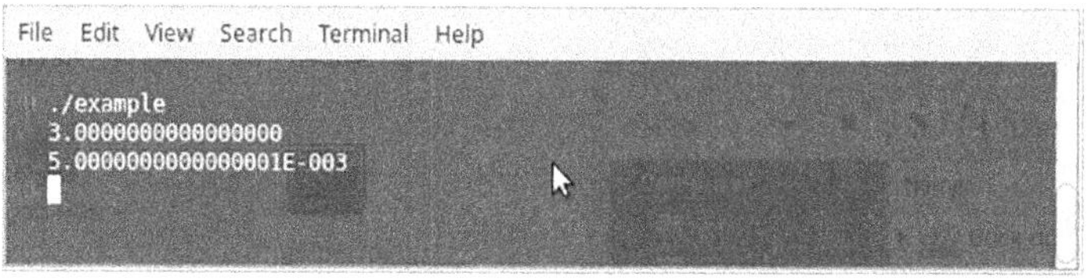

FIGURE 7.20   The execution outcomes in the Linux terminal window.

The execution outcomes for the developed software code are shown in Figure 7.20.

Taking into account that two inputs were defined as 1 and 2, and six timestep calls were defined (note that one more timestep call is needed for a time step of 0 ms) in the source code for "example.f90," the obtained outcomes (output 1 is 3 and output 2 is 5 ms) show that the developed design interface is fully workable. More detailed application of this interface in railway mechatronic investigations is described in two case studies in the text below.

7.6   CASE STUDIES

The case studies described in the following sections show examples of the implementation of co-simulation techniques for the design of a simplified wheel slip control system of a locomotive and for a study of the influence of traction control on the locomotive behavior in the train operational condition.

7.6.1   CO-SIMULATION FOR A LOCOMOTIVE TRACTION CONTROL STUDY

Most previous locomotive traction dynamics simulations have been performed using a locomotive model that has been created from the following components [3, 4]:

- Mechanical system that includes a locomotive and a track;
- Simplified traction control system implemented through a special subroutine or a control block.

However, such an approach does not work in terms of accuracy of results delivered because, while it is still appropriate to be used at the initial stages for the development of slip control algorithms, any investigation of processes at the wheel-rail interface under traction and braking should be undertaken with the application of advanced modeling [34]. The data exchange mechanism in the advanced approach is based on the TCP/IP co-simulation approach [2], as shown in Figure 7.21, and it has discrete synchronization points with a sampling time of 0.001 s for vehicle system dynamics studies [6, 19]. Ways to achieve the realization of this approach for traction studies are described in detail in [3, 4]. This approach is very good when it is required to modify some parameters in a traction control system or a slip control algorithm during a co-simulation process, but it also has one disadvantage in that it is computationally time consuming. In order to increase the computational performance, it is better to use a co-simulation approach described in Section 7.5 and where the locomotive multi-body model is integrated with a traction power system represented by a shared library generated in Simulink as shown in Figure 7.22.

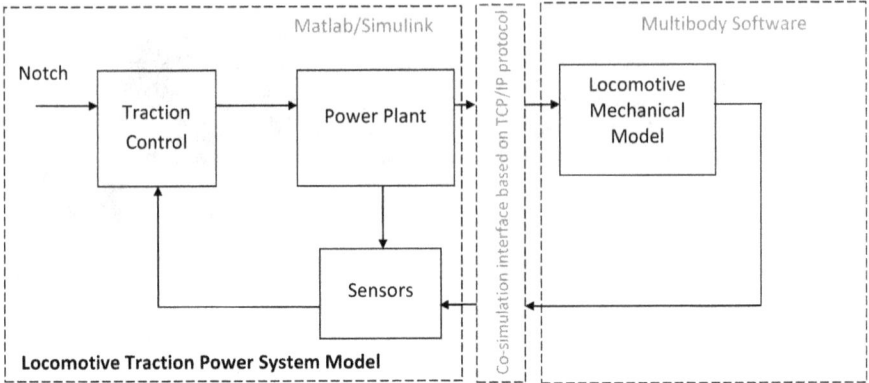

FIGURE 7.21 TCP/IP co-simulation approach for locomotive traction studies. (From [19], with permission.)

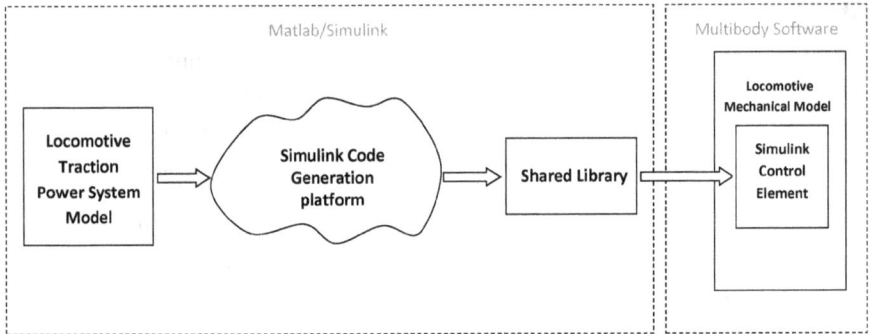

FIGURE 7.22 Shared library co-simulation approach for locomotive traction studies.

In this section, a case for the modeling of a locomotive with a bogie traction control strategy based on a simplified modeling approach is presented. The case is focused on understanding of dynamic behavior for a typical locomotive (Co-Co) with bogie traction control operated in the traction mode [2]. This allows formulating the main tasks which can be described as:

- To develop a locomotive model in Gensys
- To develop a Simulink model of a simplified traction system that can be generated into the shared library
- To perform testing of the locomotive dynamic response to variations of adhesion conditions at the wheel-rail interface

7.6.1.1 Multi-body Model of a Heavy Haul Locomotive in Gensys

A typical Australian heavy haul standard gauge locomotive has been used in this study. The locomotive has a Co-Co axle arrangement as shown in Figure 7.23, and consists

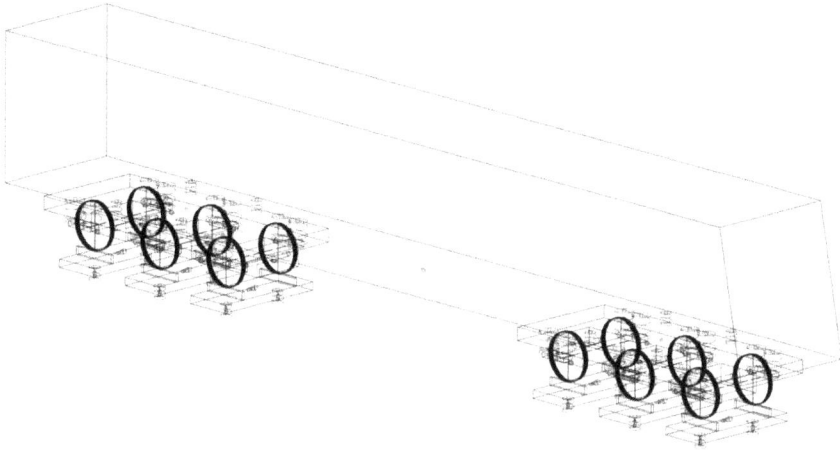

FIGURE 7.23 Locomotive model in Gensys.

of 33 bodies (1 car body, 2 bogie frames, 12 axle boxes, 6 motor housings, 6 rotors, and 6 wheelsets). All bodies have been modeled as rigid masses with 6 degrees of freedom. Some constraints are set on these bodies as listed in Table 7.1.

The connection between the car body and one bogie frame includes:

- Four rubber springs which are modeled as four coil spring elements acting in longitudinal, lateral, and vertical directions
- Two traction rods, where each is modeled as a linear spring element acting in parallel with a linear damper
- One lateral and two vertical bumpstops, modeled as non-linear spring elements acting in the corresponding directions

---

TABLE 7.1
Constraints on Bodies

	x – Longitudinal	y – Lateral	z – Vertical	f – Roll	k – Pitch	p –Yaw
Locomotive car body	√	√	√	√	√	√
Bolster	√	√	√	√	√	√
Side frame	√	√	√	√	√	√
Axle box	√	√	√	√	√, $v_k = 0$	√
Wheelset	√	√	√	√	√, $k = 0$	√
Motor housing	√	√	√	√	√	√
Rotor	√	√	√	√	√, $k = 0$	√

Note: √ = Degree considered; $k = 0$ and $v_k = 0$ refer to body pitch angle and axle box angle velocity being fixed to be equal to zero.

The primary suspension between a bogie frame and an axle box is modeled as:

- Two coil spring elements acting in longitudinal, lateral, and vertical directions
- One longitudinal traction rod modeled as a linear spring element
- One linear vertical damping element on each axle box
- One vertical bumpstop
- One lateral bumpstop with a non-linear characteristic with the middle wheelset having a different characteristic in order to model a different clearance between its wheelset and the axle box in comparison with the leading and trailing wheelsets

Constraints between wheelsets and their axle boxes are used in the model.

The traction motor is a nose suspended motor. The motor has been modeled as two bodies comprising a motor housing and a rotor based on the approaches described in [32]. Therefore, those two bodies have been constrained. The motor housing is connected to the wheelset on one end through constraints, and the spring element is used on the other end to connect the housing to the bogie frame.

For the modeling of a gearbox, a special subroutine has been developed based on kinematic relations between torques and angular velocities taking into account energy conversion efficiency. The assumption that the connection between the rotor and the wheelset is perfectly stiff has been made for the model. The processes of interaction between a gear and a pinion have not been considered for this study.

The main parameters of the locomotive represented in the simulation model are presented in Table 7.2.

The wheel and rail profiles used are for standard new S1002 wheels and new UIC60 rail. In wheel-rail contact modeling, the rails are modeled as massless bodies. Three springs normal to the wheel-rail contact surface are used in the contact subroutine. This allows modeling of three different contact surfaces in the wheel-rail contact simultaneously. The normal contact forces are also solved by consideration of these three springs. The rails are connected to the track via springs and dampers in the lateral and vertical directions. The calculations of creep forces are made using the modified Fastsim algorithm [35].

7.6.1.2 Model of a Locomotive Simplified Traction System

The bogie traction control system used in this study is based on the feedback control strategy. The proposed control system for one bogie is presented in Figure 7.24 with the following variables and parameters: T_{ref} – reference torque (torque demand); $T_{ref}*$ – corrected reference torque; T_{in} – input motor torque; T_{wheels} – traction torque applied to wheelset; ΔT – feedback torque signal; ω_1, ω_1, ω_2 – the real angular velocities of each wheelset; ω – the maximum angular velocity of wheelsets; V – the locomotive velocity; s_{est} – estimated longitudinal slip (creep); s_{ref} – the reference value of the longitudinal slip.

TABLE 7.2
Parameters for Multi-Body Model of Heavy Haul Locomotive

Car Body

Centre of gravity, vertical height above rail	1.930 m
Mass	90,510 kg
Moment of inertia, roll	132,193 kg.m^2
Moment of inertia, pitch	3,394,125 kg.m^2
Moment of inertia, yaw	3,390,553 kg.m^2

Bogie Frame

Centre of gravity, vertical height above rail	0.733 m
Mass	4903 kg
Moment of inertia, roll	3629 kg.m^2
Moment of inertia, pitch	14,453 kg.m^2
Moment of inertia, yaw	17,659 kg.m^2

Wheelset

Centre of gravity, vertical height above rail	0.5033 m
Mass	2036 kg
Moment of inertia, roll	1231 kg.m^2
Moment of inertia, pitch	255 kg.m^2
Moment of inertia, yaw	1231 kg.m^2

Axle Box

Centre of gravity, vertical height above rail	0.5033 m
Mass	239 kg
Moment of inertia, roll	50 kg.m^2
Moment of inertia, pitch	50 kg.m^2
Moment of inertia, yaw	50 kg.m^2

Motor Housing

Centre of gravity, vertical height above rail	0.5103 m
Mass	2390 kg
Moment of inertia, roll	508 kg.m^2
Moment of inertia, pitch	480 kg.m^2
Moment of inertia, yaw	453 kg.m^2

Rotor

Centre of gravity, vertical height above rail	0.5033 m
Mass	710 kg
Moment of inertia, roll	100 kg.m^2
Moment of inertia, pitch	16 kg.m^2
Moment of inertia, yaw	100 kg.m^2

Other Dimensions

Wheel spacing	1.9 m
Bogie spacing	13.7 m
Track gauge	1.435 m

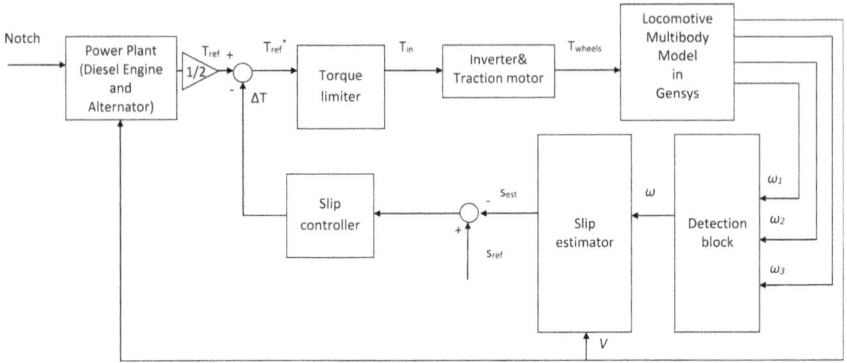

FIGURE 7.24 Simplified locomotive traction control system for a single bogie. (From [19], with permission.)

For the power plant model, the tractive effort realized by a locomotive, F_t, can be calculated as [4, 36]:

$$\text{for} \quad F_t * V < \left(N^2/64 \right) * P_{max}, \quad F_{t=} \left(N/8 \right) * Te_{max} - k_f * V \tag{7.1}$$

$$\text{else} \quad F_{t=} \left(N^2/64 \right) * P_{max} / V \tag{7.2}$$

where N is the throttle setting in notches, 0–8; P_{max} is the maximum locomotive traction horsepower; Te_{max} is the maximum locomotive traction force; and k_f is the torque reduction.

In this case, the dynamics of the diesel-alternator system can be described by means of a low-pass filter and use of a Laplace transformation, and can be written for a single wheelset as:

$$T_{ref} = \frac{1}{\tau_1 s + 1} \cdot \frac{F_t r}{n_m} \tag{7.3}$$

where τ_1 is a time constant; s is the Laplace variable; r is wheel rolling radius, m; and n_m is the number of motorized axles within the locomotive.

The inverter and motor dynamics can be described with a low-pass filter:

$$T_{in} = \frac{1}{\tau s + 1} T_{ref} \tag{7.4}$$

where τ is a time constant.

The slip estimator (see Figure 7.24) is designated to calculate the estimated slip for each axle (wheelset) which is calculated as:

$$S_{est} = \frac{wr - V}{V} \tag{7.5}$$

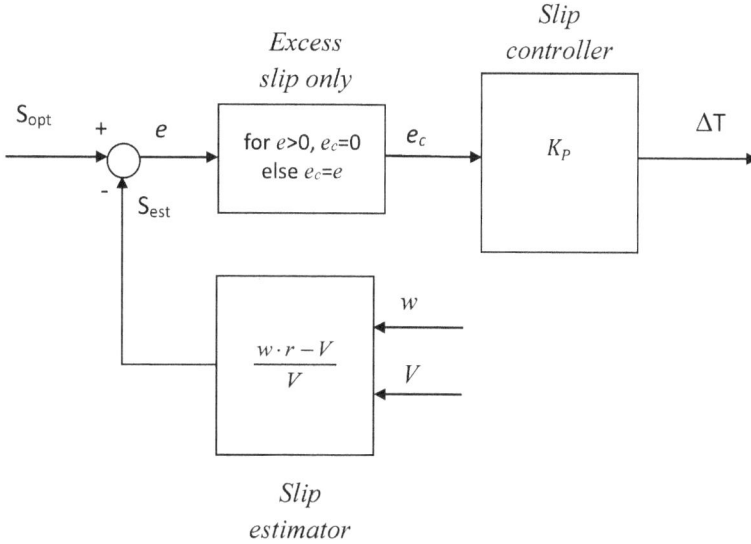

FIGURE 7.25   Wheel slip/traction control modeling approach.

The slip controller, which is part of a wheel slip/traction control system as shown in Figure 7.25, is a simple P controller which uses a slip error as the input signal to the controller. A slip error can be defined as:

$$e = s_{opt} - s_{est} \tag{7.6}$$

The slip control is active when the estimated slip value, s_{est}, is higher than the optimal slip, s_{opt}. In this case, the slip error correction, e_c, can be found as:

$$\text{for} \quad e > 0, e_c = 0 \tag{7.7}$$

$$\text{else} \quad e_c = e \tag{7.8}$$

The control law of the slip controller is represented by the following equation:

$$\Delta T = K_P \cdot e_c \tag{7.9}$$

where K_P is proportional gain which is tuned to the applied load. A more detailed description of this slip control modeling approach can be found in [4, 35].

For this model, the discrete (no continuous states) solver with a fixed time step of 10 µs was used in Simulink. The full model in Simulink used for this study is shown in Figure 7.26. The model has 7 inputs (1 – locomotive velocity; 2.6 – angular velocities of wheelsets) and 6 outputs (1.6 – traction torques applied to wheelsets).

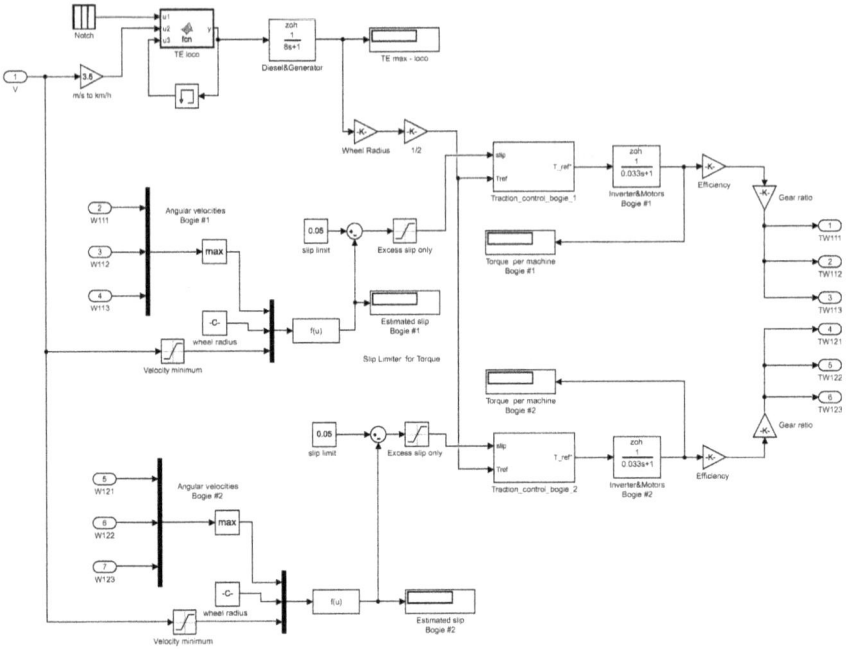

FIGURE 7.26 Simplified locomotive traction system model in Simulink.

The code for the Matlab Function, that is, for the TE block shown in Figure 7.26, is provided below:

```
function y = fcn(u1,u2,u3)
% u1=N - the throttle settings in notches, 0 to 8
% u2=V - locomotive speed, m/s
% u3=Ft - the tractive effort realized by a locomotive
(previous time step)
TEmax=600000;
N=floor(u1);
V=u2/3.6;
Pmax=2900000;
if ((u3*V)<((N*N/64)*Pmax))
    Ft=N/8*TEmax-1000*V;
else
    Ft= (N*N/64)*(Pmax/V);
end
if(N==0)
    Ft=0;
end
y = Ft;
```

The shared library for this Simulink model was built based on the procedure described in Section 7.5.1.

7.6.1.3 Dynamic Response Test to Variations of Adhesion Conditions at the Wheel-Rail Interface

This test allows checking the response of the simplified traction power system on the changes in friction conditions and on the dynamic behavior of a locomotive operating under traction mode in this case study. The main aim of this test is to be sure that the designed mechatronic system has not allowed wheelsets to exceed the slip threshold. This outcome is essential to avoid any potential damage caused by exceeding the maximum allowable traction torque applied to the wheelset. In order to create varying friction conditions and to check the performance of the developed traction control system models, three adhesion condition curves were used in this study. The parameters for the modeling of different adhesion coefficients for the modified Fastsim algorithms are presented in Table 7.3. These comprise the following parameters: k_0 – initial value of Kalker's reduction factor at creep values close to zero, $0 < k_0 \leq 1$; α_{inf} – fraction of the initial value of the Kalker's reduction factor at creep values approaching infinity, $0 \leq \alpha_{inf} \leq 1$; β – non-dimensional parameter related to the decrease of the contact stiffness with the increase of the slip area size, $0 \leq \beta$ [11]; μ_s – the maximum coefficient of friction; A – the ratio of the limit friction coefficient at infinity slip velocity to the maximum friction coefficient μ_s; and B – the coefficient of exponential friction decrease [37].

The following operational scenario was used in the co-simulation:

- Notch position is going from *idle* to 8 during the first 2 s, after which it remains at Notch 8;
- Locomotive accelerates and then runs with a constant speed of 20 km/h;
- Total simulation time is 50 s;
- Dry friction condition is used for locomotive running, except for switching to *wet friction condition* between 25 and 30 s, and to *greasy friction condition* between 40 and 45 s;
- Slip threshold for the traction control system is set to 0.05 per bogie;
- Tangent track with no track geometry irregularities.

TABLE 7.3
Adhesion Curves Parameters for Modified Fastsim Algorithm

Parameters	Adhesion Curve		
	Dry	Wet	Greasy
k_0	0.14	0.1	0.08
α_{inf}	0.025	0.02	0.015
B	0.85	0.8	0.65
μ_s	0.44	0.3	0.23
A	0.43	0.39	0.70
B (s/m)	0.72	0.17	0.07

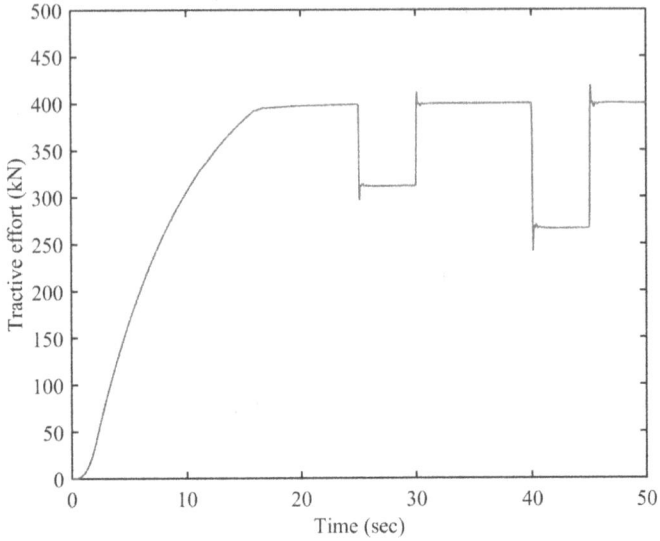

FIGURE 7.27 Locomotive tractive effort in the time-domain.

The results obtained for this locomotive model equipped with a simplified slip control, as defined in Section 7.4.1.2, are shown in Figures 7.27–7.29. Figure 7.27 shows that the locomotive tractive effort performance is as expected for this type of locomotive and it is limited by the adhesion conditions at the wheel-rail interface. Therefore, the locomotive was able to reach the slip threshold for all curves and its power is limited by a slip controller, i.e., the estimated slip values, as shown in Figure 7.28, are limited to 0.05. The simplified model as expected results in

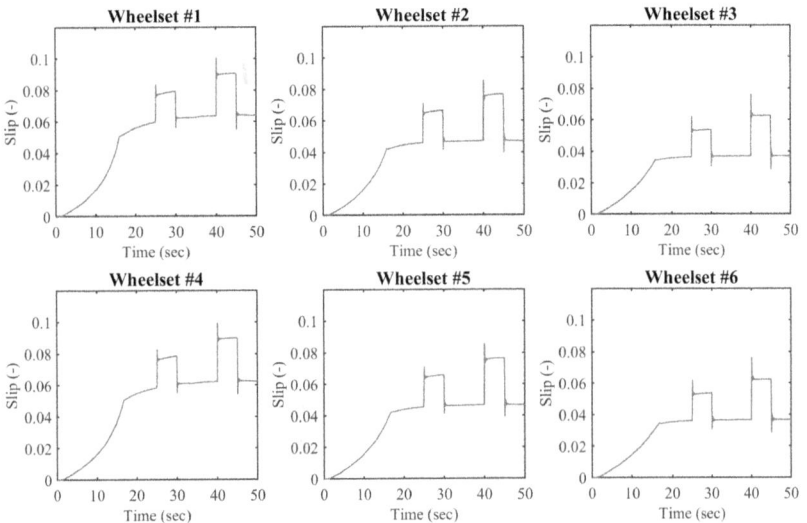

FIGURE 7.28 Longitudinal slip values in the time-domain.

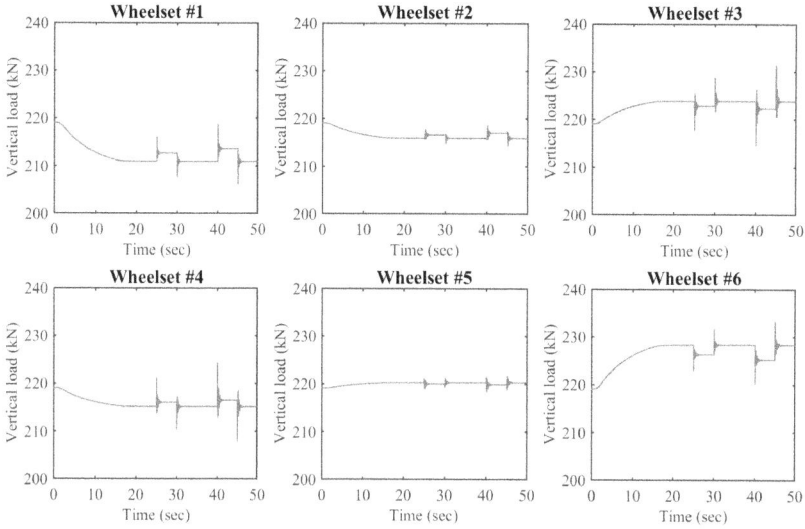

FIGURE 7.29 Vertical load values in the time-domain.

distribution of vertical loads between wheelsets as shown in Figure 7.29. Figure 7.30 shows the estimated value of the traction coefficients for each wheelset of the locomotive in the time-domain. The results confirm that the simplified modeling approach of the power traction system leads to variations in slip values between wheelsets and such variations are caused by the absence of the induction machine dynamics that forces the equalization of wheelset slip in the simplified model. However, the general performance of a full locomotive model provides adequate traction coefficient

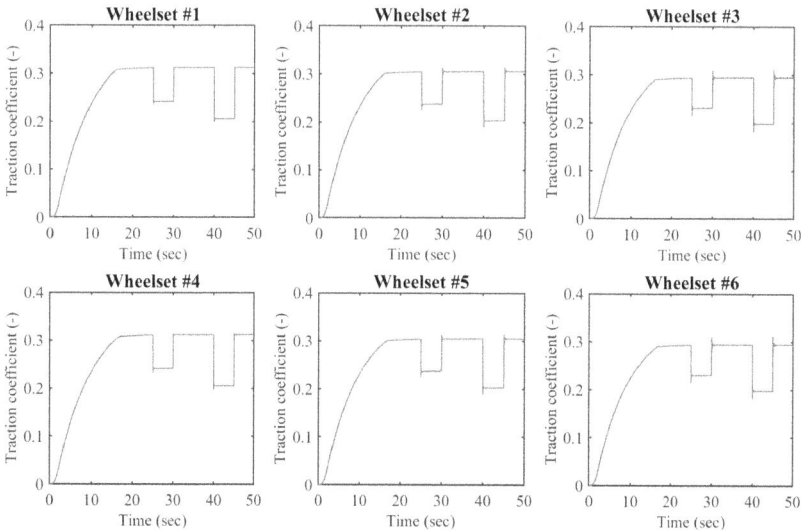

FIGURE 7.30 Traction coefficient values in the time-domain.

outputs and the test with a variation of different adhesion/friction conditions at the wheel-rail interface confirms that the developed traction control system is robust in the way it handles such changes.

The mechatronic system for the heavy haul locomotive modeled in this study is workable and the results produced by models in co-simulation between Gensys and a shared library generated from Matlab/Simulink show that this co-simulation is appropriate for railway vehicle traction studies and traction control algorithms.

7.6.2 CO-SIMULATION FOR AN ADVANCED LONGITUDINAL TRAIN DYNAMICS STUDY

The application of one-dimensional longitudinal train dynamics models cannot provide sufficient accuracy for locomotive studies because it does not consider two main components: the wheel-rail interaction and the traction control system. On the other hand, most locomotive traction/braking studies are commonly focused on the dynamics of an individual locomotive and are limited in terms of implementation of in-train forces. The individual locomotive approach has significant limitations for train dynamics studies because it does not fully depict the real behavior of locomotives in the train operational environment by not considering that in-train forces have a significant influence on locomotive dynamics and the interaction at the wheel-rail interface. For example, taking into account that the heavy haul train configuration in actual practice has up to 5 locomotives, this leads to the building of a very complex mechatronic system for the study. The multi-locomotive co-simulation approach for such a task, that also includes co-simulation between a locomotive and a traction power system for each locomotive in the train consist, is shown in Figure 7.31. A more detailed description on this architecture can be found in [19].

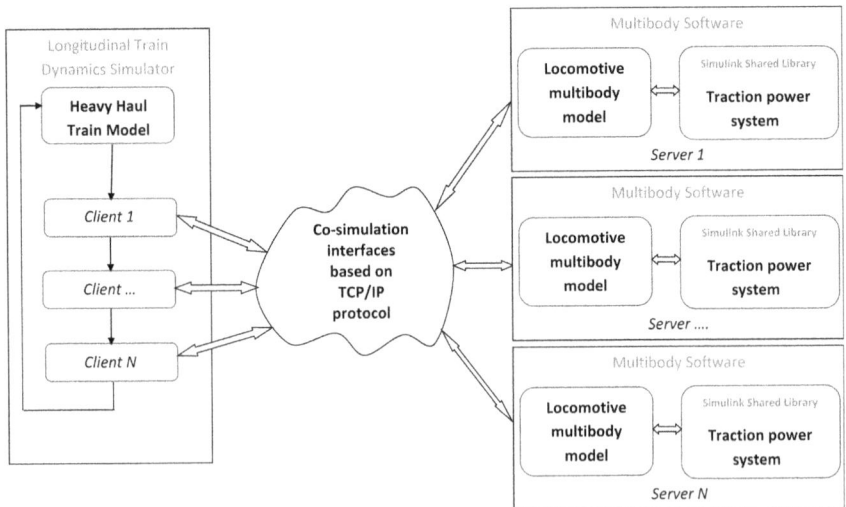

FIGURE 7.31 Architecture of the multi-locomotive co-simulation approach. (From [19], with permission.)

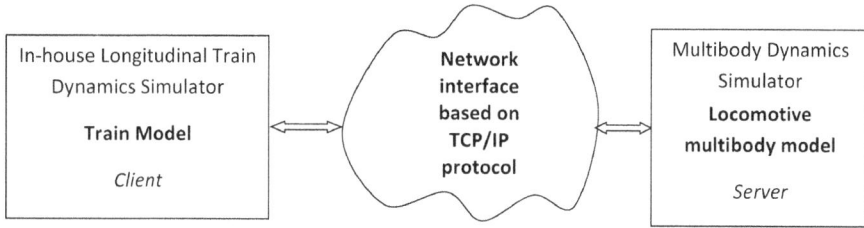

FIGURE 7.32 Architecture of the multi-locomotive co-simulation approach between LTDS and MBS.

In this study, the simplified version of this co-simulation approach between a longitudinal train dynamics simulator (LTDS) and a MBS package is considered as shown in Figure 7.32 and the main tasks are formulated as:

• To study a concept of uni-directional data exchange co-simulation approach [38, 39]
• To study a concept of bi-directional data exchange co-simulation approach [40]
• Compare results of the two locomotive dynamics studies performed with uni- and bi-directional co-simulation approaches

For this study, we consider a basic simplified concept only without describing the integration of the traction power system as shown in in Figure 7.32. This requires dividing the full complex model into two sub-models:

• Train model
• Locomotive model

The train model is based on a longitudinal train dynamics approach, which is implemented in the in-house software package [41, 42]. The locomotive model is implemented in the Gensys railway vehicle multi-body software package [43], which has a client-server architecture designated for data communication between two computers or two software packages through a TCP/IP interface.

7.6.2.1 Uni-directional Data Exchange Co-simulation Approach

The simulation of the entire train in the LTDS is required in order to obtain the speed, tractive effort and lateral coupler forces of each locomotive for use in the co-simulation in this study. The first two parameters are common outputs in longitudinal train dynamics simulations [44]. Lateral coupler forces have recently been started to be widely in use in locomotive studies [4, 45] due to their influence on the forces at the wheel-rail contact interface.

The uni-directional interface implemented for exchange of these parameters from the LTDS to MBS and a bi-directional exchange for control of the data exchange process use the following commands for data exchange between a client (LTDS) and a server (MBS, i.e., Gensys) [43]:

- *ask_iadr <string_var_name>*: this command is needed to obtain the address of a variable. If the transmission of the command was successful, Gensys will answer with 7FF0000000000001, which means that the command requires one more argument. After sending the argument (the name of the variable), Gensys replies with the address of the variable. The command calls once for each parameter and the LTDS stores this parameter in memory for the whole co-simulation process.
- *put_iadr <var_address_value>*: this command is used to overwrite the value of a variable in the Gensys model. This command is called each time step that is set for the co-simulation process. On a successful transmission of the command *put_iadr*, Gensys sends 7FF0000000000002, which means that the command requires two more arguments. The first argument is <var_address_value> that is defined with the *ask_iadr* command. After sending the first argument (the address of the variable), Gensys replies with 7FF0000000000001, which means that the command requires the second argument. After sending the second argument (the new value of the variable), Gensys replies with 0, indicating that the new value was successfully inserted in the main memory.
- *run_tout*: this command runs a simulation process for each time step. Gensys replies with the value of variable *time*, i.e., with the current calculation time.
- *run_stop*: this command stops a simulation process and closes a network connection.

If the command was executed successfully, Gensys replies with the zero value.

The following parameters are used in Gensys for the co-simulation process to control the movement and dynamic behavior of the locomotive model:

- Speed
- Distance
- Front coupler force
- Rear coupler force
- Curvature
- Front lateral coupler force
- Rear lateral coupler force
- Tractive effort
- Resistance force
- Gravitational component
- Curving resistance
- Traction force limit

The main disadvantage of this approach is that parameters inside of the LTDS are not updated during such a co-simulation process.

7.6.2.2   Bi-directional Data Exchange Co-simulation Approach

The main difference from the uni-directional data exchange co-simulation approach is that the bi-directional data exchange allows data from the MBS to be sent to the LTDS. There are only three variables that should be returned into the LTDS:

- Speed
- Distance
- Tractive effort

In order to perform such a data exchange, the additional command is in-use [43]:

- ***get_iadr <var_address>***: this command is used by the client to retrieve a value from the server. After a successful transmission, Gensys sends 7FF0000000000001, which means that the command requires one argument. After sending the argument (the address of the variable), Gensys replies with the value of the variable.

7.6.2.3   Comparison of Results Obtained with Two Data Exchange Co-simulation Approaches

The simulation scenarios for a comparison process consider the following stages:

- Development of a longitudinal train model
- Development of a locomotive model
- Co-simulation analysis

7.6.2.3.1   Train Model in LTDS

A head-end train with the configuration of 3 locomotives + 122 wagons was selected for demonstration. All locomotives are 6-axle diesel locomotives with 22.3 tons axle load (TAL); locomotive traction characteristics used in the LTDS are shown in Figure 7.33. All wagons are loaded and have 23 TAL.

A section of hypothetical track, shown in Figure 7.34, was used in a similar manner as in the studies published in [38–40].

During the simulations, the train started at the position of 100 km with zero speed. Traction notch was set to 1 at the beginning of the simulation; it was then increased by one notch position every 5 s until notch 8 was reached where it remained for the rest of the simulation.

The simulation time step inside of the LTDS was set to 1 ms.

7.6.2.3.2   Locomotive Model in Gensys

For this study, the heavy haul locomotive has similar characteristics to a locomotive description published in [46, 47]. This locomotive has semi-steering bogies and is equipped with a bogie traction control system. The locomotive weight is 134 tons and it has a Co-Co wheel arrangement.

The locomotive body, two bogies, and six wheelsets are modeled as rigid bodies and have six degrees of freedom each. The main model parameters of such a heavy haul locomotive are presented in Table 7.4.

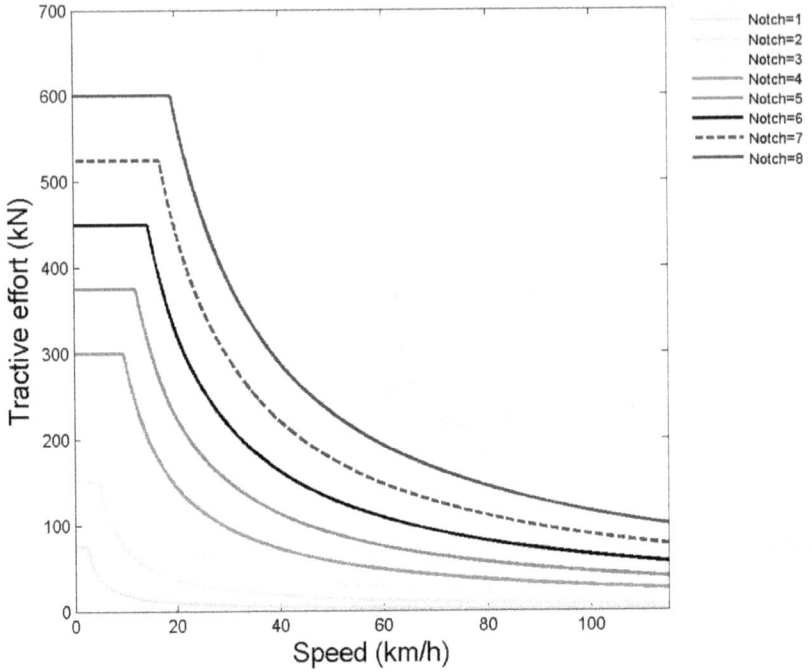

FIGURE 7.33 Tractive effort characteristics.

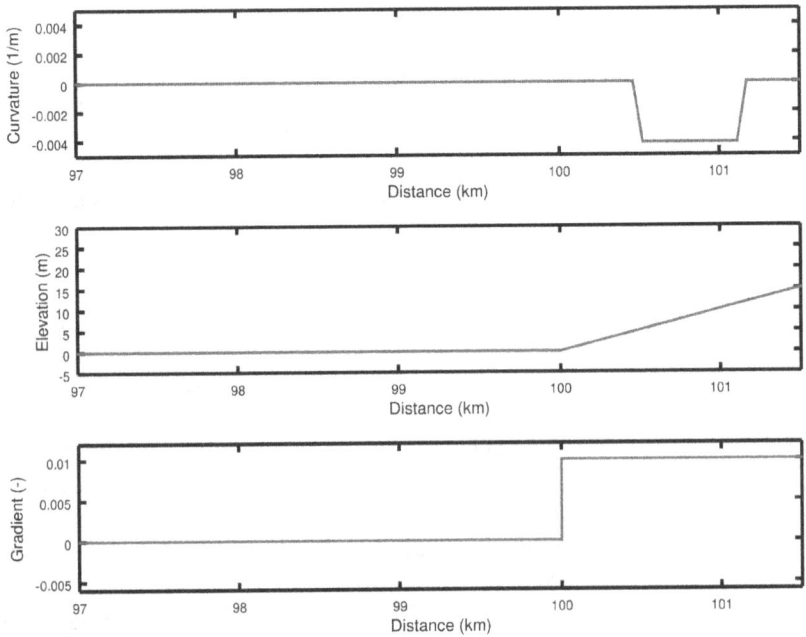

FIGURE 7.34 Track design characteristics.

TABLE 7.4
Locomotive Model Parameters [47]

Category	Parameter	Value	Units
Dimensions	Coupler longitudinal distance from car body CoG	22	m
	Nominal coupler height above rail level	0.885	m
	Bogie pivot longitudinal distance from car body CoG	7.85	m
	Bogie pivot longitudinal distance (outwards) from bogie frame CoG	0.31	m
	Bogie semi-wheelbase	1.9	m
	New wheel diameter	1.016	m
	Total mass	134,000	kg
Car body	CoG height above rail level	1.93	m
	Mass	90,000	kg
	Moment of inertia, roll	174,002	kg.m^2
	Moment of inertia, pitch	3,727,195	kg.m^2
	Moment of inertia, yaw	3,706,796	kg.m^2
Bogie	CoG height above rail level	0.733	m
frames	Mass	12,121	kg
	Moment of inertia, roll	5,318	kg.m^2
	Moment of inertia, pitch	37,007	kg.m^2
	Moment of inertia, yaw	41,029	kg.m^2
Wheelsets	CoG height above rail level = New wheel radius	0.508	m
	Mass	3,209	kg
	Moment of inertia, roll = yaw	2,014	kg.m^2
	Moment of inertia, pitch	1,351	kg.m^2
Secondary	Rubber springs – longitudinal distance from bogie frame CoG	0.95	m
suspension	(Outer) springs:		
	Lateral distance from bogie CoG	1.078	m
	Longitudinal and lateral shear stiffness	188.4	kN/m
	Vertical stiffness	7.9	MN/m
	Traction rods:		
	Stiffness	5	MN/m
	Damping coefficient	20	kN.s/m
	Lateral viscous dampers:		
	Longitudinal distance from bogie CoG	0.275	m
	Damping coefficient	40	kN.s/m
	Yaw viscous dampers:		
	Lateral distance from bogie CoG	1.24	m
	Series stiffness	45	MN/m
	Damping coefficient (at 32 mm expansion/compression)	4,600	N.s/m
	Blow-off point (at 1.032 m expansion/compression)	6,800	N.s/m
	Secondary yaw viscous dampers – Damping coefficient	40	kN.s/m

(Continued)

TABLE 7.4 (Continued)
Locomotive Model Parameters [47]

Category	Parameter	Value	Units
Primary suspension	(Axlebox) Lateral position from wheelset CoG	1.078	m
	Coil springs:		
	Longitudinal shear stiffness	45	MN/m
	Lateral shear stiffness	2.25	MN/m
	Vertical stiffness	288.5	kN/m
	Damping coefficient	10	kN.s/m
	Traction rods:		
	Stiffness	44	MN/m
	Damping coefficient	100	kN.s/m

The wheel and rail profiles used are for standard new ANZR1 wheel and new AS60 rail profiles. In the wheel-rail contact modeling, three different wheel-rail contact surfaces can be in contact simultaneously. The wheel contacts are modeled with three spring elements normal to three wheel-rail contact surfaces, which allow normal wheel-rail contact forces to be determined. The calculations of creep forces at the wheel-rail contact zone are made using the modified Fastsim algorithm [36]. The dry friction conditions have been selected for all tangent track sections and lubricated friction conditions are in use for all curves.

In order to perform studies under traction, the model includes the shared library with an advanced traction system based on the bogie traction control strategy of one inverter per bogie and the application of a PI controller [19]. The slip threshold is set to 0.07 based on the type of locomotive as explained in [18].

To ensure that the simulation environment is close to real operational conditions, the track modeled has FRA Class 5 geometry irregularities [48].

Due to the difference in modeling approaches between longitudinal train dynamics and multi-body vehicle dynamics, it is necessary in Gensys to apply a compensator (an additional resistance force that acts on the locomotive car body in the direction opposite to the locomotive movement) in addition to the usual resistance forces in order to match the locomotive speed behavior obtained in the LTDS. The compensator was implemented as a sky-hook (linear spring and damper connected in parallel) element.

A modified Heun's method, *heun_c* [43], with a variable time step is used as a solver in Gensys.

7.6.2.3.3 *Comparison Analysis*

For comparisons of results for these two co-simulation approaches, a tangent track and one curve of 240 m radius were chosen on the hypothetical track. They are positioned between 100 and 101.4 km as shown in Figures 7.35 and 7.36. The co-simulation time step between the MBS and the LTDS was set to 10 ms.

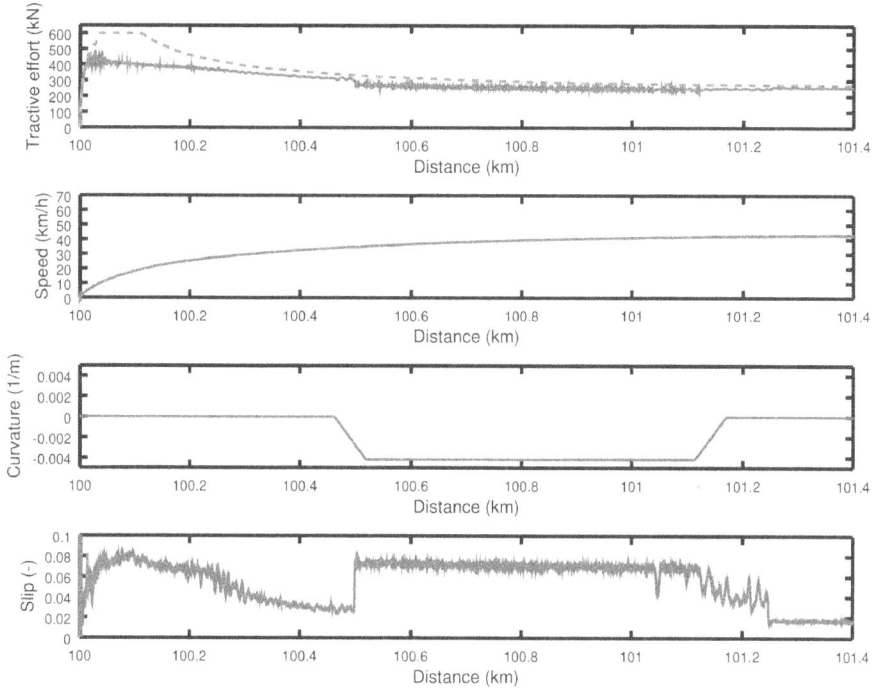

FIGURE 7.35 Tractive effort, speed, curvature, and speed versus distance for the uni-direction data exchange approach (solid line – MBS, dotted line – LTDS).

Figures 7.35 and 7.36 show the locomotive speed simulation results in Gensys that confirm that both simulation cases are reasonably well synchronized in this regard. The tractive effort results, presented in Figure 7.35, show differences between the MBS and LTDS results in the case of the uni-directional data exchange approach. This occurs due the limits imposed by adhesion and available power. The slip system, as shown in Figure 7.35, was activated at the start of simulation when a high tractive effort was generated by the locomotive, this being due to the lubricated friction conditions on the curves. As a result, the slip at the wheel-rail interface was limited by the threshold slip value of 0.07. This caused some loss of tractive effort in Gensys and the implementation of such a slip control strategy does not allow achieving the values of tractive effort delivered in the LTDS. However, in the case of the bi-directional data exchange simulation approach, very good convergence between the tractive effort results from the LTSD and Gensys has been observed as shown in Figure 7.36.

These cases confirm that the wheel-rail contact conditions have a significant influence on results delivered in LTDSs. The design of the adhesion control module that limits wheel slip, which is a part of the complex system of a locomotive/train, is also a very important part of such a simulation process. However, the main disadvantage of both these approaches is that these are time consuming in comparison with the usage of constant approximated lateral forces for each track location directly in

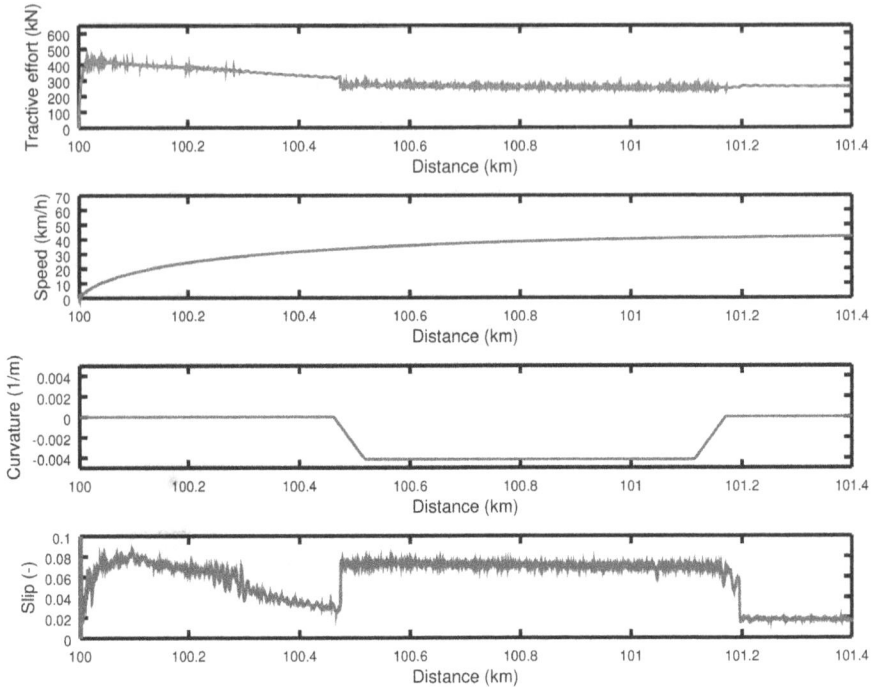

FIGURE 7.36 Tractive effort, speed, curvature, and speed versus distance for the bi-direction data exchange approach (solid line – MBS, dotted line – LTDS).

MBS that works in the non-co-simulation mode. However, this can be compensated for if a specific analysis requires obtaining more precise traction and rail damage simulation results for the dynamic behavior study with more than one locomotive in the train configuration.

REFERENCES

1. S. Bruni, R. Goodall, T. X. Mei, H. Tsunashima, Control and monitoring for railway vehicle dynamics, Vehicle System Dynamics, 45(7–8), 733–779, 2007.
2. M. Spiryagin, V. Spiryagin, Modelling of Mechatronic Systems of Running Gears for a Rail Vehicle, East Ukrainian National University, Lugansk, Ukraine, 2010. ISBN 978-966-590-871-5 (in Ukrainian).
3. M. Spiryagin, C. Cole, Y. Q. Sun, M. McClanachan, V. Spiryagin, T. McSweeney, Design and Simulation of Rail Vehicles, Ground Vehicle Engineering Series, CRC Press, Boca Raton, FL, 2014.
4. M. Spiryagin, P. Wolfs, C. Cole, V. Spiryagin, Y. Q. Sun, T. McSweeney, Design and Simulation of Heavy Haul Locomotives, Ground Vehicle Engineering Series, CRC Press, Boca Raton, FL, 2017.
5. W. Körtum, O. Vaculín, Is Multibody Simulation Software Suitable for Mechatronic Systems? Proceedings of the 5th World Congress on Computational Mechanics, Vienna, Austria, 7–12 July 2012.

6. M. Arnold, A. Carrarini, A. Heckman, G. Hippmann, Simulation Techniques for Multidisciplinary Problems in Vehicle System Dynamics, In: M. Valášek (Ed), Computational Mechanics in Vehicle System Dynamics, Supplement to Vehicle System Dynamics, Taylor & Francis, London, UK, 40, 17–36, 2004.

7. M. Arnold, B. Burgermeister, C. Führer, G. Hippmann, G. Rill, Numerical methods in vehicle system dynamics: State of the art and current developments, Vehicle System Dynamics, 49(7), 1159–1207, 2011.

8. MODELISAR, Functional mock-up interface for co-simulation, Document version 1.0, 21 October 2010.

9. MODELISAR, Functional mock-up interface for model exchange and co-simulation, Document version 2.0 Beta 4, 10 August 2012.

10. P. Häse, C. Decking, Investigation of Drive Systems Using ADAMS and MATLAB/SIMULINK, ADAMS/Rail Users' Conference 2000, Harlem, the Netherlands, 2000.

11. H. P. Kotz, Simulation of Effects Based on the Interaction of Mechanics and Electronics in Railway Vehicles, Simpack User Meeting, Freiburg, Germany, 8–9 April 2003.

12. H. P. Kotz, A Toolkit for Simulating Mechatronics in Railway Vehicles, Simpack User Meeting, Freiburg, Germany, 8–9 April 2003.

13. R. V. Kovalev, G. A. Fedyaeva, V. N. Fedyaev, Modelling of an electro-mechanical system of locomotives, Sbornik Trudov of DIIT, no. 14, 2007, 123–127 (in Russian). See: http://www.universalmechanism.com/index/download/elmechloco.pdf.

14. V. N. Fedyaev, Influence of electrical and mechanical subsystems of a shunting locomotive on the realization limit tractive efforts, Ph.D. Thesis, Bryansk State Technical University, Bryansk, Russia, 2006 (in Russian).

15. M. Spiryagin, K. S. Lee, H. H. Yoo, Control system for maximum use of adhesive forces of a railway vehicle in a tractive mode, Mechanical Systems and Signal Processing, 22(3), 709–720, 2008.

16. G. A. Fedyaeva, Forecasting of dynamic process under transient and emergency mode for traction drives with asynchronous motors, DSc Thesis, Moscow State Railway University, Moscow, Russia, 2008 (in Russian).

17. M. Spiryagin, S. Simson, C. Cole, I. Persson, Co-simulation of a mechatronic system using Gensys and Simulink, Vehicle System Dynamics, 50(3), 495–507, 2012.

18. M. Spiryagin, P. Wolfs, C. Cole, S. Stichel, M. Berg, M. Plöchl, Influence of AC system design on the realisation of tractive efforts by high adhesion locomotives, Vehicle System Dynamics, 55(8), 1241–1264, 2017.

19. M. Spiryagin, I. Persson, Q. Wu, C. Bosomworth, P. Wolfs, C. Cole, A co-simulation approach for heavy haul long distance locomotive-track simulation studies, Vehicle System Dynamics, 57(9), 1363–1380, 2019.

20. X. Liu, R. Goodall, S. Iwnicki, A Direct Control Approach for Automatic Steering and Stability of Motorized Independently Rotating Wheels, Proceedings of the 26th International Symposium on Dynamics of Vehicles on Roads and Tracks (IAVSD 2019), Gothenburg, Sweden, 12–16 August 2019.

21. GENSYS.1910 Reference Manual. Program Calc [Internet]. AB DEsolver, Östersund, Sweden. Accessed on 11/12/2019. Available at http://gensys.se/doc_html/calc.html.

22. Simulink. Developing S-functions. Matlab & Simulink 2019. Revised – September 2019. The MathWorks, Inc. Natick, MA, USA.

23. Embedded Coder® Getting Started Guide. Matlab & Simulink R2017b. Revision September 2017. The MathWorks, Inc., Natick, MA, USA.

24. Embedded Coder® User's Guide. Matlab & Simulink R2017b. Revision September 2017. The MathWorks, Inc., Natick, MA, USA.

25. Dassault Systèmes Simulia Corp. Simpack Documentation. B-14 – Multi domain interfaces/Model export. Simpack Release 2017.

26. A. C. Zolotas, J. T. Pearson, R. M. Goodall, Modelling requirements for the design of active stability control strategies for a high speed bogie, Multibody System Dynamics, 15(1), 51–66, 2006.

27. N. Bosso, A. Gugliotta, A. Somà, M. Spiryagin, Model of Scaled Test Rig for Real Time Applications, Proceedings of the 21st International Congress of Mechanical Engineering, Natal, Brazil, 24–28 October 2011.

28. S. Müller, R. Kögel, R. Schreiber, Numerical Simulation of the Drive Performance of a Locomotive on Straight Track and in Curves, 14th European ADAMS Users' Conference, Berlin, Germany, 1999.

29. S. Müller, R. Kögel, R. Schreiber, Simulation of a Locomotive as a Mechatronical System, 4th ADAMS/Rail Users Conference, Utrecht, The Netherlands, 1999.

30. O. Polach, Influence of Locomotive Tractive Effort on the Forces Between Wheel and Rail, In: Selected Papers From 20th International Congress of Theoretical and Applied Mechanics Held 28 August–1 September 2000 in Chicago, IL, Swets & Zeitlinger, Lisse, the Netherlands, Supplement to Vehicle System Dynamics, 35, 7–22, 2001.

31. J. C. Huang, J. Xiao, H. Weiss, Simulation Study on Adhesion Control of Electric Locomotives Based on Multidisciplinary Virtual Prototyping, Proceedings of the IEEE International Conference on Industrial Technology (ICIT 2008), Chengdu, China, 21–24 April 2008.

32. Vampire Help Manual – Chapter 8, Simulink Interface, Version Vampire Pro 6.00, 862–889.

33. Universal Mechanism, UM co-simulation tool – User's manual – Version 8.0, 2016.

34. Universal Mechanism, UM Control – Getting Started – Version 8.0, 2018.

35. M. Spiryagin, O. Polach, C. Cole, Creep force modelling for rail traction vehicles based on the Fastsim algorithm, Vehicle System Dynamics, 51(11), 1765–1783, 2013.

36. M. Spiryagin, P. Wolfs, F. Szanto, C. Cole, Simplified and advanced modelling of traction control systems of heavy-haul locomotives, Vehicle System Dynamics, 53(5), 672–691, 2015.

37. O. Polach, Creep forces in simulations of traction vehicles running on adhesion limit, Wear, 258(7–8), 992–1000, 2005.

38. M. Spiryagin, Q. Wu, C. Cole, F. Szanto, Advanced Studies on Locomotive Dynamics Behaviour Utilising Co-Simulation between Multibody and Train Dynamics Packages, Proceedings of Conference of Railway Excellence 2016, Melbourne, Australia, May 16–18, 2016.

39. M. Spiryagin, Q. Wu, Y. Q. Sun, C. Cole, I. Persson, Locomotive Studies Utilizing Multibody and Train Dynamics, In 2017 Joint Rail Conference (JRC2017), Philadelphia, PA, 4–7 April 2017, DOI:10.1115/JRC2017-2221.

40. M. Spiryagin, Q. Wu, Y. Q. Sun, C. Cole, I. Persson, Advanced Co-Simulation Technique for the Study of Heavy Haul Train and Locomotive Dynamics Behavior, In: W. Zhai, K. C. P. Wang (Eds), ICRT 2017: Railway Development, Operations, and Maintenance, Proceedings of the First International Conference on Rail Transportation, American Society of Civil Engineers, Reston, VA, Reston, VA, USA, 79–87, 2018, DOI:10.1061/9780784481257.010.

41. Q. Wu, S. Luo, T. Qu, X. Yang, Comparisons of draft gear damping mechanisms, Vehicle System Dynamics, 55(4), 501–516, 2017.

42. Q. Wu, X. Yang, C. Cole, S. Luo, Modelling polymer draft gears, Vehicle System Dynamics, 54(9), 1208–1225, 2016.

43. Gensys.1908 Reference Manual, Users Manual for Program CALC. Available at http://www.gensys.se.

44. Q. Wu, M. Spiryagin, C. Cole, Longitudinal train dynamics: An overview, Vehicle System Dynamics 54, 2016, 1688–1714.

45. C. Cole, M. Spiryagin, Q. Wu, Y. Q. Sun, Modelling, simulation and applications of longitudinal train dynamics, Vehicle System Dynamics, 55(10), 1498–1571, 2017.
46. M. Spiryagin, A. George, Y. Q. Sun, C. Cole, T. McSweeney, S. Simson, Investigation on the locomotive multibody modelling issues and results assessment based on the locomotive model acceptance procedure, Journal of Rail and Rapid Transit 227, 2013, 453–468.
47. A. L. George, Theoretical and numerical investigation on traction forces for high adhesion locomotives, MEng Thesis, Central Queensland University, 2015.
48. Federal Railroad Administration, Track safety standards – Classes 1 through 5, Chapter 1 in Track and Rail and Infrastructure Integrity Compliance Manual, Federal Railroad Administration, Washington, DC, 2014.

8 Microprocessor Computers and Electronics

This chapter is focused on the microprocessor computer design and application aspects for rail vehicle mechatronic systems. Possible rail vehicle application areas for microprocessor-based systems are reviewed. The design concepts of different microprocessor computer architectures are discussed. This chapter also presents a case study for the integration of a microprocessor-based system for a wagon dynamics condition monitoring device with the formulation of the power problem and the energy balance solution being provided.

8.1 INTRODUCTION

At the present time, the application of microprocessor-based systems in modern vehicular systems is a common feature for all transport modes. In the railway vehicle designs, the first reviews for such applications can be found in [1–5]. The microprocess systems described at that time were designated for rolling stock control purposes and more focused on the following application areas:

- Traction power control system
- Traction motor control
- Diagnostic applications for major components

With the development of electronic components and parts, additional emphasis was made on the following on-board microprocessor system control and diagnostic areas:

- Suspension and passenger comfort control systems
- On-board vehicle health monitoring systems
- On-board signaling systems
- Train control systems

The levels of control necessary in rolling stock on-board applications set the following areas for the application of computerized systems [6]:

- Traffic control systems are responsible for safe driving, navigation, surveillance, and signaling. This system connects to other wayside systems through specialized communication interfaces.

DOI: 10.1201/9781003028994-8

- Train control systems are responsible for the dynamic behavior of the train (i.e., all vehicles included in the train configuration), its operational procedures including traction and braking performance and also other train safety operational aspects. It is commonly controlled by a train driver (unless the train is under unmanned control) and, in a critical and emergency situation, it can be also controlled by a traffic control system through communication interfaces in order to avoid any train collisions, etc.
- Vehicle control systems are those installed on the rail vehicles and designated to control an individual vehicle's resources based on command signals obtained from train control systems if these are in use in the train configuration. The vehicle control system provides the control and synchronization of work of all computerized subsystems (e.g., traction, braking, suspension, etc.) installed on a rail vehicle.

A typical example of computerized control architecture of a heavy haul train is shown in Figure 8.1.

Further development of communication technologies has extended the application of computerized systems in the following directions:

- Remote condition monitoring
- Rolling stock information systems
- Unmanned train or vehicle control systems

To understand the complexity involved in the modern on-board computerized systems, it is necessary to examine the example of all the computerized control systems installed for a standard diesel-electric locomotive as shown in Figure 8.2.

The centralized microprocessor system presented generally controls the following sub-systems:

- Train interface control system
- Conventional driver and driver information systems
- Train protection system
- Traction and brake control system
- Battery and auxiliary power control systems
- Air conditioning, door, lighting etc. systems
- Data recorder systems

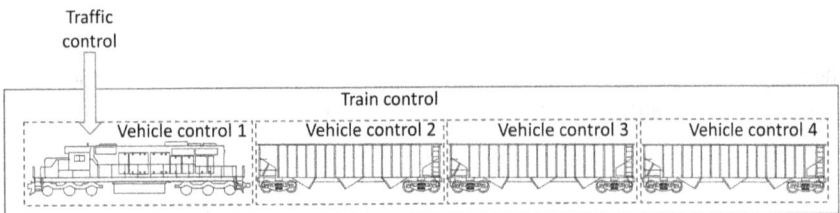

FIGURE 8.1 Levels of on-board controls of rail vehicles in a train consist.

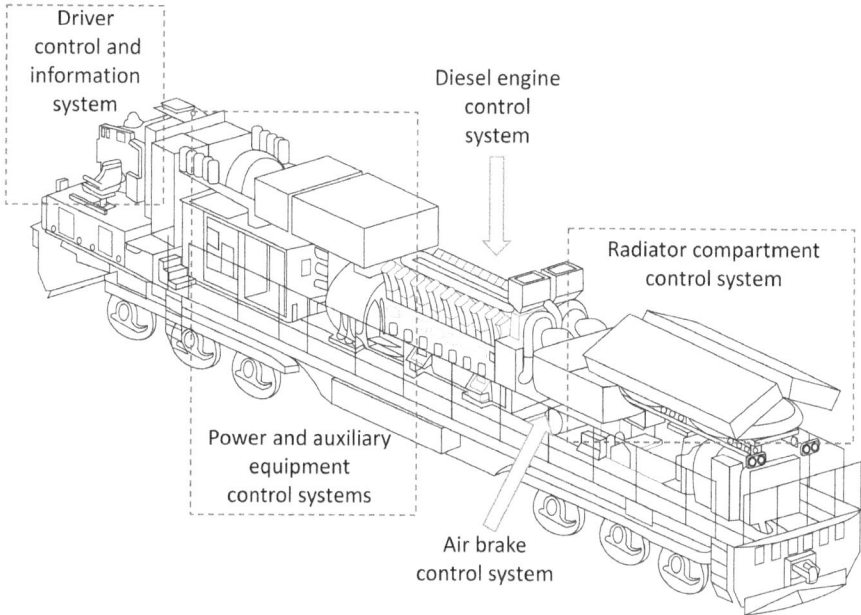

FIGURE 8.2 Example of common computerized control system unit architecture of a diesel electric locomotive.

This list above can be extended further but that is not a purpose of this chapter. This chapter is focused on the basic architecture of the rail vehicle computerized system and to consider major modules involved in the hardware design of mechatronic systems used in rail vehicles. The underlying motivation is to provide some ideas on how computer technologies that work with various subsystems are designed in order to provide control and monitoring (diagnostics) of dynamical systems of a rail vehicles.

These major modules of the computerized system can be roughly divided into two groups based on the hardware design architecture principles:

- Microprocessors (including control computers)
- Microcontrollers

The usage of these major components in terms of a rail vehicle mechatronic design can be presented as shown in Figure 8.3. The multifunction vehicle bus shown in this figure is described in Chapter 8.

The algorithms that need to be implemented in a controlled system are dependent on the tasks that should be resolved and controlled by each module. All these hardware architectures and their design purpose in rail vehicle mechatronic systems are discussed in the following sections. Also, some considerations regarding information exchange between two or more interconnected modules used in the rail vehicle computerized system are discussed.

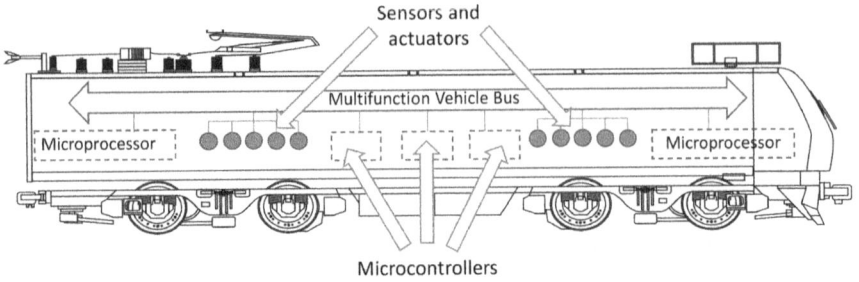

FIGURE 8.3   Example of a mechatronic rail vehicle concept equipped with various modules.

8.2   MICROPROCESSORS VERSUS MICROCONTROLLERS

The microprocessor systems shown in Figure 8.4 consist of three major components:

- A central processor unit (CPU) which is designated to recognize and run software/program instructions;
- Memory to store all hardware or program data required to run a program. There are two types of memory:
  - Permanent memory or read-only memory (ROM) that is programmed at the factory during the manufacturing process and contains information that can only be read by certain microprocessor applications. This memory cannot be erased and it is stored when power is switched off. However, in some microprocessor systems there are some possibilities of the usage of a programmable ROM type of memory such as EPROM and EEPROM that can be erased and re-written by the user under some predefined procedure. Programs which are designated for ROM are termed *firmware*.

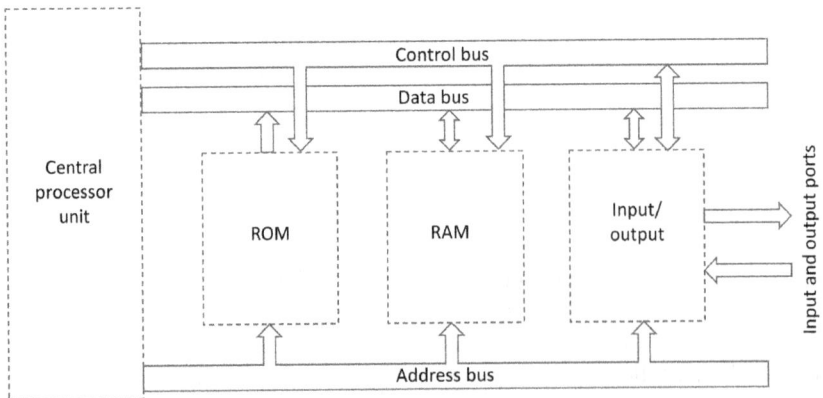

FIGURE 8.4   General form of a computerized system.

- Temporary memory or random-access memory (RAM) that allows the user to store, read, and write data which is created or generated by *software* programs, in a memory. Software is the term used for specific programs such as operating systems and their applications. The data stored is such a memory is completely erased when the system is switched off.
- Input and output interfaces which are designated for the exchange of data between the microprocessor and the peripheral devices that are commonly positioned as the external environment of essential elements linked to a computerized system.

The computerized systems used on rail vehicles are powerful computers designed for train automation purposes that allow them to carry out advanced control and diagnostic functions. The simplistic locomotive computerized system described in [7] used three different microprocessor modules that allowed reducing the number of electromechanical relays in locomotives by 70% as well as reducing electric and electronic parts by 21% in comparison with a non-computerized locomotive design.

The design of electronic components should be done as defined in the relevant standards. More information on the design requirements for such devices can be found in the BS EN 50155 and IEC60571 standards [8, 9].

8.2.1 MICROPROCESSORS

Considering all the components shown in Figure 8.4, the computerized system is usually called a microprocessor-based system where the CPU is commonly defined as the microprocessor. The first commercially available single chip 4-bit microprocessor (4004, 2300 transistors) was released by the Intel Corporation in 1971. The general architecture of a microprocessor is show in Figure 8.5. The microprocessor consists of the following major parts:

- Arithmetic and logic unit (ALU), which is designated for the data manipulation tasks.
- Registers, which are defined as memory locations and designated to hold data required to be stored for microprocessor operations. The number and functionality tasks of the register depend on the microprocessor design.
- Control unit, which defines timing and sequence of operations and allows interruption of operations if the command is a part of program operations or a result of some external event.

8.2.2 MICROCONTROLLERS

For control or monitoring purposes of mechatronic systems, it is necessary to have additional chips that allow communicating with the external processing elements through input and output ports as well as storing the involved data in the designated storage locations. The first patent for the microcontroller was issued in 1971 assigned to Texas Instruments.

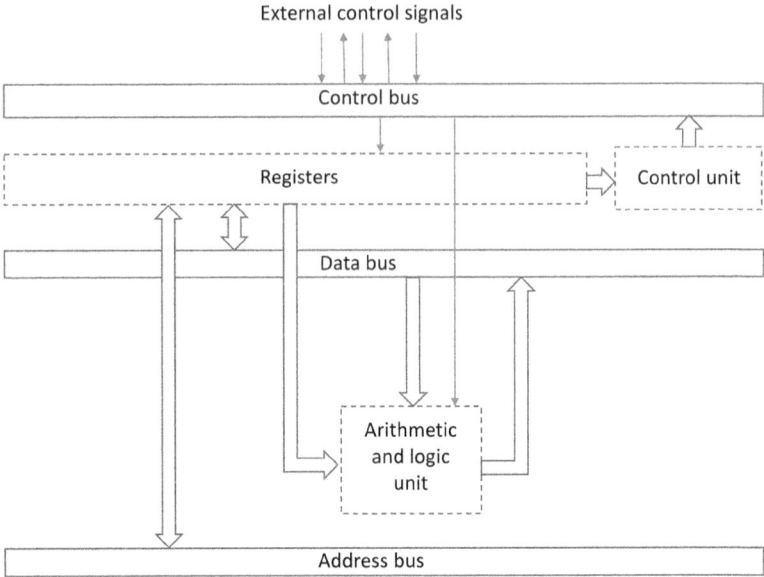

FIGURE 8.5   General architecture of a microprocessor.

A microcontroller (also called a microcomputer) is quite similar to a micropro-cessor. The main distinguishing feature is that the microcontroller contains its own software that is stored permanently in ROM which is a part of the chip unlike in microprocessors (they are programmed externally). The microcontroller rep-resents the device that integrates all components of a computerized system on a single chip as shown in Figure 8.6. The microcontrollers commonly have pins for

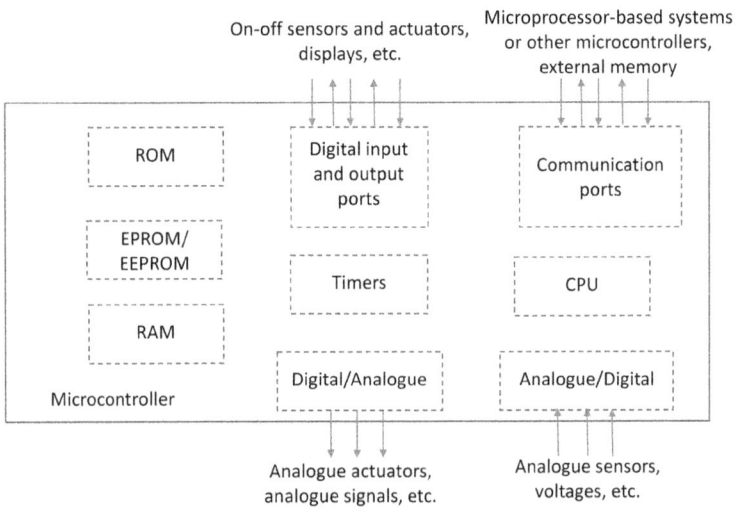

FIGURE 8.6   Example of a microcontroller design.

input and output connections with the external world including power supply, timers, and control signals.

The choice of microcontroller is a quite difficult task because a lot of design factors should be taken into consideration. One such example is the choice of the microcontroller for the on-board sensor node device which is needed to detect a wheel flat on rail vehicles as per the concept design described in publications [10, 11]. Such a system requires the sensor nodes built on the Adafruit ItsyBitsy M0 Express microcontrollers [12] shown in Figure 8.7. The choice of this microcontroller was made based on the following parameters: low cost, miniature form-factor, extensive documentation, the sufficient number of input and output ports, and ultra-low power for the required MHz performance. The configuration of such a microcontroller is ATSAMD21G18 32-bit Cortex M0+ with 256KB Flash and 32 KB RAM, 3.3V logic, 48 MHz, 32-bit processor, 2 MB SPI FLASH chip, and 23 Input/Output pins.

Some other well-known manufacturers of microcontrollers that can be considered in the design solutions are:

- **Texas Instruments:** https://www.ti.com
- **Analog Devices:** https://www.analog.com
- **Microchip:** https://www.microchip.com
- **NXP:** https://www.nxp.com
- **STMicroelectronics:** https://www.st.com

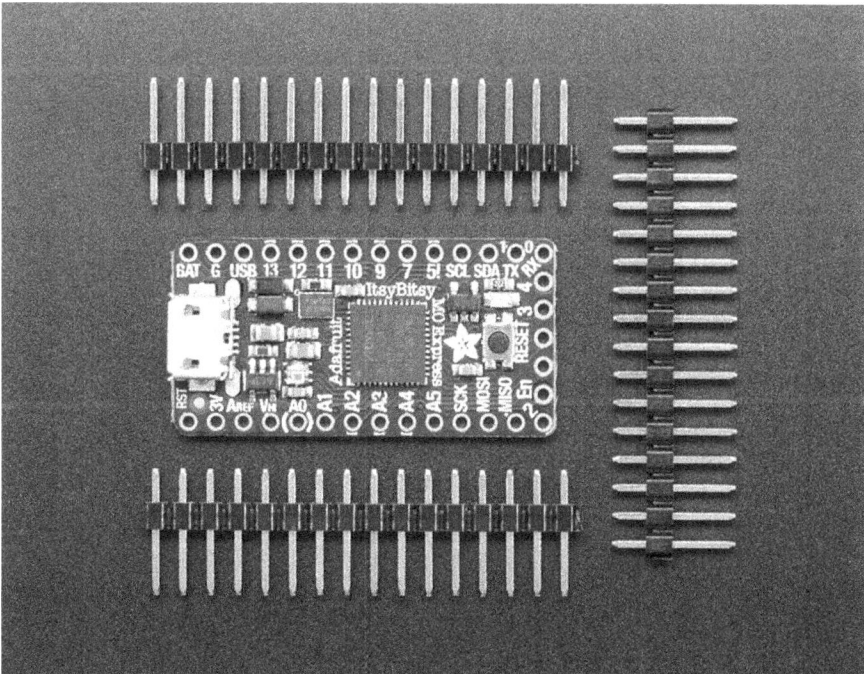

FIGURE 8.7 Microcontroller used for the detection of wheel flats.

8.3 CONTROL COMPUTERS

The control computer is a type of microprocessor-based system that commands, manages, and directs other control system devices and even other computerized systems in case of the high complexity of mechatronic system design. Their application on rollingstock assumes that such computers should be ruggedized and completely adapted for the control of mechatronic system processes, i.e., to control dynamic behavior and work of the mechatronic systems of rail vehicles.

The control computers should be designed considering the safety integrity level as defined in IEC 61508 [13]. In order to meet railway specific requirements, the EN 50126, EN 50128, and EN 50129 standards [14–16] were derived from the IEC 61508. The standards prescribe that the control systems should be deigned considering four safety integrity levels from SIL-1 to SIL-4 depending on the areas of application. The differentiation between the levels is based on the probability of failure/hour and risk reduction factor parameters. The SIL-4 level is defined as the highest risk reduction factor and is reserved for highly critical safety functions that may cause significant casualties that should be prevented at all costs. Common train control systems should be designed considering SIL-1 and SIL-2 requirements. For example, the control systems that are oriented toward preventing derailments and focused on the monitoring of lateral acceleration from lateral movement of wheels should be designed as SIL-2, providing a probabilistic risk reduction factor from 10^{-6} to 10^{-7}.

The two common types of control computers used in rail applications are described in the next two sub-sections.

8.3.1 PROGRAMMABLE LOGIC CONTROLLERS

A programmable logic controller (PLC) or programmable controller is a specialized industrial microprocessor-based device that is designated for interfacing to and controlling/monitoring of analogue and digital devices. In other words, the PLCs are easily programmable controllers, but they have been designed to replace hard-wired relay logic systems. The idea and component design aspects of PLC devices are similar to microcontrollers; however, their functions usually require their operations to occur in harsh operational conditions (including, e.g., high levels of pollution, dust, vibration and impact, power fluctuations, extreme environment conditions) and this dictates that their robustness and reliability properties must be at the highest level.

The form-factor of PLCs can start from small modular devices with a small number of inputs and outputs and finish as a large rack-mounted modular system that includes a great number of PLC devices with many arrangements of input and output ports. Those devices are commonly connected through special designed network interfaces as well as using such network interfaces to connect them to the main computerized system if such a design solution is required.

The software programs to control mechatronic equipment are typically stored in a special memory that is backed-up with a battery installed inside of the PLC.

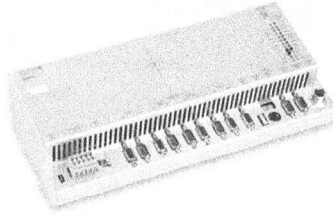

FIGURE 8.8 PLC unit for rail application manufactured by Lütze Transportation (see [17]).

The PLC design requires the application of real-time systems that should allow them to produce output results within a limited time despite the variations in input conditions.

In a general classification, a PLC can be classified as an industrial microprocessor-based controller with programmable memory that allows storing all required program instructions and permits its functionality to be extended with various functions depending on technical requirements of the designed system. A multi-functional PLC device with x86 CPU for rail application [17] is shown in Figure 8.8. The design and testing of this unit were performed based on the EN50155, EN50121, EN50124, and EN61373 standards [18–20].

The descriptions of some possible application areas of PLC devices in the control systems of rail vehicles can also be found in [21–23].

8.3.2 Field Programmable Gate Arrays

Unlike a traditional PLC which is designed as a specialized device for applications where that specific microprocessor architecture requires just to read sensor inputs and then to act with the predefined control strategy and produce control outputs, a Field Programmable Gate Array (FPGA) is a more general device that uses parallel processes and incorporates large resources of logic gates and RAM blocks to implement complex digital computations. The first FPGA was invented in 1985 and it comprises programmable silicon chips that combine processor-based systems with application-specific integrated circuits. The FPGA is commonly considered as a successor of the Complex Programmable Logic Device (CPLD). The difference between the CPLD and FPGA is architectural, where a CPLD has a quite restrictive structure that consists of one or more programmable logic arrays with a small number of clocked registers. The FPGA design is more flexible and complex, and it contains adders, multipliers, memory, and serializers/de-serializers. One more distinguishing feature is that the memory used in CPLDs is embedded while FPGAs work with external memory. As a result, the FPGA is more applicable for solving a task that requires computational capabilities.

The descriptions of some possible application areas of FPGAs in the control and monitoring systems of rail vehicles can also be found in [24–28].

8.4 MULTI-MODULE STRUCTURES FOR MICROPROCESSOR-BASED CONTROL SYSTEMS

The multi-processor and multi-controller structure that is currently in use on modern on-board computerized systems can be used on one vehicle as well as on multiple vehicles consists. In this latter case, one control unit in such a structure should be operated as a master unit while the other units should be operated as slave units. The master unit is focused on the coordination of work of all rail vehicle subsystems through the relevant slave units via a computer network. The slave units are designed to perform specific designated tasks and might require feedback instructions from the master unit. All communication interfaces between such subsystems are further discussed in Chapter 8.

A good example of the development of modular control systems is the family of SIBAS (Siemens Bahn Automatisierungs System) automation systems. The first generation was introduced in 1983 [29], called SIBAS 16 and it was based on the 16-bit Intel 8086 processor. The system was designated for all logic and control duties in traction applications. The system included two microprocessor-based units – a traction control unit (TCU) and the central control unit (CCU). The main design objective of SIBAS 16 was to establish proper control of a rail vehicle by using standard hardware equipment with software modules that can easily be extended and modified.

Considering further trends in the design of rail vehicle control systems and requirements to control everything from the electrical equipment to the entire train, Siemens introduced SIBAS 32 [30] in 2007 and then SIBAS 32C in 2010 [31] which incorporates vehicle control units (VCU), TCU, brake control units (BCU), and auxiliary power supplies (APS). Powerful microprocessors are used in the design of this system for a high volume of real time calculation and for processing a large amount of data obtained from all the functional units including the co-ordination tasks for work of all subsystems and for providing open-loop and closed-loop control (see Chapter 4) for on-board equipment installed on the entire train.

8.5 CASE STUDY: MICROCONTROLLER IN MONITORING SYSTEM

The rail industry is a rugged environment for the integration of modern technology, particularly in freight and heavy haul where suspension designs are simple and primitive resulting in a number of steel-on-steel contacts that can produce moments of high shock and vibration. Like automotive applications, there are also significant temperature and humidity considerations as well. From the 1990s onwards, there has been a steady migration of microcontroller-based systems in locomotives. Modern locomotives are now almost completely microcontroller driven from braking to traction, driver information panels and auto-haul variants.

Due to their high volumes and simplicity, freight and heavy haul wagons/trains have not experienced the same level of integration with current systems limited to Electronically Controlled Pneumatic Brake hardware and condition monitoring systems (including early warning systems). The condition monitoring hardware must be low cost and easy to install due to the aforementioned volume issue. The types of

wagon condition monitoring systems include areas such as wheel bearing temperatures, acoustic wheel squeal, and wagon dynamics. Of these, the wagon dynamics
monitor is the most common type of device. For completeness it should also be noted
that these devices not only serve as wagon condition monitoring systems but assess
track condition as well.

8.5.1 DESIGN

So, what does a wagon dynamics condition monitoring device look like? Due to the
lack of available power on freight and heavy haul wagons, the units must be autonomous in terms of both power and communication, resulting in ultra-low-power
devices. Figure 8.9 shows the typical layout comprising 5 major components, microcontroller, inertial measurement unit (IMU), GPS, communications interface, and
energy storage (and harvesting). An example of this type of hardware arrangement
can be seen in US Patent 7853412, Estimation of Wheel-Rail Forces [32]. In this
device, the three-axis accelerometer measures wagon body accelerations and uses
an inverse physics model to calculate the approximate wheel-rail contact forces of
one or more wheelsets. If the algorithm determines that a derailment condition may
be imminent in the wagon, an alert notification is transmitted to the driver through
a Zigbee wireless network.

A brief general discussion of the different components and specific examples of
how they were most recently applied to the above device are listed below.

a. **Microcontroller:** contains the algorithm for measurement and signal processing of the wagon dynamics metric/s under observation. For our example
as described above, this is the inverse physics model. The microcontroller
also manages the peripheral interface to the IMU, GPS, and communication
interface. Typically, these peripheral interfaces are now all digital through
SCI, SPI, or I2C, however, IMU devices, e.g., accelerometers, may still be
acquired through an analog interface in some circumstances.

b. **IMU:** this is usually a 6 degree of freedoms device or subset thereof,
comprises three-axis accelerometers and gyroscopes. Nowadays these are
usually Micro-Electromechanical Systems (MEMS) devices, however,
piezoelectric accelerometers are still used. There is a wide array of manufacturers of these devices with common types being the Bosch BMI160
and ST LSMxxx series of chips. Mobile phone development has driven the
development of ultra-low power varieties of these devices where continuous
power consumption of <5 mW is possible.

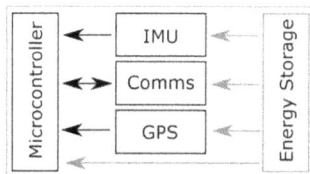

FIGURE 8.9   Rollingstock autonomous device topology.

c. **GPS:** enables spatial tagging of conditioning monitoring data if required. This may be an optional device depending on the application. The active mode power usage of GPS is also quite high in comparison to other peripherals and the Microcontroller. For example, the MTK3329 chipset used by the example consumes between 120 and 160 mW to acquire and track satellites. Methods for low power utilization of GPS are an important consideration in wagon condition monitoring devices.

d. **Communications Interface:** early versions of the presented example use Zigbee (a common mesh network topology) communication to hop a message up a train. There are now a number of different low power communication technologies available such as LoRA, LTE, and SigFox. In all cases the power required to transmit is still quite high, for example LoRa can be up to 100 mW, so the amount of data to be transferred is an important design consideration. For the example discussed, data transmission is restricted to derailment only events, and this event mechanism is a common approach in this field to conserve power.

e. **Energy storage and harvesting:** the most common battery storage medium is lithium and its derivatives. Of the lithium types, the iron phosphate is the most temperature stable available, however, compared to practical field temperatures, these battery types still experience charge threshold cut-offs, i.e., threshold is 60°C for LiFEPo4. The most common harvesting method is solar. Vibration energy harvesting via piezo devices is used commercially, however, it has serious limitations in the quantity of energy harvestable, usually only in the order of microwatts. It does have the benefit of being maintenance free though.

Depending on the installation location of these devices, it is common for axle and side-frame mounted devices to be cast completely in flexible epoxy to prevent shock and vibration from destroying the device. Wagon body mounted devices experience less vibration, however care must still be taken on any connection points in the device's enclosure, specifically screw mounts, cabling, and ingress protection.

8.5.2 PROBLEM FORMULATION

A modern ultra-low power microcontroller must be selected for an instantiation of the derailment detection device described previously. The device design considerations are shown in Table 8.1. It is necessary to calculate the remaining power for the microcontroller and identify a microcontroller with the required performance and power.

8.5.3 SOLUTION

The microcontroller power available is the solution of the energy balance problem. The total energy available for a 24-hour period is:

$$2\text{W Solar Panel} \times 3 \text{ h/day} = 6000 \text{ mWh}$$

TABLE 8.1
Design Parameters of the Microcontroller Device

Parameter	Value
Required Microcontroller Performance	>40 MHz
Solar Panel	2 W
Available sunlight	3 h/day
Communication:	
• Power	100 mW
• Packet size	256 bytes
• Data rate	1 kbps
Expected Impact Events per day	10
Communications Packets per Impact Events	1
GPS	
• Active mode (Tracking satellites) (assume continuous usage)	150 mW
6 DoF IMU	5 mW

Energy out for the other components is:

150 mW GPS × 24 h = 3600 mWh
10 events/day × 256 byte packet × 8 bits/byte = 20480 bits transmission
20480 bits/1000 bits/second = 20.48 s air time
(20.48 s air time/3600 s/h) × 100 mW = 0.568 mWh
5 mW IMU × 24 h = 120 mWh
Total Energy Out = 3600 mWh + 0.568 mWh + 120 mWh = 3720.568 mWh

Remaining energy for microcontroller:

$$(6000 \text{ mWh} - 3720.568 \text{ mWh}) \div 24 \text{ h} = 94.97 \text{ mWh}$$

There are a number of microcontroller manufacturers offering ultra-low power performance microcontrollers utilizing the ARM Cortex-Mx architecture. Some examples are the Atmel SAMD51 series operating at 120 MHz with FPU support and 65 μA/MHz current consumption and the Texas Instruments MSP432 which operates at 48 MHz with FPU support and 80 μA/MHz current consumption. The supply voltage input range on these chips is typically 1.8–3.3 VDC.

REFERENCES

1. S. Sone, I. Okumura, The Microprocessor in Railway Control Systems, In: S. G. Tzafestas (Ed), Microprocessors in Signal Processing, Measurement and Control, Springer, Dordrecht, the Netherlands, 411–436, 1983. ISBN 978 94 009 7007 6.
2. D. J. Mitchell, Overview of Microprocessor-Based Controls in Transit and Concerns About Their Introduction, Conference on Light Rail Transit, Pittsburgh, PA, 8–10 May 1985. Available at: http://onlinepubs.trb.org/Onlinepubs/state-of-the-art/2/2-026.pdf

3. C. R. Avery, Living with the Microprocessor on Traction – the British Rail Experience, IEE Colloquium on Living with Microprocessor Controlled Traction Systems, London, UK, 3 April 1990.

4. R. J. Bevan, Evolution of Microprocessor Control (for Railway Applications), IEE Colloquium on Living with Microprocessor Controlled Traction Systems, London, UK, 3 April 1990.

5. P. Aström, Control Electronics of Rail Vehicles, Proceedings of the ASME/IEEE Spring Joint Railroad Conference, Atlanta, GA, 107–116, 31 March to 2 April 1992.

6. H. Schneider, J. Vitins, MICAS-32 distributed traction control for motive power units, ABB Review, 5, 11–20, 1995.

7. G. Kaplan, T. A. Adam, Transportation: AC propulsion and the use of microprocessors for locomotive control and diagnostics may help an ailing railroad industry retrack itself, IEEE Spectrum, 23(1), 73–75, 1986.

8. BS EN 50155:2017, Railway Applications –Rolling Stock –Electronic Equipment, British Standards Institution, London, 2017.

9. IEC 60571:2012, Railway Applications –Electronic Equipment Used on Rolling Stock, International Electrotechnical Commission, Geneva, Switzerland, September 2012.

10. E. Bernal, M. Spiryagin, C. Cole, Wheel flat detectability for Y25 railway freight wagon using vehicle component acceleration signals, Vehicle System Dynamics, 58(12), 18931913, 2019.

11. E. Bernal, M. Spiryagin, C. Cole, Ultra-low power sensor node for on-board railway wagon monitoring, IEEE Sensors Journal, 20(24), 15185–15192, 2020.

12. Adafruit Industries, Development Boards – Adafruit ItsyBitsy M0 Express – for CircuitPython & Arduino IDE. Accessible on 29 March 2021. Available at: https://www. adafruit.com/product/3727.

13. IEC 61508:2010, Functional Safety of Electrical/Electronic/Programmable Electronic Safety-Related Systems, International Electrotechnical Commission, Geneva, Switzerland, 2010.

14. BS EN 50126:2017, Railway Applications – the Specification and Demonstration of Reliability, Availability, Maintainability and Safety (RAMS), British Standards Institution, London, 2017.

15. BS EN 50128:2011, Railway Applications – Communication, Signalling and Processing Systems – Software for Railway Control and Protection Systems, British Standards Institution, London, 2011.

16. BS EN 50129:2018, Railway Applications, Communication, Signalling and Processing Systems – Safety Related Electronic Systems for Signalling, British Standards Institution, London, 2018.

17. Lütze Transportation GmbH, Vehicle control unit data sheet "DIORAIL PC2 powerful PLC for rail vehicles", 18.02.2019 – Part-No. 746024. Accessible on 29th March 2021 at: https://www.luetze-transportation.com/fileadmin/tx_luetzeasimfetch/746024-vehicle-control-unit-en-data-sheet-011720580.pdf.

18. BS EN 50121:2017, Railway Applications – Electromagnetic Compatibility, British Standards Institution, London, 2017.

19. BS EN 50124:2017, Railway Applications – Insulation Coordination, British Standards Institution, London, 2017.

20. BS EN 61373:2010, Railway Applications, – Rolling Stock Equipment – Shock and Vibration Tests, British Standards Institution, London, 2010.

21. P. Q. Acuin, J. Q. Puerto, R. O. Tamayo, G. L. Abulencia, R. F. Ibuig, Development of a Functionally-Tested Hybrid Electric Train, In Proceedings of 2019 IEEE 11th International Conference on Humanoid, Nanotechnology, Information Technology, Communication and Control, Environment, and Management (HNICEM), IEEE, Laoag, Philippines, 29 November to 1 December, 2019.

22. US Patent 9102309, System and method for detecting wheel slip and skid in a locomotive, 20 December 2016.
23. US Patent 9522687, System and method for remotely operating locomotives, 15 October 2020.
24. M. Lehtla, Microprocessor Control Systems of Light Rail Vehicle Traction Drives, PhD thesis, Faculty of Power Engineering, Tallinn University of Technology, 2006. Available at: https://digikogu.taltech.ee/et/Download/5a6a478d-59d5-447d-9299-37621a79e224.
25. Y. J. Zhou, T. X. Mei, S. Freear, Field programmable gate array implementation of wheel-rail contact laws, IET Control Theory and Applications, 4(2), 303–313, 2010.
26. T. X. Mei, Y. J. Zhou, Systems-on-chip approach for real-time simulation of wheel–rail contact laws, Vehicle System Dynamics, 51(4), 542–553, 2013.
27. D. Macii, M. Avancini, L. Benciolini, S. Dalpez, M. Corrà, R. Passerone, Design of a Redundant FPGA-Based Safety System for Railroad Vehicles, 17th Euromicro Conference on Digital System Design, IEEE Computer Society, Verona, Italy, 27–29 August 2014. ISBN 978 1 4799 5793 4.
28. US Patent 9026282, Two-tiered hierarchically distributed locomotive control system, 5 May 2015.
29. N. C. Jones, J. H. McNeil, S. C. S. Royle, An Electronic Freight Locomotive, International Conference on Main Line Railway Electrification, IET, New York, UK, 25–28 September 1989.
30. SIEMENS, SIBAS 32 Das Steuerungssystem für alle Schienenfahrzeuge, Siemens AG. Transportation Systems, 2007.
31. SIEMENS, SIBAS – the optimal control for all types of trains. Available at: https://www.mobility.siemens.com/global/en/portfolio/rail/rolling-stock/components-and-systems/traction-converters.html. Accessed on 29 March 2021.
32. US Patent 7853412 B2, Estimation of Wheel-Rail Forces, 14 December 2010.

9 Communications, Networks, and Data Exchange Protocols

This chapter is focused on the description of communication principles, architectures, and technologies used in individual rail vehicles and train consists. This chapter describes major communication architectures that are currently in use in rail vehicles and contains a review of related railway standards used for enabling communication technologies in rail vehicle design. The case study presented in this chapter looks at the advancement in freight train braking possible through the introduction of the end of train device and the development of a robust train communication network.

9.1 INTRODUCTION

At the present time, the introduction of digital and communication technologies in rail vehicle design has provided a great opportunity for the establishment of train communication networks (TCNs) for different types of train configurations despite the different communication equipment and components that are in use in individual rail vehicles. However, it was not a case when these technologies started to be implemented. The implementation of communication digital technologies can be divided into four stages [1]:

- **Stage 1 (1980–1990):** Each rail vehicle has its own communication network which connects various microprocessor computers that control propulsion, brakes etc. in a single vehicle control system.
- **Stage 2 (1990–2000):** Implementation of new functionalities for control and diagnosis of the performance of the whole train and for providing passenger information inside of train consists.
- **Stage 3 (2000–2005):** Implementation of communication connections between on-board train and ground based wayside communication systems for the improvement of train operation and maintenance purposes.
- **Stage 4 (2005–present):** The implementation of distributed system approaches that allow connecting computers through wireless network connections inside of the train consists and between train and ground communication systems.

Considering the mechatronic background of this book and particularly the microprocessor system design and application principles previously described in Chapter 8,

DOI: 10.1201/9781003028994-9

this Chapter focuses on the three main communication architectures that are in use in rolling stock mechatronic systems:

- Intra-car communication architecture
- Inter-car communication architecture
- Train-to-ground communication architecture

9.1.1 INTRA-CAR COMMUNICATION ARCHITECTURE

Intra-car communication technologies represent on-board communication networks that are designated to transfer information commonly through wired connections between different control and computerized equipment installed in trains. The term *intra-car communication* refers to all communications that occur within a rail vehicle. The standardization of these networks was facilitated by the introduction of the IEEE Standard 1473–1999 (superseded by IEEE Standard 1473–2010 [2]) for protocols related to the Train Control and Management System (TCMS) covering the two segments of the on-board TCN:

- Multi-functional Vehicle Bus (MVB)
- Wire Train Bus (WTB)

Similarly, the definition of a complete network specification for a TCN also has its origins in 1999 with the adoption of the international standard which can now be found in IEC 61375-1 [3–5]. It is necessary to understand the difference between TCMS and TCN:

- TCMS is a system that is designated to provide a single point of control for all computerized systems installed on the train.
- TCN is the infrastructure that enables the information and data exchange throughout the train.

While the MVB is designated to connect standard onboard equipment as shown in Figure 9.1, the WTB is designated to inter-connect rail vehicles in train configurations. In the freight train configurations, the WTB can also be referred to as the train line.

In Figure 9.1, the following units are connected to the WTB and the MVB:

- Head-End Unit (HEU, this unit controls the major train operational activities)
- Automatic Train Control Unit (ATC)
- Auxiliary Power Controller (APC)
- Traction Control Unit (TCU)
- Brake Control Unit (BCU)
- Brake Compressor Control Unit (BCC or BCCU)
- Equipment Control Unit (ECU, this unit controls the on-board equipment such as doors, air conditioner, etc.)

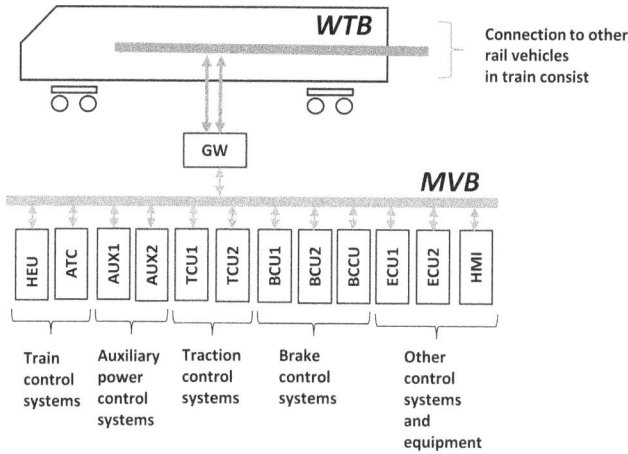

FIGURE 9.1 Example of the intra-car architecture.

- Human Machine Interface (HMI)
- Gateway (GW, this device, which is also known as a node, is an advanced train communication computer that allows establishing interfaces between train and vehicle bus technologies)

The TCN has a hierarchical structure that allows fast data transmission due to the separation of data streams between MVBs and WTB and the load reduction on the WTB [6]. Traditionally, the combination of the WTB and MVB technologies is considered as a vehicle bus. However, other technologies, for example, Ethernet, CAN, and Serial Links, are also widely used as vehicle buses.

9.1.2 INTER-CAR COMMUNICATION ARCHITECTURE

The term *inter-car communication* is designated for all communications between vehicles or between vehicles and sensors installed at various locations in the train set. The example of a TCN which represents the inter-car communication architecture and consists of a combination of one WTB and five MVBs is shown in Figure 9.2. This figure also includes one more unit in comparison with Figure 9.1 that is defined as the Car Control Unit (CCU). It is necessary to mention that the HEU and CCU definitions can be replaced by the Vehicle Control Unit (VCU) (for an example, see [7]). The inter car communication architecture is commonly built based on wired, optical fiber cabling, or wireless technologies. This architecture is also referred to as Vehicle-to-Vehicle (V2V) communications [8]. Practical examples on the development of the train and consist wireless networks can be found in [9].

9.1.3 TRAIN-TO-GROUND COMMUNICATION ARCHITECTURE

The train to ground communication requires using wayside transponders. Train to/ from ground communication is commonly built based on wireless technologies that

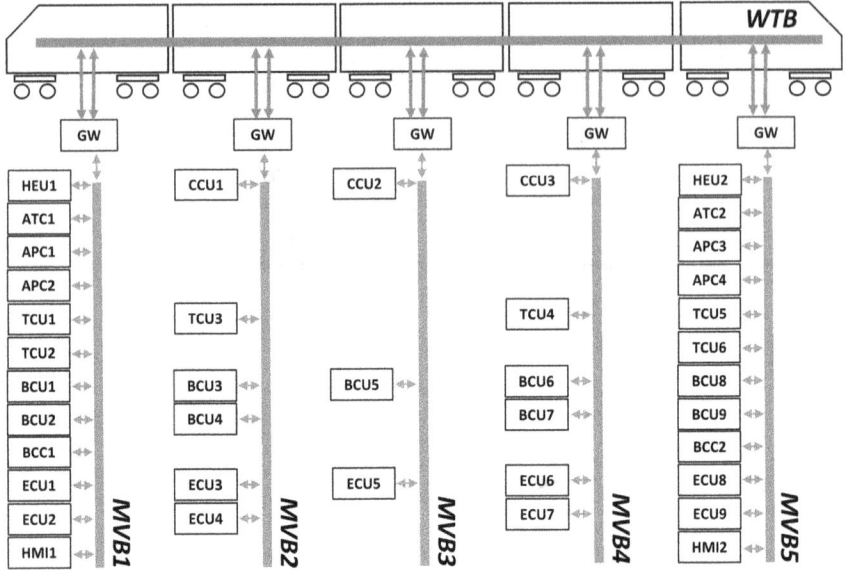

FIGURE 9.2   Example of the inter-car architecture.

allow data to be broadcast to trains or to individual rail vehicles [10]. This architecture is also referred to as Vehicle-to-Infrastructure (V2I) communications [8]. Practical examples on the development of the architecture for the train to ground communication can be found in [11].

9.2   COMMON TYPES OF NETWORKS

Several types of network configurations are applicable for implementation in rail vehicles. The choice of a network type should be decided considering the network requirements defined in task specifications for the specific mechatronic system. In the following subsections, the commonly used network technologies for the establishment of connections between different control units or sensor equipment are described.

9.2.1   WIRED NETWORKS

A wired network is commonly defined as a network configuration that is built based on electric or optic cables which establish connections between units and/or other devices on the networks they are installed on and are connected through MVBs and WTB. Wired networks have a lot of advantages in comparison with other types of networks:

* Fast data transfer (e.g., a Faster TCN built based on a standard Ethernet with a data rate of 100 Mbps or even on a Gigabit Ethernet with a data rate of 1 Gbps);

- Standard equipment that is widely used by other industries;
- High level of security and the protection against unauthorized access.

From the other hand, the disadvantages of this type of networks are:

- Extensive cabling and wiring in the design of complex systems;
- High repair costs in the case of equipment failure;
- Wired networks may lack of flexibility in terms of installation and operation mobility.

It is a common practical solution to establish a Local Area Network (LAN) or Control Area Network (CAN) based on the wired network topologies. This is a reason why there are a great number of wired network technologies that have found worldwide application on rolling stock such as a Real-Time Ethernet (e.g., Ethernet Train Backbone (ETB) [12] or Ethernet Consist Network defined in [13]), CANOpen (see CanOpen Consist network in [14]), LonWorks, Profibus, Power Line Communication, Leaky Coaxial Cable (LCX), Train Communication Network, and WorldFIP.

9.2.2 WIRELESS NETWORKS

A wireless network is a configuration that allows establishing a connection between units/other devices on the network by means of radio frequency or microwave signals. Such technologies are working with wireless-enabled devices that communicate through wireless local area networks (WLANs), wireless sensors networks, cellular networks, or even via the Internet. The main advantage of the wireless network is that it can reduce cabling costs and ease the addition of new equipment (including sensors) in the existing rolling stock where the installation of new wiring cannot be performed for some technical reasons and other issues [15]. The main disadvantages of such networks are associated with the high network security requirements (in terms of their practical realizations and related costs), the strength and reliability of the radio signals, and the protection from unauthorized access.

The wireless technologies applicable for rolling stock applications can be divided into the following groups:

- WLAN is defined in the IEEE 802.11 standard and it is aimed for short-range communication. This standard is related to well-known Wi-Fi technologies working in the 2.4GHz and 5 GHz bands. The experimentation performed for most Wi-Fi standards showed that this technology has a limitation for train application because it does not support communication at speeds above 80 km/h [1]. The approved amendment to the IEEE 802.11 standard, referenced as IEEE 802.11p, is designated to provide wireless access in vehicular environments in the 5.9 GHz band at speeds of up to 250 km/h.
- Worldwide Interoperability for Microwave Access (WIMAX) is defined in IEEE 802.16 and it aims to provide coverage out to 30 km in the band range from 2 to 11 GHZ.

- Cellular networks are widely in use in transportation because such technologies are well distributed in many locations and provide a good coverage in metropolitan areas and around main transportation networks. The cellular network technologies and their applications in the railway industry are currently covered by the standard *Global System for Mobile Communications – Railway*, also called GSM-Railway or GSM-R. This standard is an international wireless communications standard for railway communication and applications. The wireless network systems built on this standard's requirements guarantee performance at speeds up to 500 km/h with a high level of communication reliability and a peak data rate up to 172 Kbps. The forecasted obsolescence of GSM-R technology might happen in 2030. In order to replace this standard in the future, the new technology called LTE for Railways (LTE-R) was proposed in 2011. This technology presents a next-generation communication network designated to work at speeds up to 500 km/h with a peak data rate up to 50 Mbps.

More information on wireless technologies including their Industrial Internet of Things (IIoT) applications can be found in [16].

9.2.3 MIXED NETWORKS

There always exists a lack of connectivity and interoperability between different subsystems in train configurations. Mixed networks are a part of a TCN that is commonly established by joining wired and wireless network technologies used in an individual rail vehicle or among vehicles when they are required to be operated as a single train consist and are connected through so-called virtual couplings. The latter solution is still under development and more information can be found in [17].

9.3   COMMON COMMUNICATION PROTOCOLS

A data exchange communication protocol represents a system of rules that allows two or more devices to transmit information through the buses or wireless networks. A great number of protocols exist and they are fully dependent on the architecture (including software and hardware) in use in train configuration. Each protocol defines its own rules, structure (e.g., syntax and semantics), synchronization of communication, collision avoidance, and possible error recovery methods.

For mechatronic systems, it is important that the communication is done through real time protocols. The protocols can be divided into two categories:

- Specially designed for rolling stock (Mux G, Tornad, TCN, etc.)
- Adapted from standard technologies (CAN, LonWorks, WorldFIP, etc.)

Three of them were standardized – TCN [3], CANOpen [14], and LonWorks [2] for train control applications. In the TCN architecture, all busses work with the same real-time protocols which share two communication services, namely the process data (variables) refreshed through broadcasting as a distributed real-time database and the message data (messages) which are transmitted on demand [18] through the master-slave architecture to ensure that time-critical data are transferred in real time.

The CAN and LonWorks protocols are both based on a Carrier Send Medium Access (CSMA) technique, but they are two different CSMA protocols. The CAN protocol is based on the non-destructive collision protocol (CSMA/BA) that provides the highest priority message by the way of Bitwise Arbitration. The LonWorks (also known as Type L) protocol was developed as a general-purpose control networking protocol that connects computerized systems with equipment with associated sensors and actuators and it was submitted to American National Standards Institute (ANSI) and accepted as a standard for control networking (ANSI/CEA-709.1-B). The LonWorks protocol is also a non-destructive collision protocol that provides collision avoidance by means of random delays in the transmission, where each delay is computed at node level.

More detail on related network technologies and communications protocols used for inter-car and intra-car data communications between mechatronic subsystems can be found in [2, 12–14, 19].

9.4 CASE STUDY: ELECTRONICALLY CONTROLLED PNEUMATIC BRAKES COMMUNICATION NETWORK

This case study looks at the advancement in freight train braking possible through a robust train communication network. For nearly 150 years freight train braking has utilized a pneumatic brake pipe system extending across the length of the train. This brake pipe performed two functions, one as a supply of air to each wagon where locally installed reservoirs store air for braking applications, and the second to propagate the required brake application or release through the driver activating a drop or rise in brake pipe pressure respectively. There has been iterative development of the system over the years including an additional pipe dedicated to wagon reservoir charging and individual wagon modifications such as accelerated release of brake cylinder pressure. The introduction of one- and two-way End of Train (EOT) devices helped provide additional monitoring information for the performance of the brake pipe at the rear of the train, however, drivers largely relied on how the train qualitatively felt under braking to understand its performance. Some of the drawbacks of the system are:

- No feedback on the performance of individual wagons, both in terms of pressure measurements and mechanical actuation. Faults in wagons must be identified through regular routine testing of the brake system prior to the train departing, and only catastrophic faults are typically identified in service.
- As trains continue to become longer and heavier, there is an increase of in-train forces under braking, as the propagation of the brake application always begins at the lead locomotive. Distributed power trains have reduced the effect of this by actuating braking at their remote locomotive locations in conjunction with the front locomotive/s, however, distributed power train consists also come with other operational considerations.
- Brake pipe viscous friction effects limit the distance pneumatic brake signals can propagate before becoming attenuated to the point of not actuating the wagon brake equipment as expected. This is due to conventional triple valves relying not only a pressure level, but the rate at which it drops and replenishes (for release case).

9.4.1 Inception of Electronically Controlled Pneumatic Brakes

To overcome these issues with train braking and to enable longer and heavier trains to be developed, the Association of American Railroads (AAR) started the development of a new train braking methodology in the early 1990s. The emphasis of this new system was to continue utilizing the pneumatic system as a supply only, integrating local electro-pneumatic control of the rolling stock. As a result, the modern freight train railway braking system called Electronically Controlled Pneumatic Brakes (ECP or ECPB) was developed. A power and communication train-line runs down the length of the train connecting to each wagon mounted electro-pneumatic unit. Unlike typical network communication where the cable only carries communication signals, the ECPB train-line injects the communications over the power-line negating the requirement for an additional network communication cable. A similar analog is the use of Power-Over-Ethernet (POE) for devices such as CCTV cameras.

9.4.2 Network Communication

The network standard adopted for ECPB communications consists of ANSI/EIA 709.1-A Control Network Protocol Specification and ANSI/EIA 709.2-A Control Network Power Line, also known as the LonWorks CAN. It is half duplex and operates on a single carrier frequency with a 132 kHz center band with a resultant theoretical throughput of approximately 5.4 kbps. Each device on the network must have a transceiver for communication and they are configured with a 40-byte message payload (total packet size including header, of 64 bytes), the default for LonWorks networks.

9.4.3 Device Types

The ECPB network comprises four device types, HEU, Car Control Device (CCD), EOT device, and Power Supply Controller (PSC). The integration of these devices in a train configuration is shown in Figure 9.3.

The HEU in the lead locomotive is the primary controller of the network and manages the operating mode of the network, the Train Brake Command (TBC), and feedback from devices. In normal operation, it commands the activation and deactivation of train-line power and continuously monitors for train-line continuity.

Each wagon unit has a CCD that interfaces with the train-line and actuates commands received from the Lead HEU. The CCD continually monitors wagon braking specific parameters such as brake cylinder pressure, reservoir pressures, brake pipe pressure, and network communication continuity. If it detects a fault, depending

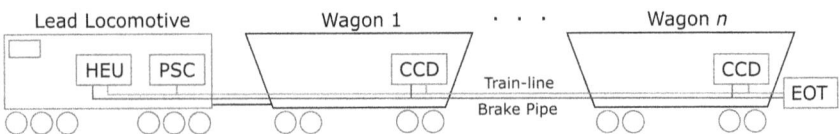

FIGURE 9.3 ECP block diagram.

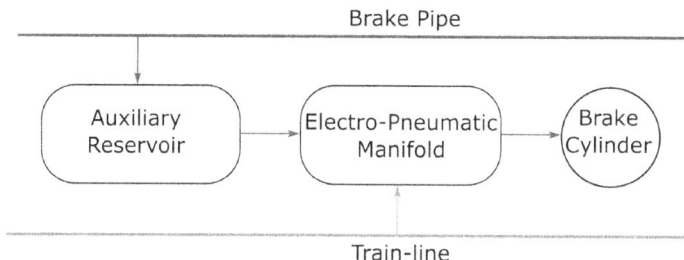

FIGURE 9.4 CCD block diagram showing electronic and pneumatic connectivity.

on the type and severity of the fault it may transmit the fault directly to the HEU and, under significantly degraded performance, autonomously actuate an emergency brake application. A block diagram of the CCD is shown in Figure 9.4.

As its name suggests, the EOT device is the last connected device on the ECP network. It continually monitors the brake pipe pressure and train-line continuity. Like a conventional pneumatic EOT it provides a flashing light and movement detection, however, has no braking capability.

The PSC powers the train-line when commanded to do so by the Lead HEU. It monitors train-line continuity and voltage and is able to autonomously power-down the train-line if a fault is detected.

9.4.4 PROBLEM FORMULATION

For drawbacks 1 and 2 of the conventional pneumatic system, identify potential methods using the ECP network to solve the drawbacks.

Things to consider in identifying potential methods:

- Packet size versus throughput. What are the limitations on the network performance?
- Half duplex network topology
- Communication topologies
  - Command/response mechanisms, i.e., polling
  - Event driven from the nodes
  - Broadcasting

9.4.5 SOLUTION – DRAWBACK 1

With a packet size of 64 bytes (~500 bits) and throughput of 5.4 kbps, the LonWorks network is quite limited in the amount of data it can transfer, therefore the ECP standard uses a per-second periodic polling mechanism initiated by the Lead HEU for querying the status of devices attached to the network, specifically the CCDs and PSCs. The transmission of the Lead HEU message and response from the device results in a <20% utilization of the network in normal operation, leaving capacity for other network communications.

In the scenario where a device experiences a fault, depending on the fault type, it may be permitted by the protocol to transmit the fault directly to the Lead HEU. It is considered that, in normal operation, fault occurrences are low and therefore uncontrolled utilization of the network, i.e., not initiated from the Lead HEU, is permitted.

9.4.6 SOLUTION – DRAWBACK 2

The reduction of in-train forces is solved by mandating that the brake application is actuated by all CCDs in parallel. This minimizes the opportunity for run-ins that occur through a head of train-initiated brake application. The standard stipulates that a CCD must begin actuation of a brake application within one second of receiving the command from the Lead HEU. To make this possible, the Lead HEU broadcasts a per second beacon message containing the current Train Braking Command. To reduce network utilization, the broadcast beacon message also contains the address of the polled device described in the first solution.

REFERENCES

1. B. Bouchez, L. de Coen, Communication Systems for Railway Applications, Chapter 4, In: M. Emmelmann, B. Bochow, C. C. Kellum (Eds), Vehicular Networking – Automotive Applications and Beyond, John Wiley & Sons Ltd, Chichester, UK, 83–104, 2010.
2. IEEE Standard 1473–2010, IEEE Standard for communications protocol aboard passenger trains, IEEE Vehicular Technology Society, New York, NY, 2011. DOI: 10.1109/IEEESTD.2011.5724313.
3. IEC 61375-1:2012, Electronic railway equipment – Train communication network (TCN) – Part 1: General architecture, International Electrotechnical Commission, Geneva, Switzerland, 2012.
4. IEC 61375-3-1:2012, Electronic railway equipment – Train communication network (TCN) – Part 3-1: Multifunction vehicle bus (MVB), International Electrotechnical Commission, Geneva, Switzerland, 2012.
5. IEC 61375-2-1:2012, Electronic railway equipment – Train communication network (TCN) – Part 2-1: Wire train bus (WTB), International Electrotechnical Commission, Geneva, Switzerland, 2012.
6. K. J. Gemmeke, Experiences with the implementation of the train communication network, WIT Press Transactions on the Built Environment, 6, 253–267, 1994. DOI: 10.2495/CR940321.
7. C.-H. Cho, J.-D. Lee, J.-H. Lee, K.-H. Kim, Y.-J. Kim, Design of the Train Network Simulator Based on Train Communication Network, International Symposium on Industrial Electronics Proceedings (Cat. No.01TH8570), IEEE, 343–347, 2001. DOI: 10.1109/ISIE.2001.931811.
8. Vehicular Communications and Networks – Architectures, protocols, operation and deployment, Woodhead Publishing Series in Electronic and Optical Materials: Number 72, Edited by W. Chen. Elsevier, Cambridge, UK, 2015.
9. Roll2Rail, Report R2R-T2.5-D-BTD-003-18, New dependable rolling stock for a more sustainable, intelligent and comfortable rail transportation in Europe, D2.5 – Architecture for the train and consist wireless networks, 2017. Available at http://www.roll2rail.eu/Page.aspx?CAT=DELIVERABLES&IdPage=45291e18-8d8f-4fd6-99f8-5d4b7a519b9c.

10. IEC 61375-2-6:2018, Electronic railway equipment – Train communication network (TCN) – Part 2–6: On-board to ground communication, 2018.
11. Roll2 Rail. Report R2R-T2.6-D-SIE-096-21. New dependable rolling stock for a more sustainable, intelligent and comfortable rail transportation in Europe. D2.6 – Architecture and interface definition for the train to ground communication. 29 August 2017. Available at http://www.roll2rail.eu/Page.aspx?CAT=DELIVERABLES&IdPage =45291e18-8d8f-4fd6-99f8-5d4b7a519b9c.
12. IEC 61375-2-5:2014, Electronic railway equipment – Train communication network (TCN) – Part 2–5: Ethernet train backbone, International Electrotechnical Commission, Geneva, Switzerland, 2014.
13. IEC 61375-3-4:2014, Electronic railway equipment – Train communication network (TCN) – Part 3–4: Ethernet consist network (ECN), International Electrotechnical Commission, Geneva, Switzerland, 2014.
14. IEC 61375-3-3:2012, Electronic railway equipment – Train communication network (TCN) – Part 3–3: CANopen consist network (CCN), International Electrotechnical Commission, Geneva, Switzerland, 2012.
15. Shift2Rail, Report CTA2-T1.1-D-CAF-005-10, Contributing to Shift2Rail's next generation of high capable and safe TCMS, Phase II, D1.1 – Specification of evolved wireless TCMS, 2020. Available at https://projects.shift2rail.org/download.aspx? id=3aa6a599-6f01-4049-b5b6-b2ffeebcb293.
16. P. Fraga-Lamas, T.M. Fernández-Caramés, L. Castedo, Towards the internet of smart trains: A review on industrial IoT-connected railways, Sensors, 17(6), 1457, 2017. Available at https://doi.org/10.3390/s17061457.
17. Shift2Rail, MOVINGRAIL project, Deliverable D3.1, Virtual coupling communication solutions analysis, Revision 1.1, 2020. Available at https://projects.shift2rail.org/download.aspx?id=c8411334-22f0-4658-9fd9-5909db7b2c05.
18. H. Kirrmann, P. A. Zuber, The IEC/IEEE train communication network, IEEE Micro, 21(2), 81–92, 2001. DOI: 10.1109/40.918005.
19. M. Wahl et al., Survey of Railway Embedded Network Solutions – Toward the Use of Industrial Ethernet Technologies, Les Collections de INRETS, IFSTTAR, Marne-la-Vallée, France, 2010. Available at https://www.ifsttar.fr/fileadmin/user_upload/editions/inrets/Syntheses/Syntheses_INRETS_S61.pdf.

10 Data Acquisition and Data Processing Techniques

10.1 INTRODUCTION

There are two main reasons why data acquisition and data processing techniques are relevant to a mechatronic system. The first stems from the growing popularity of digital controllers in mechatronic system, and the second is that mechatronic systems are also often provided with condition monitoring/health monitoring capabilities.

In modern applications of mechatronics, digital controllers have largely replaced analog controllers. The general scheme of a digital control implementation is shown in Figure 10.1. Assuming a multi-input multi-output (MIMO) control system, a number n of controlled variables from the controlled system (the plant) are measured by suitable sensors and processed by the digital controller (a microprocessor or industrial computer) to define the commands to m actuators producing the desired control actions on the plant. Given that most sensors are analog transducers and most actuators are driven by analog command signals, an analog to digital converter (ADC) is required to transform the analog signals in output from the sensors to digital series that can be processed by the digital controller, and a digital to analog converter (DAC) is required to convert the digital commands in output from the controller to the analog commands in input to the actuators. Note the scheme in Figure 10.1 does not include some components such as the sensor conditioning system, anti-aliasing filters, and analog filters in output from the DAC. The function of these additional components is described in Sections 10.3–10.5 of this chapter.

The processing of the measured signals performed by the digital controller typically involves filtering the signals and sometimes may require more complex treatment like state estimation or frequency domain analysis of the signal. All these tasks can be efficiently performed on the digital signals in output from the ADC, whereas they would either be impossible or result in a less efficient implementation if performed on analog signals.

The second reason that makes data acquisition and data processing relevant to mechatronic systems is that these systems are often provided with a condition monitoring unit, which means sensors are also used to acquire information on the system's health state. This again requires that measured signals are acquired in a digital format, processed appropriately (possibly entailing advanced data treatment such as spectral analysis or state estimation), and stored on a mass storage unit. In railway vehicles, the mass storage is often located in a wayside server, so the transfer of data from the onboard monitoring system to the wayside server must be performed.

DOI: 10.1201/9781003028994-10

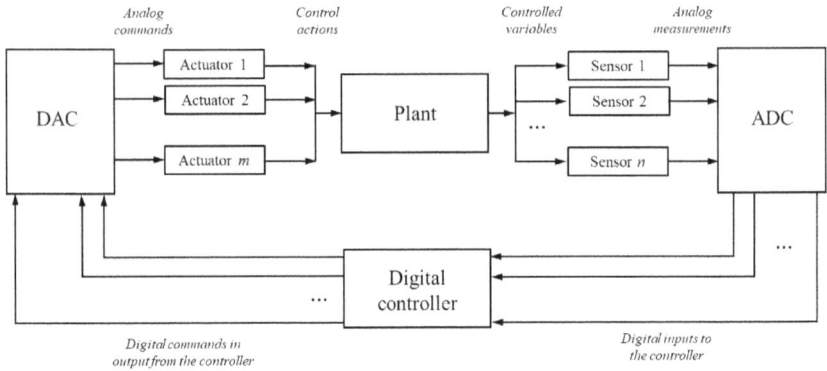

FIGURE 10.1 Configuration of a multi-input multi-output digital control implementation.

In this chapter, we focus on the main relevant issues related with data acquisition, introducing the general layout of a data acquisition/data processing system with its main hardware components. We describe the basic principles of analog-to-digital and digital-to-analog conversion and we introduce some fundamental digital data processing techniques, namely digital filtering and frequency analysis for digital signals. The description of sensing technologies for different types of sensors is outside the scope of this chapter (for this topic, the reader is referred to Chapter 6 and [1]), but we provide in Section 10.3 a general description of the issues related to signal conditioning, which is typically part of the data acquisition system.

10.2   GENERAL LAYOUT OF A DATA ACQUISITION AND DATA PROCESSING SYSTEM

A more detailed break-down of the data acquisition and data processing system is shown in Figure 10.2, delineating hardware and processing modules. The sensor signal conditioning phase is directed by the type of sensor and capability of the ADC conversion module and is implemented as analog hardware circuitry. A common example of analog signal conditioning is an anti-aliasing filter. Refer to Section 10.3 for more information on signal conditioning. Nearly all modern microcontrollers

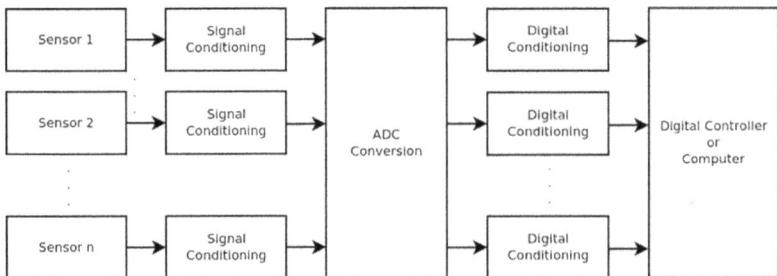

FIGURE 10.2   General layout of a data acquisition and data processing system.

offer some type of internal ADC module and analog reference voltage. A separate ADC may be employed if insufficient performance is available from the host microcontroller internal ADC or the channel count is insufficient. An inter-chip interface such as SPI, I2C, or Serial is used for communication between the host microcontroller and external ADC.

It is now common for a digital conditioning (filtering) phase to be performed after the conversion and prior to usage by the digital controller or computer. This enables a level of abstraction and re-programmability to the module rather than tying all the signal conditioning in analog circuitry prior to the ADC conversion step.

10.3   SIGNAL CONDITIONING

The pre- and post-ADC signal conditioning functions are an important step in ensuring a clean reliable signal is available to the host microcontroller application layer. The pre (analog) signal conditioning phase can be one or more of the following types:

- **Anti-Aliasing filter:** ensures that the waveform frequency converted by the ADC meets Nyquist sampling theorem.
- **Amplification/Scaling Signal Transducer ranges to ADC input range:** in practice it is highly likely that the sensor output range does not match the ADC input range. For example, a strain gauge might only have a 2.5 mV range, connected to the 3V ADC input. Analog circuitry is required to amplify or scale the sensor range to maximize the available dynamic range of the measurement.
- **Multiplexing:** this enables multiple analog channels to be acquired by a single ADC channel by sequentially iterating (switching) through the channels and converting them.

The post (digital) Signal Conditioning phase can be one or more of the following types:

- **Decimation/Oversampling:** a method of reducing the ADC conversion rate without inducing aliasing in the resultant signal. This is conceptually achieved through various accumulation techniques. Refer to Section 10.4.3 for a detailed discussion.
- **General purpose DSP filter:** this captures all other filtering types that are performed on a microcontroller such as high pass and band pass/rejection filtering.
- **Frequency Domain Analysis:** translation of the temporal data series to the frequency domain.
- **Digital Error Correction:** removal of known non-linearities in the ADC converter.
- **Calibration and unit conversion:** The ADC converted data is provided in the resolution of the converter, such as 10, 12, 16-bit signed or unsigned values. This step uses specific knowledge of the sensor type stored or downloaded to the host microcontroller that enables it to apply a functional translation of the raw value to specific units for use in the application layer.

10.4 ANALOG-TO-DIGITAL CONVERSION

Most sensors are analog transducers and an ADC is therefore required to transform the analog signals in output from the sensors and their conditioning system in digital signals that that can be treated by a processor and stored on a memory storage device.

There are various architectures of ADC devices available on the market, and it is not the scope of this chapter to describe their operating principles; for this topic, the reader is referred to [2, 3]. For the sake of clarifying the issues of analog-to-digital conversion in relation to the scope of this chapter, it is sufficient to consider the ADC as composed of two blocks working in series: a sample and hold (S&H) device and a quantizer. The S&H translates the value of the input analog signal (generally a voltage) at a given sampling time into a constant analog signal which is maintained until a new sample is acquired. The S&H stage is required to make sure that the sampled value from the input signal will not change while the quantization process is performed.

10.4.1 QUANTIZATION AND QUANTIZATION ERROR

The quantizer translates the analog value out of the S&H into a digital value expressed by a finite number of bits, the *word*. The quantization step logically consists of iteratively comparing the sampled analog value to fixed levels corresponding to a uniform division of the range spanning from the minimum value V_{min} to the maximum one V_{max} that can be accepted by the quantizer. The digital value in output of the DAC is then the best approximation of the sampled analog value, for the given word size and range of the quantizer.

Clearly, there is a quantization error involved in this operation, and this error depends on the full-scale range of inputs accepted by the ADC and on the number of bits in the word representing a single digital value. For instance, an 8-bit ADC working on an input range of $0 \div 10$ V will be able to represent $2^8 = 256$ discrete levels, so the value associated to the least-significant bit (LSB) will be:

$$\text{LSB} = \frac{10 \text{ V}}{256} = 39.1 \text{ mV}$$

and the (ideal) quantization error will be $\pm\frac{1}{2}$ LSB $= \pm 19.5$ mV. However, a 12-bit ADC working in the same range will have a LSB equal to 2.4 mV and an ideal quantization error of ± 1.2 mV.

The quantization error can be treated as a noise component affecting the output of the DAC. According to the IEEE 1241-2010 standard [4], the signal-to-noise-and-distortion ratio (SINAD) of a DAC is "the ratio of root-mean-square (rms) signal to rms noise and distortion (NAD)." In an ideal DAC, not subject to distortion or noise sources, the SINAD is only depending on the quantization error and can be shown to be expressed in dB according to the following expression:

$$\text{SINAD} = (6.02 \text{ n} + 1.76) \text{ dB} \qquad (10.1)$$

with n the number of bits of the converter. Of course, a large word size provides a higher SINAD and is desirable, but there is a trade-off with both the sampling rate and the cost of the ADC.

In a real ADC device, there are other sources of error beyond quantization, particularly an offset of the zero value, a deviation of the actual ADC gain from its nominal value, and a distortion caused by non-linearity, the latter source of error meaning that the discrete values to which the sampled value is compared in the quantization process are not exactly equally spaced [5]. Offset and gain errors can be compensated at least partially, but linearity errors are more difficult to mitigate. As a result of these errors, the SINAD of a real ADC device will be lower than the value predicted by Equation (10.1). The performance of a real ADC device is therefore often expressed using the effective number of bits (ENOB). According to IEEE 1241-2010 [4], "For an input sine wave of specified frequency and amplitude, after correction for gain and offset, the effective number of bits (ENOB) is the number of bits of an ideal ADC for which the rms quantization error is equal to the rms noise and distortion of the ADC under test." For instance, a 12-bit ADC might have an ENOB of 10.5 and this latter number is more representative of the accuracy of the analog-to-digital conversion performed than the *theoretical* number of bits. Note that the ENOB depends on the amplitude and frequency of the signal in input so, at least in principle, the ENOB value should be accompanied by a specification of the amplitude and frequency of the harmonic signal for which this value was estimated [4].

10.4.2   SAMPLING FREQUENCY AND ALIASING

In the analog-to-digital conversion process, the analog signal in input is transformed into a finite sequence of values sampled at discrete time values. In most cases, a constant size of the sampling interval Δt is used, and the sampling frequency f_s is defined as the reciprocal of the sampling interval. The choice of the sampling frequency is pivotal to preserving in the digital sample the information contained in the input analog signal. The well-known Nyquist sampling theorem states that the minimum sampling frequency required to correctly sample an analog signal is two times the highest frequency component in the signal. We exemplify this requirement using an example. In Figure 10.3, a harmonic analog signal with frequency of 10 Hz is sampled using a sampling frequency equal to 100 Hz (upper subplot) and 8.33 Hz (lower subplot). The samples (neglecting the quantization error) are shown by dots, and a dashed line is used in the lower subplot to connect the sampled values, to help with the visualization of the result. The original signal can be correctly reconstructed from the digital signal sampled at 100 Hz, but if the lower sampling rate of 8.33 Hz is used, the most straightforward reconstruction of the analog signal from the digital one, by linear interpolation for example, is a sine function having frequency 1.67 Hz.

This result exemplifies the well-known phenomenon called aliasing: a harmonic signal sampled at a sampling frequency not fulfilling the requirement of the Nyquist theorem will be reconstructed to a harmonic signal having a frequency lower or equal to $f_s/2$. Aliasing can be explained based on the properties of harmonic functions: the

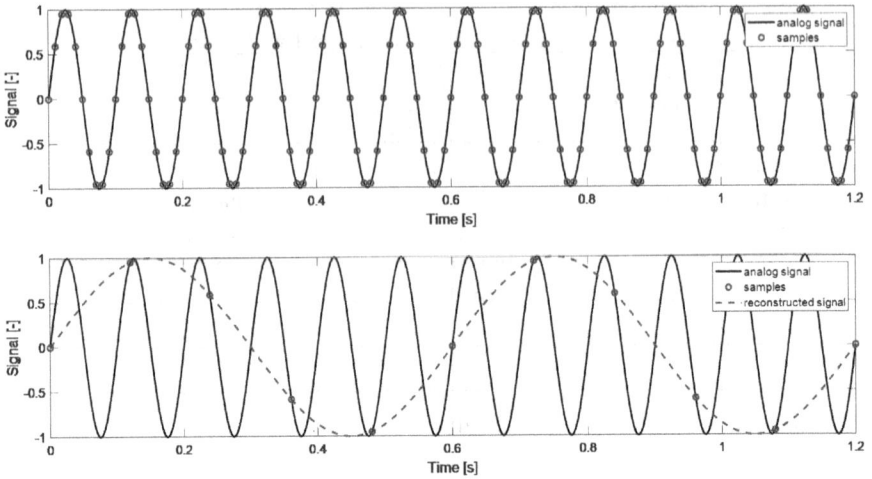

FIGURE 10.3 Sine signal at 10 Hz frequency sampled at at 100 Hz (top) and 8.33 Hz (bottom).

sequence of discrete values produced by the sampling at frequency f_s of a sinusoidal signal with frequency f is:

$$x_k = \sin\left(2\pi f k\Delta t\right) = \sin\left(2\pi f \frac{k}{f_s}\right) = \sin\left(2\pi f \frac{k}{f_s} + N2\pi\right) = \sin\left(2\pi\left(f + Nf_s\right)\frac{k}{f_s}\right)$$

(10.2)

with N having any integer value. Hence, after sampling the signal, it is impossible to distinguish the signal with frequency f from a signal having frequency $f + Nf_s$. For instance, signals having frequency $1.3\,f_s$, $2.3\,f_s$, ... will produce the same sampled sequence of a signal having frequency $f = 0.3\,f_s$. Furthermore, the following relationship holds:

$$x_k = \sin\left(2\pi f \frac{k}{f_s}\right) = -\sin\left(2\pi f \frac{k}{f_s} + N2\pi + \pi\right) = -\sin\left(2\pi\left(f + \left(N + \frac{1}{2}\right)f_s\right)\frac{k}{f_s}\right)$$

(10.3)

Therefore, apart from a phase difference, the sine signal with frequency f cannot be distinguished from a signal having frequency $f + \frac{N}{2}f_s$. This means signals having frequency corresponding to any integer value of N such as $0.8\,f_s$ ($N = 1$) or $1.3\,f_s$ ($N = 2$), will produce a sequence of samples that could be interpreted as being produced by a sinusoidal signal with frequency $f = 0.3\,f_s$. The effect expressed by Equation (10.2) together with Equation (10.3) is known as *folding* and is well represented by the diagram in Figure 10.4, where the frequency of the signal after analog-to-digital conversion f_{ADC} is plotted as function of the ratio of the frequency of the

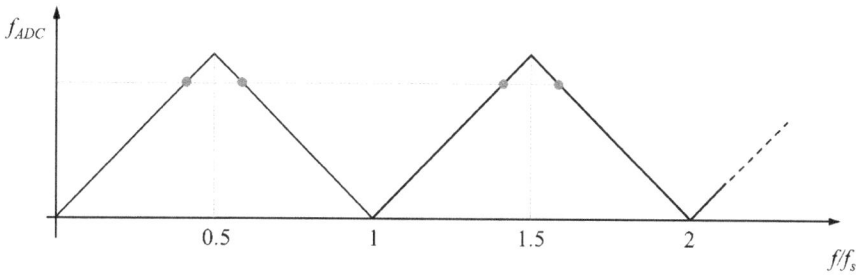

FIGURE 10.4   Frequency folding produced by aliasing.

analog signal f over the sampling frequency f_s. The diagram shows clearly that a one-to-one correspondence between the actual frequency of the signal and the frequency of the sampled signal can be ensured only for frequencies up to $\frac{1}{2}f_s$, as stated by Nyquist theorem.

10.4.3   ANTI-ALIASING FILTERS AND OVERSAMPLING

In Section 10.4.2 we have considered the effect of aliasing on a single harmonic signal. In a general case, the signal in input to the ADC consists of various harmonic components. Any harmonic component at a frequency above one half of the sampling frequency will be subjected to aliasing and will fold in the 0-$\frac{1}{2}f_s$ frequency range as shown in Figure 10.4. After the sampling of the signal is performed, it is not possible to correct this error, as the contribution to the spectrum of the sampled signal coming from the high-frequency harmonics of the analog signal subject to aliasing cannot be separated from the contribution of low-frequency harmonics in the input signal. Hence, the only way to avoid aliasing is to apply a low-pass filter to the analog signal before it is input to the ADC. This filter is referred to as the anti-aliasing filter and is part of the signal conditioning system described in Section 10.3. The cut-off frequency of the anti-aliasing filter depends on the filter's roll-off (see Section 10.5) and must be sufficiently lower than the Nyquist frequency $f_N = \frac{1}{2}f_s$.

Modern ADC devices often allow the use of a sampling rate which is much in excess of the upper frequency of interest in the analog signal being acquired. In this case, it is possible to relax the requirements posed to the design of the analog anti-aliasing filter, choosing deliberately to sample the signal at a much higher frequency than would be required for the scope of data acquisition. The Nyquist frequency f_N is then much higher than the maximum frequency of interest in the signal and hence a simple, low-order anti-aliasing filter can be used to efficiently remove from the analog signal any content at frequencies higher than f_N prior to sampling The digital signal is then decimated to reduce its size, so the signal in output from the DAC is transformed in a new digital signal sampled at a significantly lower frequency consistent with the actual maximum frequency of interest. Decimation cannot consist of a mere down-sampling of the signal, as this would cause aliasing in the decimated signal caused by harmonic components at frequencies higher than the new Nyquist frequency which were not efficiently removed by the analog anti-aliasing filter. Therefore, before decimation the digital signal in output from the ADC is passed through a digital low-pass filter having a cut-off frequency consistent with the new Nyquist frequency and a steep roll-off. The two steps of digital filtering and decimation can be performed together, thereby improving the computational efficiency of the process and the use of memory [6]. This technique, known as *oversampling*, takes advantage from the fact that it is much easier to realize a high-performance digital filter with steep roll-off, instead of its analog equivalent.

10.5   DIGITAL-TO-ANALOG CONVERSION

A DAC transforms a digital signal into its analog counterpart. A DAC is not required in a data acquisition system, but it may be required in a digital control system as the actuators are often driven by analog reference signals. Hence, the digital reference elaborated by the digital control system must be converted to an analog equivalent.

Similar to ADCs, we are not interested in describing the various possible architectures of a DAC's hardware (for this topic, the reader is referred to [2]), but rather to provide a simple description of how a continuous signal in time can be generated from a finite sequence of samples defined at constant time intervals, and to mention the main issues involved with this process.

The simplest model of the digital-to-analog conversion process is the zero-order hold (ZOH): each sample of the digital signal is converted into a fixed-level analog signal which is maintained over one sampling interval and is then updated to a new fixed-level signal corresponding to the new sample of the digital signal, ideally resulting in a sequence of steps. Of course, the transition from one level to the next one of the output signal takes place over a finite time according to the dynamic response of the DAC device, so the actual signal generated will be more complex than a mere sequence of steps. Alternatively, the digital-to-analog conversion can be performed by a first-order hold (FOH), which transforms the digital signal into a piecewise linear analog signal through a linear interpolation of the samples. Both ZOH and FOH lead to high-frequency components in the reconstructed analog signals, which are integer multiples of the sampling frequency. These spurious harmonic components must be removed from the reconstructed signal by means of an analog low-pass filter with cut-off frequency lower than the Nyquist frequency.

10.6   DIGITAL FILTERS

Filtering is the modification of a signal that is used to modify its harmonic components, attenuating unwanted features in the signal. In particular, band-pass filters are used to maintain unaltered the frequency components falling in a specified frequency range, i.e., the filter's *passband*, while reducing the other components of the signal falling in the *stopband*. Typical examples are low-pass filters, used to smooth out high-frequency components, and high-pass filters, used to remove the DC component and drift from the signal. In the first case, the passband is the frequency range

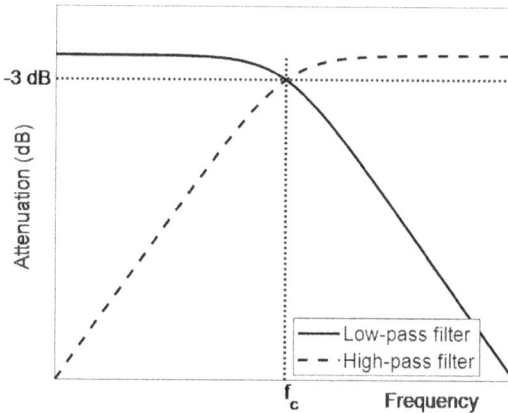

FIGURE 10.5   Attenuation versus frequency for a low-pass and a high-pass filter.

below a cut-off frequency which is a design parameter of the filter. In the second case, the passband is the frequency range above the cut-off frequency. The ratio of the filtered signal to the original signal at a given frequency is called *attenuation*; Figure 10.5 shows the attenuation curve of a low-pass and of a high-pass filter as a function of frequency. *Stopband rejection* is the property of the filter to provide the desired level of attenuation in the stopband and the *roll-off* of the filter is the steepness of the attenuation curve in the transition from the passband to the stopband. These two parameters, together with the cut-off frequency, are typically the main targets in the design of the filter.

Filtering can be performed on both analog signals and digital signals. Analog filters are electronic circuits made up of analog components such as resistors, capacitors, inductors, etc., whereas digital filters are defined as a convolution involving the digital signal in input and the coefficients defining the digital filter. Figure 10.6 shows the process of producing a filtered digitalized signal from an unfiltered analog signal using analog and digital filters.

There are several advantages with using digital filters instead of analog filters. First of all, analog filters as any electronic circuit are sensitive to noise caused, e.g., by electromagnetic disturbance or temperature variations, whereas digital filters are

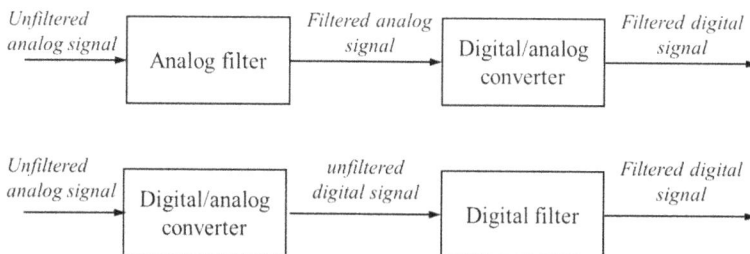

FIGURE 10.6   Filtering of a signal using analog or digital filters.

totally insensitive to these effects. Secondly, analog filters are a hardware component and, as such, are subject to failure. Of course, digital filters also need to be run using a hardware processor, but normally the processor is in charge of performing several tasks, so adding a digital filter will not have an impact on reliability, although it will impact the demand of computing effort. Furthermore, the complexity of analog filters rapidly increases with the required performance in terms of stopband rejection and roll-off, while the performance of digital filters can be improved with relative ease at the expense of a higher demand of computational effort. Finally, digital filters can be easily re-configured as the filter's frequency response is defined by the coefficients used in a convolution, hence different digital filters can be stored on the computer hardware and used depending on specific needs.

In the same way as analog filters, digital filters cause a delay of the filtered signal with respect to the unfiltered one: this is a consequence of the *causality* of the filter and cannot be avoided when the filter is applied in real time, i.e., the filter is applied to the input signal as soon as new samples are available from the ADC. The delay of the filter can be eliminated only in the case of deferred-time processing of the input signal.

There are different types of digital filters, but a main distinction can be made between Finite Impulse Response (FIR) and Infinite Impulse Response (IIR) filters. In a FIR filter, the k-th sample of the filtered signal y_k is defined as a linear combination of the last $N + 1$ available samples of the unfiltered signal $x_k, x_{k-1}, x_{k-2}, \ldots, x_{k-N}$, according to:

$$y_k = \sum_{i=0}^{N} b_i x_{k-i} \tag{10.4}$$

with N representing the order of the filter and $b_0, b_1, \ldots, b_N$ the coefficients defining the frequency response function of the filter.

For an IIR filter, the input-output relationship is instead defined as:

$$y_k = \sum_{i=1}^{N} a_i y_{k-i} + \sum_{i=0}^{M} b_i x_{k-i} \tag{10.5}$$

and the order of the filter is the largest value between N and M.

We see therefore that the main difference between FIR and IIR filters is that there is feedback of previous filtered samples on the new sample of the filtered signal in an IIR filter, while this is not included in a FIR filter. The consequence is that, when a FIR receives as input a digital impulse, i.e., a signal having a single non-zero sample, the filtered signal will become zero after a finite number of steps of application of the filter. However, in an IIR, the output produced by a digital impulse in input will remain non-zero at any subsequent time, due to the feedback component of the filter.

A simple example of a FIR filter is a moving average filter. For instance, the unweighted moving average of the last $N+1$ values from the input signal is a FIR filter with coefficients $a_i = \frac{1}{N+1}$.

The frequency response of a FIR filter is defined by its order N and by the coefficients $a_0, a_1, \ldots, a_N$, and the quality of the FIR filter performance in terms of roll-off and of stopband rejection is increasing with the order of the filter, so there is a trade-off between filter performance and demand of computational effort placed on the processor.

One popular way to design IIR filters is based on mapping the poles and zeros of an analog filter having the desired frequency response to the z-variable plane and, out of this, deriving the coefficients $a_1, a_2, \ldots, a_N, b_0, b_1, \ldots, b_M$ of the filter [7]. In particular, Butterworth filters are frequently used as they provide a frequency response having maximally flat magnitude. Therefore, IIR filters are frequently obtained as digital realizations of an analog Butterworth filter.

An IIR filter can provide a steeper transition from the passband to the stopband and improved stopband rejection compared to a FIR filter having the same order; this is the main reason for using IIR filters, but there are also some advantages with using FIR filters. Namely, FIR filters are unconditionally stable, their design is simple and they have a linear phase, which means there will be no phase distortion in the filtered signal.

To illustrate the performance of FIR and IIR digital filters, we consider an example where we want to apply a low-pass filter with a cut-off frequency of 50 Hz to a digital signal defined as:

$$x_k = \sin\left(2\pi f_1 k\Delta t\right) + 0.5\sin\left(2\pi f_2 k\Delta t\right) \qquad (10.6)$$

with $f_1 = 10$ Hz, $f_2 = 70$ Hz, and $\Delta t = 2.5$ ms corresponding to a sampling frequency of 400 Hz. After applying the filter, we expect the first harmonic component in the signal to be preserved as its frequency is lower than the cut-off frequency and the second one to be attenuated as it falls in the stopband of the filter. We consider three alternatives for the digital filter: a 4th order FIR, an 8th order FIR, and a 4th order IIR Butterworth filter.

The original and filtered signals are shown in Figure 10.7: the upper subplot shows the original signal (solid line), and the filtered signals from the 4th order and 8th order FIR filters. A larger attenuation of the harmonic component at 70 Hz is observed for the 8th order filter. In the lower subplot the unfiltered signal is compared with the filtered signal produced by the 4th order FIR filter and by the 4th order Butterworth filter. The filtered signals from the Butterworth filter show a higher attenuation of the high-frequency harmonic component, comparable to the one provided by the 8th order FIR filter, confirming the superior performance of an IIR filter in terms of rejection of harmonic components falling in the stopband compared to a FIR filter of the same order.

As expected, the filtered signals in both subplots show a delay with respect to the unfiltered signal and the delay produced by the 4th order Butterworth filter is higher than for the 4th order FIR filter and comparable to the one produced by the 8th order FIR filter.

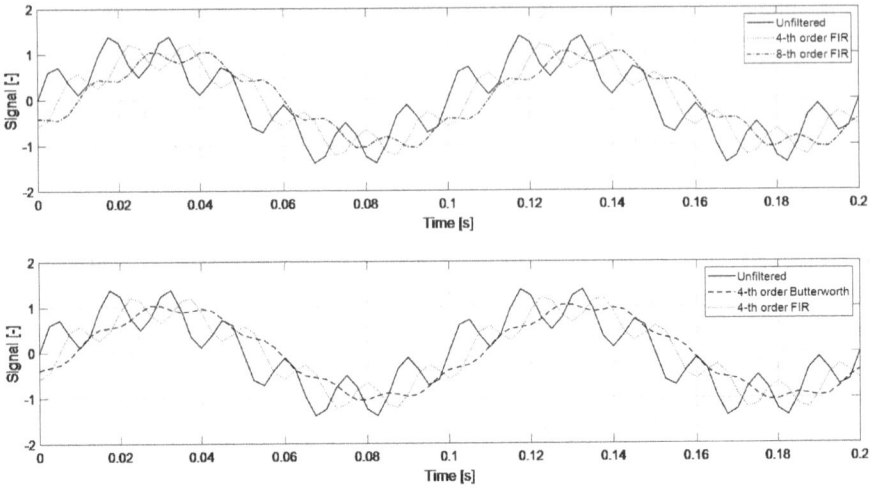

FIGURE 10.7   Example of performance of FIR low-pass filters with different orders.

10.7   FREQUENCY ANALYSIS FOR DISCRETE SIGNALS

After the measured signals are converted to digital signals and filtered according to the scopes of the measurement, it may be useful to analyze their spectral contents, i.e., obtain a representation of the different harmonic components in the signal. This can be done through the equivalent of the Fourier transform for digital signals.

Considering a digital signal consisting of n_s samples equally spaced in time:

$$x_k = x(k\Delta t) \quad k = 1, 2, \dots, n_s \tag{10.7}$$

the digital form of the Fourier transform is known as the Discrete Fourier Transform (DFT) and transforms the sequence of n_s real-valued samples x_k representing the digital signal into a sequence of n_s complex-valued coefficients defined as:

$$X_m = \sum_{k=1}^{n_s} x_k e^{-j\frac{2\pi mk}{n_s}} \quad m = 0, 1, n_s - 1 \tag{10.8}$$

In Equation (10.8), j is the imaginary unit and m is the index denoting the single term in the sequence. The inverse transformation of Equation (10.8) is:

$$x_k = \frac{1}{n_s} \sum_{m=0}^{n_s-1} X_m e^{j\frac{2\pi mk}{n_s}} \quad k = 0, 1, n_s - 1 \tag{10.9}$$

We note that, in Equation (10.8), the DFT value X_{n_s-m} is the complex conjugate of X_m, hence:

$$X_{n_s-m} = \sum_{k=0}^{n_s-1} x_k e^{-j\frac{2\pi(n_s-m)k}{n_s}} = \sum_{k=0}^{n_s-1} x_k e^{-j2\pi k} e^{+j\frac{2\pi mk}{n_s}} = \sum_{k=0}^{N-1} x_k e^{+j\frac{2\pi mk}{n_s}} = X_m^*$$

where the star denotes the complex conjugate. Using this property in Equation (10.9) we get:

$$x_k = \frac{1}{n_s} \sum_{m=0}^{\frac{n_s}{2}-1} \left(X_m e^{j\frac{2\pi mk}{n_s}} + X_m^* e^{j\frac{2\pi(n_s-m)k}{n_s}} \right) = \frac{2}{n_s} \sum_{m=0}^{\frac{n_s}{2}-1} |X_m| \cos\left(\frac{2\pi mk}{n_s} + \angle X_m \right)$$

$$= \frac{2}{n_s} \sum_{m=0}^{\frac{n_s}{2}-1} |X_m| \cos\left(\frac{2\pi m}{T} k\Delta t + \angle X_m \right)$$

(10.10)

with $|X_m|$ and $\angle X_m$ representing the magnitude and phase of the m-th term in the DFT of the signal and $T = n_s\Delta t$ being the total sampling time. From this result, we see that the digital signal x_k can be interpreted as a single period of a periodic function sampled at discrete times $k\Delta t$, with the magnitude of the first $N/2$ values in the DFT representing, after normalization by a factor $2/n_s$, the magnitude of the harmonic components of the signal, and with the phases of the same terms in the DFT representing the phase lag of each harmonic component. Hence, a spectral representation of the digital signal can be directly obtained from the DFT.

For stationary random signals, spectral contents shall be analyzed using the Power Spectral Density (PSD) function, also known as the auto-spectrum of the signal. In cases when more than one measured signal is considered, it may also be of interest to consider the cross-spectrum of signals. Auto-spectra and cross-spectra can also be obtained from digital signals, using the equivalents of the definition of the auto-spectrum and cross spectrum for analog signals, see [8, 9].

REFERENCES

1. E. Doeblin, Measurement Systems: Application and Design, 5th Edition, McGraw-Hill Science, ISBN: 978-0072922011, 2003.
2. D. Stranneby, W. Walker, Digital Signal Processing and Applications, Elsevier Ltd, ISBN 978-0750663441, 2004.
3. R. van de Plassche, CMOS Integrated Analog-to-Digital and Digital-to-Analog Converters, The Springer International Series in Engineering and Computer Science, ISBN 978-1402075001, 2003.
4. IEEE 1241-2010, Terminology and Test Methods for Analog-to-Digital Converters, IEEE, New York, NY, 2010.
5. A. D. Björsell, D. A. Björsell, Conversion, In: A. Ferrero, D. Petri, D., P. Carbone, M. Catelani (Eds), Modern Measurements: Fundamentals and Applications, Wiley-IEEE Press, ISBN 978-1118171318, 2015.

6. M. Parker, Digital Signal Processing 101: Everything You Need to Know to Get Started, Elsevier, ISBN 0128114533, 2017.

7. S. Tantaratana, C. Charoenlarpnopparut, P. Boonyanant, Y. C. Lim, IIR Filters, In: WK. Chen (Ed), The Circuits and Filters Handbook, Third Edition, CRC Press, ISBN 9781315222387, 2009.

8. J. S. Bendat, A. G. Piersol, Random Data: Analysis and Measurement Procedures, Fourth Edition, Wiley, ISBN 978-04-7024-877-5, 2010.

9. A. G. Piersol, Signal Processing, In: M. J. Croker (Ed), Handbook of Noise and Vibration Control, John Wiley & Sons, Inc., ISBN 9780471395997, 2007.

11 Mechatronic Suspensions

11.1 INTRODUCTION

Mechatronic suspensions are a very important application of general control and actuation techniques for mechanical systems and consist of replacing or complementing parts of a vehicle's passive suspension with full-active or semi-active devices with the aim of improving the vehicle's behavior under different operational conditions.

A detailed discussion of the scope of using mechatronic suspensions in a railway vehicle is provided in Section 1.2 of this book but, in short, the functions that can be assigned to a mechatronic suspension incorporated in a railway vehicle are typically:

- To improve the vehicle's running behavior in terms of guidance, stability, and curving behavior
- To enhance ride comfort, or to provide the same comfort level for higher vehicle speed/worse track quality

Active suspensions can also be used to enable a simplified mechanical design of the vehicle: for instance, two-axle vehicles with a single stage of suspension have a very simple and lightweight design, thus offering substantial advantages in terms of either increased payload or reduced mass, energy consumption and track damage. However, these vehicles often provide quite poor curving behavior and ride quality. With the use of active suspensions, the performance of two-axle vehicles can be enhanced, compensating for the simple mechanical architecture and opening new perspectives in the design of future generations of railway vehicles, especially for mass-transit applications.

It is important to point out that there are several considerations under which the running behavior and ride comfort shall be evaluated for a railway vehicle and that different issues may arise depending on some fundamental design features of the vehicle, e.g., the use of solid axles versus independently rotating wheels (IRWs) and the design of mechatronic suspensions for powered or non-powered running gear. Furthermore, the running dynamics and ride comfort requirements are very much different depending on the different field of application, such as high-speed trains, mass transit, light rail, locomotives, etc. Owing to these reasons, a large variety of mechatronic suspension solutions have been proposed and, in some cases, have reached the stage of mature technology proven in service, while other mechatronic suspension concepts are still in developmental stages and this explains the wide variety of concepts that will be reviewed in this chapter. The reader is also referred to [1–3] for classical and recent reviews of the State-of-the-Art in this field.

DOI: 10.1201/9781003028994-11

As introduced in Chapter 2, there are often two stages of suspensions in a railway vehicle: *primary suspensions* connecting the un-sprung masses (wheelsets) to a bogie frame and *secondary suspensions* connecting the bogie frame to the car body. Mechatronic suspensions can be implemented in both suspension stages, as active primary/ active secondary suspensions. A special case of active secondary suspensions is represented by car body tilting systems, which have been successfully in use over the past 30 years and now represent the most mature example of implementation in service of active suspensions for railway vehicles.

The aim of this chapter is to review functions, configurations, and control strategies for both active primary and active secondary suspensions, including an outline of car body tilting systems and of the use of mechatronic suspensions in two-axle vehicles without bogies.

11.2 ACTIVE PRIMARY SUSPENSIONS

Active primary suspensions are part of the vehicle's running gear, i.e., the ensemble of components in the railway vehicle responsible for its running behavior, and are used to improve the vehicle's running behavior. The benefits that can be achieved through the use of active primary suspensions are substantial but, at the same time, the correct functioning of primary suspensions is pivotal to the vehicle's safety, so that any active device included in the primary suspension needs to be designed to be fully safe under any possible fault condition. Safety concerns have so far hindered the use of active suspensions on vehicles in service despite the very large benefits offered by this technology. Active primary suspensions are covered extensively in this chapter in view of their extremely high potential to reduce maintenance costs for the rolling stock and infrastructure (especially in terms of wear) and also reduce energy consumption and other externalities such as noise. However, this will require further improvements with respect to the present State-of-the-Art in terms of sensor and actuation technology, design of fault-tolerant configurations, and advancements in standards for the certification of railway vehicles with mechatronic suspensions.

Aims and functions for active primary suspensions are introduced in Section 11.2.1, practical configurations will be introduced in Section 11.2.2, while sensed variables and control strategies will be discussed in Section 11.2.3.

11.2.1 ACTIVE PRIMARY SUSPENSION FUNCTIONS

The role of active primary suspensions is to improve the vehicle's running behavior under different respects:

- Active steering to improve the vehicle's curving performance
- Active running gear stabilization
- Active guidance of wheelsets with IRWs

We do not consider here the use of active/semi-active primary dampers to improve ride comfort as this particular application will be covered in Section 11.3 where ride comfort issues are treated.

11.2.1.1   Active Steering

Active steering is used to reduce the creepages and creep forces arising at the wheel/rail contact interface while the vehicle runs through a curve and may lead to substantial reduction of wear and damage, especially in tight (i.e., relatively small) radius curves. In Chapter 2, it was shown that in order to optimize its curving behavior, each wheelset should take a radial direction if centrifugal forces arising due to track curvature are exactly compensated by track cant, or a nearly radial position in the case of non-zero cant deficiency. This is not possible with passive primary suspensions because a relatively high yaw primary stiffness is required to provide the vehicle with a sufficient degree of stability, but it can be achieved through active control. An active steering vehicle will therefore include an actuation system capable of generating the desired yaw angle for the wheelset, based on sensing some kinematic quantities that can be related to the vehicle's curving behavior. Among different applications of active primary suspensions, active steering is the one for which the largest number of practical implementations are presently available.

11.2.1.2   Active Running Gear Stabilization

The stability of a railway vehicle was introduced in Chapter 2 and is a typical issue in the design of the running gear. Active suspensions can be used to raise the critical speed of a vehicle, thereby extending the range of speeds at which the vehicle can be operated in service. Stability is mainly related to the motion of the running gear in the horizontal plane, so active stabilization can be pursued by introducing yaw or lateral actuators in the primary suspension.

As explained in the book by Wickens [4], the hunting instability of a single wheelset can be described as arising from a natural feedback loop involving the wheelset's lateral and yaw motion, with the creep forces acting as the coupling terms that give rise to the closed loop. Active control can affect the feedback loop by altering either the lateral force or the yaw torque applied to the wheelset. Practical ways to implement this concept will be introduced later in this chapter.

11.2.1.3   Active Guidance

Most railway vehicles use solid wheelsets, i.e., a pair of wheels rigidly attached to a common axle. This means the rolling speed of the two wheels is the same, due to the rigid connection provided by the solid axle. However, some other vehicle designs (especially but not exclusively for trams and other light rail vehicles) use IRWs. In this case, the wheels on the two sides of the vehicle are attached to stub axles and can rotate independently one from the other. Apart from practical advantages which are not relevant here, the yaw torque produced by longitudinal creep forces is virtually reduced to zero for IRWs, thus breaking the natural feedback loop that is the cause of hunting instability. This means hunting instability cannot take place for a vehicle with IRWs but, at the same time, a vehicle with IRWs will not show inherent guidance, i.e., the ability to follow the track which is typical of solid wheelsets with conical wheel profiles. Poor guidance means the wheels may be subject to an erratic lateral motion in response to track irregularities, will often run in flange contact even when lateral forces are low, and the transition from tangent track to curve will often produce abrupt flange contacts. Active control can be used to mitigate these issues, and this can be implemented in a relatively simple way when the independent wheels are driven by motors. In this case an obvious implementation, known as the *electric shaft*, is to apply differential torques on the left and right wheels which are proportional to the differential rotation of the two wheels.

11.2.2 ACTIVE PRIMARY SUSPENSION CONFIGURATIONS

Active primary suspensions can be implemented according to various configurations [2, 3], and the differences between different implementations are mostly in terms of:

- Principle of actuation
- Direction in which actuation is applied
- Body/bodies to which control forces are applied

The three main configurations are actuated solid wheelset (ASW), actuated independently rotating wheels (AIRW), and driven independently rotating wheels (DIRW). These concepts are described below. It shall be mentioned that a fourth configuration, named directly steered wheels (DSW) is also considered in [2, 3] but so far this concept has been only the subject of theoretical studies and laboratory experiments, so it is not further covered here.

11.2.2.1 Actuated Solid Wheelset

The ASW configuration is shown in Figure 11.1. It consists of a standard two-axle bogie (or vehicle) in which passive primary suspensions are partly replaced or complemented by actuators applying on each wheelset either a controlled yaw torque (scheme on the left) or a lateral force (scheme on the right). Previous studies have shown that yaw actuation is preferable to lateral actuation both in terms of requiring a lower actuation force and of providing better ride quality for a two-axle vehicle

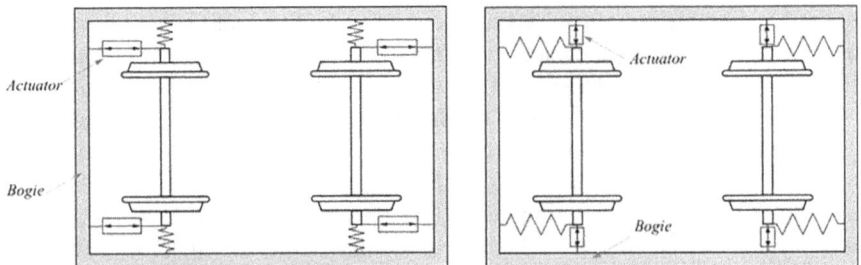

FIGURE 11.1 Primary active suspensions: actuated solid wheelset configuration with yaw actuation (left) and lateral actuation (right).

FIGURE 11.2 A practical realization of actuated solid wheelset control: the Bombardier mechatronic bogie. (From [6].)

running on a random track [5]. Additionally, lateral actuation is not well suited to providing active steering. For these reasons, practical implementations of this configuration have been proposed only considering yaw actuation.

An example of practical realization of this configuration is the Bombardier mechatronic bogie [6] shown in Figure 11.2. This is a modified version of the Bombardier VT612 passive steering bogie. Active control is applied by means of two electric motors controlling the motion of the steering linkages so that the yaw rotations of the two wheelsets relative to the bogie frame are independently actuated. The intended scope of ASW in this application was to ensure bogie stability despite removing yaw dampers from the bogie and, at the same time, improve curving performance. The prototype bogie was tested on the DB's roller rig and showed stable running at speeds in excess of 300 km/h.

A different implementation of ASW is presented in [7] and again is a modified version of a passive steering bogie design. In this case, the steering actuator is attached to the bolster and a lever-and-link mechanism is used to produce a steering angle between the two wheelsets in the bogie. This actuation scheme requires the use of just one actuator but does not allow the independent control of the yaw angle of the two wheelsets. It is therefore suited for active steering but not for stability control.

11.2.2.2 Actuated Independently Rotating Wheels

The AIRW configuration is shown in Figure 11.3. It is similar to the ASW configuration with yaw actuation introduced in Section 11.2.2.1 but using axles with IRWs instead of solid wheelsets. Compared to ASW, the advantage of this configuration is that the magnitude of the actuation forces is much lower, thanks to the fact that IRWs develop near-zero longitudinal creep forces. The disadvantage of this configuration is that IRWs do not have inherent guidance, so active guidance must be provided.

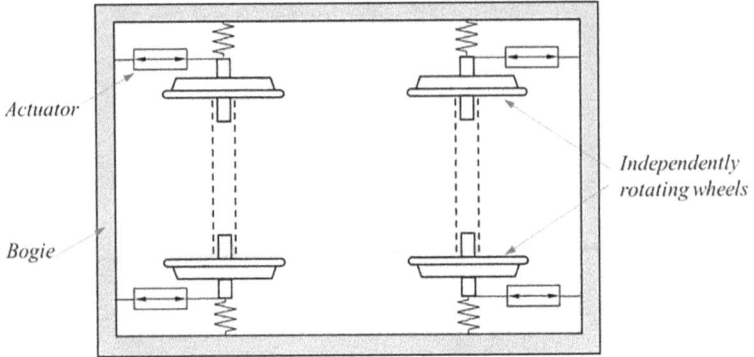

FIGURE 11.3 Primary active suspensions: actuated independently rotating wheelset configuration.

11.2.2.3 Driven Independently Rotating Wheels

A third configuration of active primary suspension is DIRW (see Figure 11.4). It consists of a standard bogie (or two-axle vehicle) with independently rotating wheels driven by pairs of separate traction motors. Active control is achieved by applying controlled torques to the wheels and can be used to provide active guidance together with active steering, provided the yaw stiffness of the passive primary suspension is low enough to allow the traction/braking forces generated by the motors to affect the yaw angle of the axle.

Compared to the other configurations described above, one advantage of DIRW is that there is no need to incorporate actuators in the primary suspensions. This is important because one weak point with active primary suspension is with proving the fault tolerance of the system, and it is certainly easier to manage a fault occurring in a traction motor rather than one occurring in a linear or rotary actuator like the ones included in other active primary suspension configurations. The disadvantage of this configuration is obviously that a larger number of traction motors is required, although the size of the motors will be smaller on average.

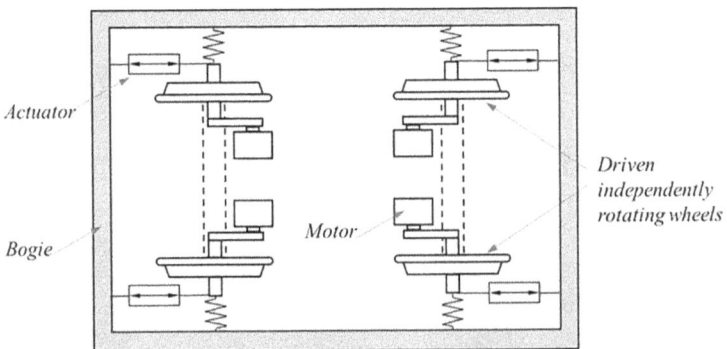

FIGURE 11.4 Primary active suspensions: driven independently rotating wheels configuration.

The DIRW configuration has been initially introduced for low floor urban vehicles, but recently it is being considered for an advanced concept of high-speed train in the *Next Generation Train* project [8].

11.2.3 Control Strategies for Active Primary Suspensions

Depending on the active primary suspension configuration chosen for a given vehicle and on the targets of the active control application, the control strategies adopted may set a focus on active steering, active stabilization, or active guidance. In some cases, more than one of these objectives are addressed together by a single control strategy, but each one is separately reviewed below.

11.2.3.1 Strategies for Active Steering

Control strategies for active steering are concerned with reducing unnecessary creep forces arising while the vehicle runs through a curve, see Section 1.2.1.2 in Chapter 1 of this book. For a vehicle running in a curve at zero cant deficiency, the optimal curving condition is obtained when the two wheelsets in the same bogie or in a two-axle vehicle are both aligned along the radial direction. In this condition, the angle of attack and steady-state lateral creep forces are eliminated. However, if the vehicle runs through the curve at non-zero cant deficiency, a small angle of attack will be required to balance non-compensated lateral forces [9].

In this case, the target for active steering is to achieve an equal angle of attack on the two wheelsets, as exemplified by Figure 11.5. If this steering condition is realized, the minimum amount of lateral creep forces required to balance the centrifugal

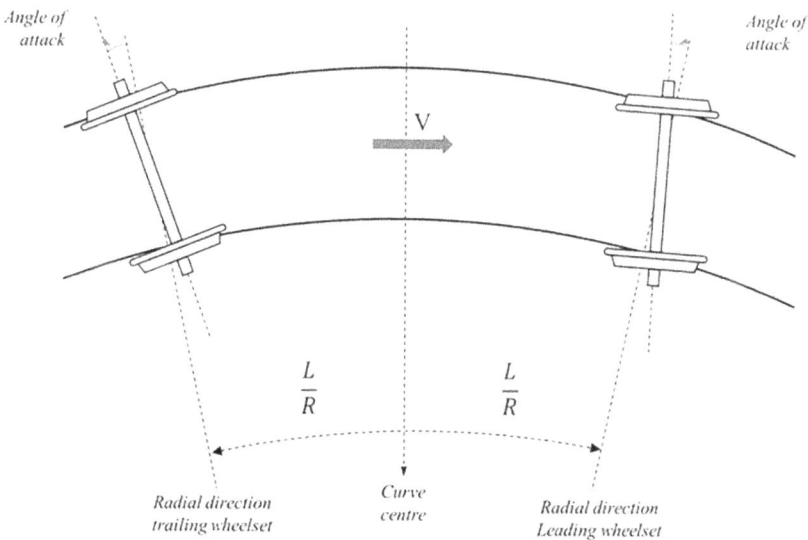

FIGURE 11.5 Perfect steering for a two-axle bogie or vehicle. Direction of motion is to the right, positive rotations counterclockwise.

forces caused by cant deficiency is established at the two wheelsets, damage and wear on the four wheels is minimized, flange contact is avoided and track shifting forces on the two axles are equalized. This can be implemented by controlling the yaw angles of the wheelsets with respect to the bogie, as in this condition the bogie will naturally settle to a nearly zero angle of attack. The required yaw angles for the two wheelsets relative to the bogie are (sign conventions are stated in the caption of Figure 11.5):

$$
\sigma_{\text{leading}} = -\sin^{-1}\left(\frac{L}{R}\right) - \sin^{-1}\left(\frac{\left(\frac{1}{2}m_b a_{nc}\right)}{f_{22}}\right) \approx -\left(\frac{L}{R}\right) - \left(\frac{\left(\frac{1}{2}m_b a_{nc}\right)}{2 f_{22}}\right)
$$

$$
\sigma_{\text{trailing}} = \sin^{-1}\left(\frac{L}{R}\right) - \sin^{-1}\left(\frac{\left(\frac{1}{2}m_b a_{nc}\right)}{f_{22}}\right) \approx \left(\frac{L}{R}\right) - \left(\frac{\left(\frac{1}{2}m_b a_{nc}\right)}{2 f_{22}}\right)
$$

(11.1)

with L the semi-wheelbase of the bogie, R the radius of the curve, m_b the sum of the bogie mass plus the portion of car body mass applied on the bogie, a_{nc} the non-compensated lateral acceleration, and f_{22} the coefficient of Kalker's linear theory relating the lateral creep force to the lateral creepage (see Section 2.3.3.1 in Chapter 2 of this book). Note in Equation (11.1) that the same value of the f_{22} coefficient is assumed for all four wheels in the bogie, but differences may occur, especially between the outer and inner wheels, due to different normal loads and different contact patch shapes.

In the right-hand side of Equation (11.1), the first term is the yaw angle required to align the wheelsets radially, while the second term is the equal attack angle on the two wheelsets producing the creep forces required to balance non-compensated centrifugal effects.

In principle, a control scheme based on Equation (11.1) is possible but requires that the f_{22} coefficient is estimated, which is impractical. A good approximation consists of neglecting the second term in Equation (11.1), i.e., producing a relative yaw angle between the two wheelsets that compensates for the difference between the radial directions at the two wheelsets (see Figure 11.5). In this way, the two wheelsets are forced to take the same angle of attack and the bogie will naturally find itself a way to realize the correct angle of attack required to balance non-compensated centrifugal forces.

An alternative control strategy is known as *yaw relaxation* [10], and the objective of this control strategy is to provide active cancellation of the primary yaw stiffness so that the wheelsets in the bogie are able to take *naturally* the optimal attack angle. The problem with this strategy is that primary yaw stiffness must be preserved at relatively high frequency corresponding to the hunting mode of the bogie, so that the stability of the vehicle is not compromised. Therefore, the control strategy is designed so that active cancellation only occurs in a low-frequency range corresponding to the time required for the vehicle to negotiate curve entry/exit transitions.

11.2.3.2 Strategies for Active Stabilization

The hunting of a solid wheelset can be suppressed by applying feedback control separately on each wheelset. The control strategy shall be defined in a way that improves the stability of the vehicle but does not negatively affect its curving behavior. Among different schemes proposed, it is worth mentioning *active yaw damping*, which uses a yaw torque applied to the wheelset by longitudinal actuators (see Figure 11.1 left for a solid wheelset and Figure 11.3 for an axle with IRWs) proportional to the lateral velocity of the axle [5]. Alternatively, the yaw angle of the wheelset can be used as the feedback variable, a technique known under the name of *skyhook stiffness* control [11]. Finally, a lateral force can be used as control action while using as the feedback variable the yaw velocity of the wheelset, a technique known as *active lateral damping* [5].

Figure 11.6 exemplifies the three control strategies in the form of a flowchart, using the 2-DOFs model of a single wheelset with primary suspensions introduced in

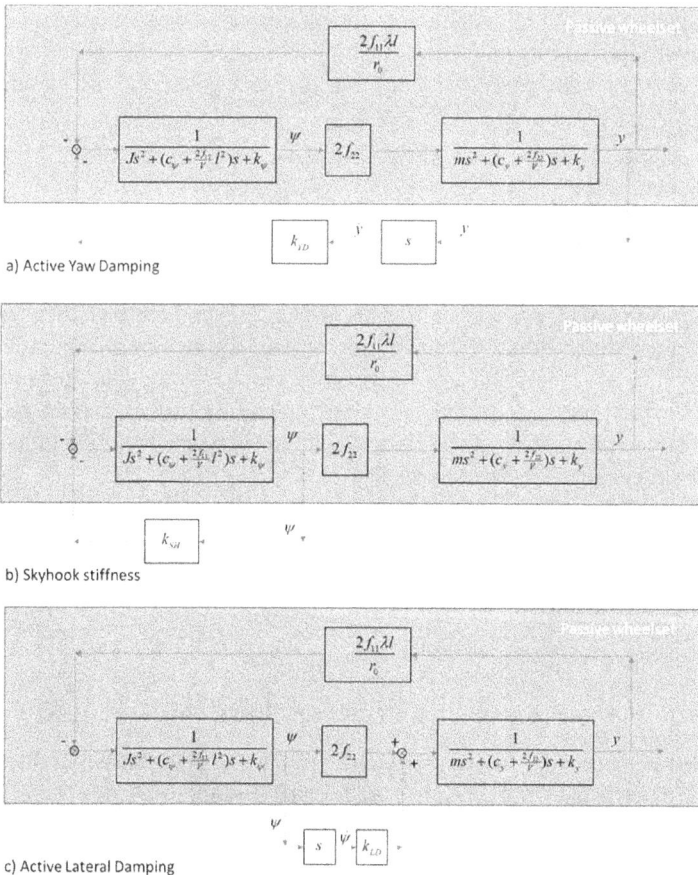

FIGURE 11.6 Active control strategies exemplified for a single wheelset with primary suspensions: (a) active yaw damping, (b) sky-hook stiffness, (c) active lateral damping.

Chapter 14 (Case A). In Figure 11.6, K_{YD}, K_{SH}, and K_{LD} are the control gains respectively for active yaw damping, skyhook stiffness, and active lateral damping. For the meaning of other symbols, refer to Section 14.1 in Chapter 14 of this book.

11.2.3.3 Strategies for Active Guidance

Railway vehicles equipped with solid wheelsets with coned wheels benefit from a *natural* ability to follow the track called guidance. However, mechanical configurations using IRWs need guidance to be provided by active control.

For the DIRW configuration, a simple way to provide guidance is to apply traction/braking torques on the two wheels proportional to the difference of their rotation (the *electric shaft* concept).

For DSW vehicles, active guidance can be provided using a feedback control nulling the lateral wheel/rail displacement, but this control has to be carefully designed in terms of its bandwidth to avoid excessive sensitivity to high-frequency disturbance from track irregularities. Another issue with this control strategy is concerned with measuring or estimating the feedback signal. One practical solution is to estimate the lateral wheel/rail displacement from the difference of the rolling speed of the left and right wheels, but this method is sensitive to the conicity and hence to unknown parameters such as deviations from the nominal track gauge and rail inclination and wear of wheel and rail profiles.

11.3 ACTIVE AND SEMI-ACTIVE SECONDARY SUSPENSIONS

The use of mechatronic components in the secondary suspensions of railway vehicle has achieved at present a more mature status of implementation compared to the use of mechatronic primary suspensions. This is due to several reasons, the most important one being that a failure occurring in the primary suspension often has severe effects on safety, while secondary suspensions are more easily designed to be fault tolerant. Furthermore, components in the primary suspensions may be required to generate large actuation forces and are subject to high levels of vibration, resulting in more challenging design requirements.

Following the same order of presentation used for active primary suspensions, the functions realized by active secondary suspensions are first introduced in Section 11.3.1. Then, configurations of active and semi-active suspensions are presented in Section 11.3.2. Finally, the control strategies are introduced in Section 11.3.3.

11.3.1 Active and Semi-Active Secondary Suspension Functions

Secondary suspensions of railway vehicles are mostly designed to provide isolation of the car body from disturbance caused by track irregularities. It follows that most concepts for active or semi-active secondary suspensions address the improvement of ride comfort for the passengers. Yet, some concepts have been proposed where active secondary suspensions are aimed to improve the vehicle's curving behavior and/or to increase running stability.

11.3.1.1 Improvement of Ride Comfort

The use of mechatronic suspensions to minimize car body acceleration in the lateral and vertical direction is widely used in recent generations of railway vehicles, especially for high-speed application where the use of advanced suspensions enables to maintain satisfactory ride quality even at very high speed and can also be used to compensate for a lower geometric quality of the track.

In general, separate active or semi-active suspensions are used to improve ride comfort in lateral and vertical direction, as the control forces required have to be applied along different directions. Active lateral suspensions are used more widely compared to active vertical suspensions as they can effectively improve ride comfort in curves and also deal with specific issues such as filtering out the effect of yaw motion of the running gear arising at high running speed.

A common approach is to have the mechatronic secondary suspension in parallel to a passive secondary suspension, so that the large static force coming from the weight of the car body is borne by the passive suspension while the mechatronic suspension will only be in charge of generating a dynamic force with low or even zero mean value. This arrangement in particular makes it possible to use semi-active components instead of full-active ones, resulting in a simpler design of the mechatronic suspension thanks to the fact that no power supply is required. In this way, both the impact on the overall vehicle's design and the cost of the mechatronic suspension can be reduced substantially. It should be noted, however, that most implementations of mechatronic lateral suspensions are designed to bear the non-compensated part of the centrifugal force acting on the car body in curves (see below), so in this case the suspension has to include full-active actuators.

Active secondary suspensions for ride comfort improvement can also be categorized based on a distinction between *low-bandwidth* and *high-bandwidth* systems. The aim of low-bandwidth systems is to mitigate car body motion arising from deterministic features of the track, especially curves. A typical example is represented by the so-called *hold-off* system, whose target is to maintain the car body in a centered position with respect to the bogies, so that the bumpstops limiting the lateral displacement of the car body relative to the bogies will not enter into contact, as this contact would cause a stiff connection between the body and the bogies, eventually causing increased transmission of vibration [12].

High-bandwidth systems aim instead to reduce car body vibration in the frequency range relevant to ride quality, i.e., approximately 0.5–30 Hz. It is worth noting that this frequency range affects the resonances of the car body corresponding to rigid modal shapes (bounce, sway, roll, pitch, and yaw modes) and some of the resonances corresponding to non-rigid modal shapes such as the 1st and 2nd bending mode and the first torsional mode.

11.3.1.2 Improvement of Running Behavior (Stability and Curving)

In Section 11.2 it was shown how the running behavior of a railway vehicle can be improved using different concepts for active primary suspensions. Despite the fact that these concepts may theoretically lead to substantial benefits in terms of, e.g., reducing creep forces and wheel wear in curve or improving vehicle stability, their implementation requires radical changes to the design of the running gear and also may raise concerns about the capability of active primary suspensions to tolerate faults without leading to potentially unsafe running conditions.

Active secondary suspensions are less effective compared to active primary suspensions when it comes to improving the curving behavior and stability of a railway vehicle, as they cannot affect directly the creepages at wheel/rail contact. However, their implementation is much easier compared to primary active suspensions because they can be designed as replacements for passive components of the secondary suspension (typically yaw dampers) and also because they pose less challenges in terms of fault-tolerant design.

Active/semi-active secondary suspensions to improve the vehicle's running behavior typically consist of actuators or controlled dampers installed in the longitudinal direction, connecting the car body to the bogie frame with at least one actuator per side. These actuators are used to generate a yaw torque exchanged between the bogie and the car body; a concept called *secondary yaw control* (SYC).

11.3.2 Configurations and Hardware

In this section, configurations and hardware of mechatronic secondary suspensions are reviewed, considering three main applications:

  i. Active/semi-active lateral suspensions
 ii. Active/semi-active vertical suspensions
iii. Active/semi-active SYC

11.3.2.1 Active/Semi-Active Lateral Suspensions

Active lateral suspensions consist of one or more actuators generating a controlled force in the lateral direction between the car body and the bogie. Depending on the bandwidth of the actuator and on the control strategies adopted, the actuator can be operated as a hold-off system (see Section 11.3.1.1) and/or as a *smart damper* replacing or complementing passive lateral dampers and providing improved isolation of vibration for the lateral sway, yaw, and roll modes of the car body. In some cases, the two modes of operation are realized by the same physical device.

For low-bandwidth systems, pneumatic actuators are often used which provide a limited pass-band of actuation, consistent with the scope of the active suspension, facilitating the rejection of high-frequency disturbance from track irregularities. Pneumatic actuators are also easily incorporated in the secondary suspension, thanks to the availability of pressurized air in the vehicle used for other suspension components (air springs) and auxiliaries (doors and others). Figure 11.7 shows the active lateral suspension of Shinkansen trains series E2 and E3, which is realized by a pneumatic actuator.

To increase the bandwidth of actuation, other implementations make use of electrohydraulic actuators, such as the active lateral suspension tested in Sweden in the framework of the *Green Train* project [13] and now used in service on the Italian ETR1000 high-speed trains. The use of electro-magnetic actuators has also been

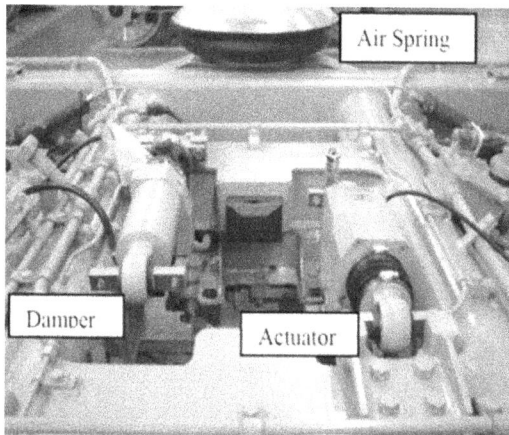

FIGURE 11.7   The active lateral suspension of Shinkansen trains series E2 and E3. (From [2].)

proposed [14, 15], but the drawback of this technology is represented by the large ratio of the actuator's volume to the maximum force that can be generated.

An alternative option for the lateral suspension is to use semi-active dampers. In this case, the implementation is easier as no power unit is required, but the hold-off function cannot be realized because semi-active dampers cannot generate a steady force. For this reason, semi-active dampers can be set in parallel to an active hold-off device.

A mature technology for semi-active dampers in secondary suspensions consists of electronically controlled hydraulic dampers [12], while the use of magneto-rheologic dampers is being considered from a research point of view but seems to be still quite far from practical application [16].

11.3.2.2   Active/Semi-Active Vertical Suspensions

The design of an active or semi-active vertical suspension system is more challenging compared to a lateral suspension as, in vertical direction, both rigid modes and flexible modes of the car body have to be controlled which means a higher bandwidth is required. Furthermore, it is sometimes difficult to fit vertical actuators or semi-active dampers in the secondary suspension. An experimental set-up incorporating active hydraulic actuators and realizing an active vertical suspension was subjected to line tests as part of the Green Train project [17] but no implementations have been applied on vehicles in service so far.

Semi-active vertical suspensions are a promising alternative and have been the subject of intensive research in past years. The concept for a semi-active vertical secondary suspension allowing the simplest implementation is to control vertical damping through a variable striction orifice installed in the connection between the air spring and surge reservoir [18, 19] (see Figure 11.8). In this way, the dissipative force generated by the air spring can be varied between a minimum and maximum value, enabling the implementation of a semi-active control scheme as

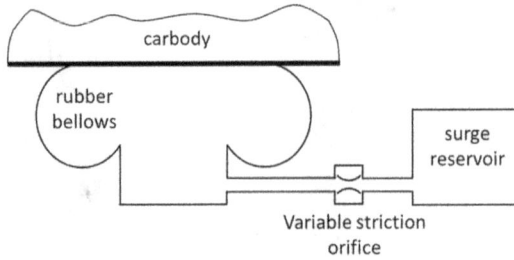

FIGURE 11.8 Air spring with variable striction orifice.

described in Section 11.3.3. The drawback of this solution is that only rigid car body modes can be controlled due to the limited bandwidth of the air spring suspension. The results of line tests performed on a Shinkansen vehicle equipped with air springs having a variable striction orifice show that the power spectral density of vertical car body acceleration can be approximately halved compared to the passive vehicle in the frequency range around 1.5 Hz which is dominated by the resonances of the car body rigid modes [19].

A second possible application of semi-active vertical suspensions consists of the use of electronically controlled hydraulic dampers. This type of semi-active damper can be designed to have a bandwidth sufficient to control the resonance of the first bending mode of the car body, which is a well-known issue affecting ride quality in high-speed trains.

A final concept worthy of being mentioned concerning semi-active vertical suspensions consists of incorporating semi-active vertical dampers in the primary suspension of the vehicle. While this solution should be categorized under the topics covered by Section 11.2, it is described here because it is the only known concept for mechatronic primary suspensions having as its objective the improvement of ride quality. This concept is described in [19] and is suitable in cases when fitting high-bandwidth controlled dampers in the secondary suspension is not possible or unwanted for some reason. The combined use of semi-active hydraulic primary dampers and of air springs with variable orifices is demonstrated to be successful with reducing car body vibration related to both the rigid and first flexible modes of the car body.

11.3.2.3 Active/Semi-Active Secondary Yaw Control

SYC is a configuration of active/semi-active secondary suspensions intended to improve the vehicle's stability and/or curving behavior. Actuators or semi-active dampers are applied between the car body and the bogie frame in the longitudinal direction as shown in Figure 11.9, replacing conventional yaw dampers, and are used to apply a controlled yaw torque to the bogie. The concept was originally proposed by Diana and co-workers [20, 21] using electromechanical actuators to improve the stability and curving behavior of an ETR470 (Pendolino) vehicle.

As far as the improvement of vehicle stability is concerned, SYC can be used to mimic the behavior of passive yaw dampers, but with superior performance compared

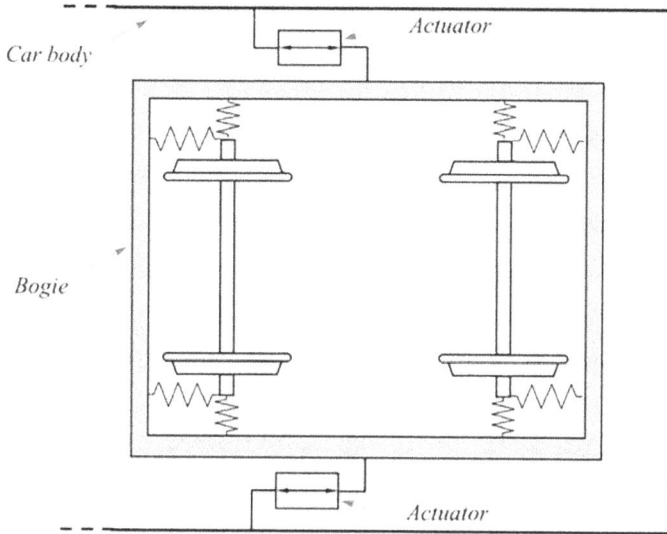

FIGURE 11.9 Scheme of secondary yaw control.

to conventional passive dampers thanks to the use of large bandwidth actuators, and also to the use of a derivative component in the control scheme to recover the effect of delays in the control and actuation chain [21].

SYC can also be used to improve curving behavior, as the actuators can be used to steer the bogie frame during curve negotiation, as proposed by Liebherr with their ADD system (ADD stands for *Aktiver Drehdämpfer*, i.e., active yaw damper). Compared to the wheelset steering systems presented in Section 11.2.2, this configuration is less effective in reducing unwanted creep forces and related wear effects because it is not capable of directly controlling the angle of attack of each single wheelset. However, as mentioned above, it is seen as a viable trade-off between performance improvement and ease of implementation.

Recently, SYC was revisited using semi-active hydraulic dampers in place of a full-active system. In this case, a careful design of the semi-active damper is required to actually provide an improvement of damping capability compared to a conventional damper. The main advantage of this unit is therefore the ability to adapt the damping force to the vehicle's running condition so that the maximum damping capability is developed only when required, reducing the wear and tear of the damper and also reducing the transmission of forces to the car body [22].

For Co-Co locomotives (vehicles with two bogies having three wheelsets each), Simson and Cole proposed a set-up similar to SYC but also including a passive linkage which is used to steer the outer wheelsets in the bogie (see Figure 11.10) [23–25]. The longitudinal actuators on the two sides of the bogie are used to steer the bogie but, at the same time, through the effect of the linkage, produce the alignment of the leading and trailing wheelsets to the local radial direction. This configuration is mainly intended to facilitate curving for three-axle bogies and to improve the tractive effort of the locomotive in a curve.

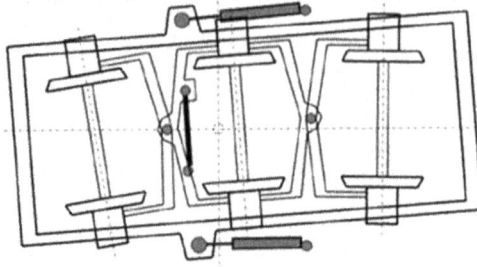

FIGURE 11.10   Actuated yaw forced steering (AY-FS) bogie. (From [25].)

11.3.3   Control Strategies for Active and Semi-Active Secondary Suspensions

In this section, an outline is provided of control strategies used for active and semi-active secondary suspensions.

11.3.3.1   Low-Bandwidth Control for Ride Comfort

Low-bandwidth control strategies are intended to improve the vehicle's behavior in response to low-frequency deterministic track inputs, mainly represented by curves. The classical example is the hold-off system, whose aim is to maintain the car body in a laterally centered position avoiding contact of the lateral bumpstops. To achieve this goal, a lateral control force is applied to the car body to balance the effect of non-compensated lateral forces. The control force can be defined according to a feed-forward or to a feed-back scheme. In the feed-forward scheme, the control force F_c is defined as proportional to the non-compensated lateral acceleration a_{nc} through a gain M_c corresponding to the portion of car body mass resting on the bogie (for a vehicle with two bogies, M_c will be one half of the total car body mass):

$$F_c = M_c a_{nc} \tag{11.2}$$

In this way, the control forces applied by the lateral actuators balance the non-compensated centrifugal force on the car body which is then kept in a centered position while the vehicle negotiates the curve. Of course, deviations from this *perfect* behavior arise from uncertainties related to the actual mass of the car body (which is varying with the payload). However, the small residual lateral force is balanced by the stiffness of the passive lateral suspension with just a limited lateral displacement so that the ultimate goal is achieved.

Alternatively, a feed-back control scheme can be used. In this case, the control force is defined as:

$$F_c = K_P \left(y_c - y_b \right) + K_I \int \left(y_c - y_b \right) dt \tag{11.3}$$

where K_P and K_I are the proportional and integral gains of the controller and $y_c - y_b$ is the observed variable representing the lateral displacement of the car body relative to the bogie measured across the active suspension. This latter quantity can be measured by an internal displacement sensor embedded in the actuator. Given the low bandwidth of this control scheme, a low value of the proportional gain compared to the integral gain should be used, avoiding a deterioration of ride comfort due to a stiffening of the suspension caused by active control.

11.3.3.2 Skyhook Control

Skyhook control is a classical control strategy intended to provide isolation of the vehicle body from disturbance from track irregularities. This control strategy was originally defined for road vehicles but is now widely applied to railway vehicles as well. Skyhook control can be realized with both full-active and semi-active suspensions. The basic idea is shown in Figure 11.11: the secondary damper in the passive suspension is replaced by a control force proportional to the absolute speed of the car body, thus mimicking the effect of a viscous damper hooked to the sky.

The rationale behind this control scheme is explained by the transmissibility diagram, i.e., the ratio of the amplitude of the sprung mass displacement over the amplitude of the base excitation. This non-dimensional quantity is a function of frequency and is shown in Figure 11.12a for a passive suspension: it is observed that, at frequencies higher than $\sqrt{2}f_c$ with f_c the natural frequency of the car body over the secondary suspension, the transmissibility is lower for lower non-dimensional damping of the suspension so, in this frequency range, removing the secondary damper is beneficial to ride comfort. At the same time, some damping of car body motion is required to reduce the transmissibility in the resonance region and to reduce the effect of transient oscillation arising in response to deterministic track inputs or other excitation (e.g., wind forces). However, skyhook damping is not detrimental to the transmissibility function in the high frequency range, because (at least in principle) the damper is not connected to the base (i.e., to the un-sprung mass). Therefore, a satisfactory isolation of the body from vibration in the bogie is achieved in the entire

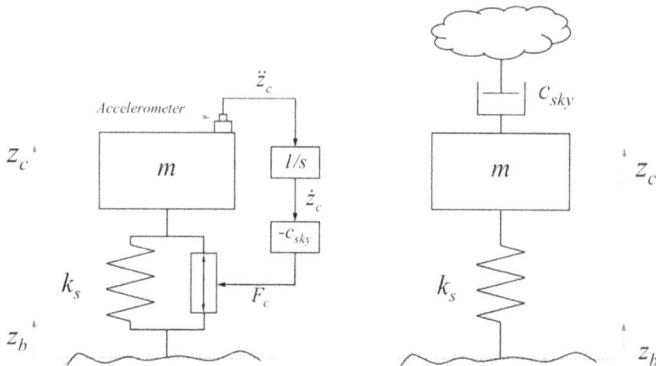

FIGURE 11.11 Principle of skyhook control.

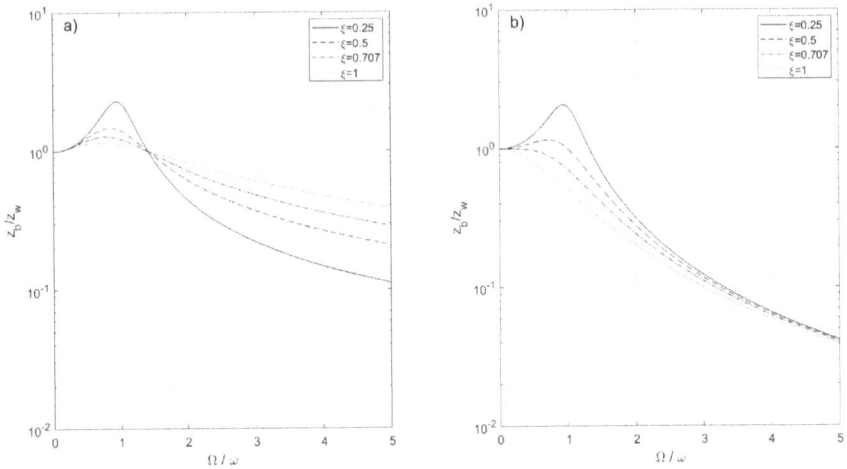

FIGURE 11.12 Transmissibility diagram for a passive suspension (a) and for ideal skyhook control (b).

frequency range, as shown in Figure 11.12b where the transmissibility diagram is plotted for an ideal skyhook control like to one shown in Figure 11.11.

Using an active suspension, skyhook control can be applied through a control force F_c defined as:

$$F_c = -c_{sky}\dot{z}_c \qquad (11.4)$$

with c_{sky} a gain representing the damping coefficient of the skyhook damper and $\dot{z}_c$ the absolute velocity of vibration of the car body (this can be either a vertical or lateral velocity component, depending on whether a lateral or vertical active suspension is considered). The absolute velocity signal is obtained from the integration of an acceleration signal which is high-pass filtered to remove low-frequency components related to deterministic track features, e.g., the quasi-static lateral acceleration in a curve. This control scheme is schematically shown in Figure 11.11 on the left.

Skyhook control for a semi-active suspension is slightly more complicated. In this case, the control force is generated by a controllable damper connecting the car body and the bogie and hence is defined as:

$$F_c = -c_v(t)\dot{z}_{def} \qquad (11.5)$$

where $c_v(t)$ is the variable damping coefficient of the semi-active damper and $\dot{z}_{def}$ is the suspension deflection velocity, i.e., the velocity of the car body relative to the bogie:

$$\dot{z}_{def} = \dot{z}_c - \dot{z}_b \qquad (11.6)$$

Therefore, to realize the control force defined by Equation (11.4), the variable damping realized by the semi-active damper should be:

$$c_v(t) = c_{sky} \frac{\dot{z}_c}{\dot{z}_{def}} = c_{sky} \frac{\dot{z}_c}{\dot{z}_c - \dot{z}_b} \tag{11.7}$$

Depending on the relative sign of the absolute car body velocity $\dot{z}_c$ and suspension deflection velocity $\dot{z}_{def}$, Equation (11.7) may lead to negative values for parameter c_v which cannot be realized by the semi-active device as this would entail a positive mechanical power provided by the semi-active component. The variable damping needs then to be saturated as follows:

$$c_v(t) = \begin{cases} c_{min} & \text{if } c_{sky} \dfrac{\dot{z}_c}{\dot{z}_{def}} < c_{min} \\[2mm] c_{max} & \text{if } c_{sky} \dfrac{\dot{z}_c}{\dot{z}_{def}} > c_{max} \\[2mm] c_{sky} \dfrac{\dot{z}_c}{\dot{z}_{def}} & \text{otherwise} \end{cases} \tag{11.8}$$

In some cases, semi-active dampers are designed to switch between two states providing a minimum and maximum damping, respectively, instead of providing continuous regulation of the damping coefficient in a min-max range. By just switching between the two extreme damping values, semi-active dampers can be designed to be faster (i.e., have a shorter switching time) or can be simplified leading to a lower cost of the hardware. The on/off skyhook control strategy for a two-states semi-active damper is then further simplified to:

$$c_v(t) = \begin{cases} c_{min} & \text{if } \dot{z}_c \dot{z}_{def} \leq 0 \\[2mm] c_{max} & \text{if } \dot{z}_c \dot{z}_{def} > 0 \end{cases} \tag{11.9}$$

11.3.3.3 Local versus Modal Control

In Sections 11.3.3.1 and 11.3.3.2, different control strategies for active/semi-active control of car body vibration have been introduced. The simplest way to implement any of these strategies is to use local control, i.e., the control strategy is implemented independently in the suspensions connecting the front and rear bogies (we assume here the common case of a vehicle with two bogies). In this case, the control force generated by the active/semi-active suspension in one bogie will be defined based on one single feedback signal coming from a sensor located in the same bogie. This implementation is simple and straightforward but the control forces generated at the two bogies will both affect multiple modes of vibration of the car body, each one having a different resonance frequency, damping, and modal shape so that the definition of controller gains will inevitably involve a trade-off.

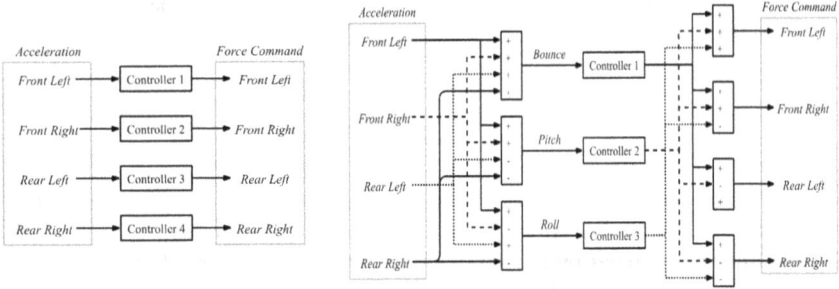

FIGURE 11.13 Comparison of local (left) and modal (right) control schemes. (From [2], reproduced with permission.)

An alternative approach is represented by the so-called Modal Control Approach [14]; in this case, the signals from sensors installed at the two bogies are combined to obtain new feedback signals, each one representative of a single car body mode of vibration. Pass-band filters can also be used to select signal components in specified frequency ranges characteristic of different modes of vibration, e.g., to separate the effect of bounce and first bending modes of the car body. Having available feedback signals that only bear information regarding one car body mode allows to design de-coupled controllers for each mode, so that the gains of each controller and, potentially, also the strategy of each single controller can be tailored to the specific features of the mode being controlled.

After appropriate control actions are defined for each mode, the control forces to be applied at each bogie are defined, these are transformed into the reference forces for actuators mounted in the front and rear bogies. Figure 11.13 clarifies the different structure of local and modal control schemes.

11.3.3.4 Control Strategies for Secondary Yaw Actuation

When used to improve the vehicle's running stability, SYC is operated to mimic the behavior of yaw dampers. In this case, the control force is defined as:

$$F_c = c_v v_{rel} + m_v a_x \tag{11.10}$$

with v_{rel} the speed of elongation of the actuator (measured by an internal sensor), a_x the longitudinal acceleration of the bogie frame due to yaw motion (obtained from one or two bogie-mounted accelerometers), and c_v and m_v are gain coefficients, the first one providing a component of control force proportional to the speed of elongation, like in an ideal viscous damper, and the second being tuned to compensate delays from sensors, control, and actuation [21].

SYC can also be used to improve vehicle curving. In this case, some simple strategies have been applied in pilot studies and demonstrations. One example is Liebherr's ADD: this is a double acting cylinder controlled by a three-state valve. As long as the length of the device measured by an internal sensor remains within pre-set limits, the valve is operated to set the two chambers of the cylinder in connection through a lamination valve and the actuator behaves as a passive damper. When the limits on the measured actuator length are exceeded, which means the bogie is negotiating a tight radius curve, the state of the valve is changed so that the high-pressure branch of the hydraulic circuit is connected to one chamber of the cylinder and the actuator produces a force with pre-defined constant intensity to steer the bogie through the curve. This application is intended to reduce wear in tight radius curves and was successfully tested at ÖBB on Siemens locomotives [26].

For the actuated yaw forced steering system proposed by Simson and Cole for locomotives, control strategies deal at the same time with vehicle stability and curving and may be either intended to minimize creep forces or to control yaw misalignment of the bogie with respect to a target value determined from the known or estimated track curvature [25].

11.3.3.5 Modern Control

The classical control strategies summarized above work well in a number of applications, but are not fully suited to deal with cases where the controller should be designed to improve the vehicle's performance with regard to more than one concern, e.g., improve running stability and curving performance at the same time or reduce car body vibration without exceeding a maximum allowed suspension deflection. Furthermore, classical controllers assume unlimited capability of the actuators to generate the required control force and therefore saturation effects related to force limits in the actuation system are not considered in the design of the control scheme and can only be handled by means of trial-and-error tuning. Finally, classical control strategies are, to some extent, not robust to uncertainty of parameters such as conicity, friction, mass of payload, etc.

Some of the modern model-based control strategies introduced in Chapter 4 of this book can be applied to overcome the issues related to the use of classical control. In particular, Linear Quadratic (LQ) optimal control allows the synthesis of a regulator as the solution of a minimization problem involving a cost function J weighting the performance of the active/semi-active suspension and the required actuation. A simple cost function example relevant to the design of an active suspension is:

$$J = \int_0^\infty \left(Q_a \ddot{z}^2 + Q_d z_{\text{def}}^2 + R F_a^2 \right) d\tau \tag{11.11}$$

with $\ddot{z}$ (car body acceleration) and z_{def} (suspension deflection) as the performance indexes, F_a the actuation force and Q_a, Q_d and R being three coefficients weighting the importance in the overall cost function of the two performance indexes and of the actuation force. Of course, more complex formulations of the cost function are possible in those cases where multiple actuators are used.

LQ control assumes full-state feedback, i.e., the actuation forces are defined as a function of the entire set of state variables describing the dynamics of the vehicle. Because a direct measure of the entire state is not feasible, in practical implementations LQG control is used, consisting of a combination of LQ and state estimation to reconstruct the vehicle's state from a restricted set of measurements.

Another state-based control technique widely applied to mechatronic suspensions is H_∞ control which results in a robust controller capable of handling unmodeled dynamics of the system and model uncertainties. This is especially useful in those cases when the mechatronic suspension is intended to affect vehicle's stability and curving behavior, because these aspects are strongly influenced by wheel-rail contact forces which are non-linearly depending on the vehicle's state while in many cases linear models are used for the synthesis of the regulator and also because the conicity and the coefficients describing the creepage/creep force relationship may be subject to large uncertainties. Examples of application of H_∞ control to railway vehicle suspensions can be found in [27] and [28].

It is finally worth mentioning that state feedback control strategies such as LQG and H_∞ are suitable also for semi-active suspensions. In this case, a simple implementation is to apply a saturation to the control force defined by the regulator so that the force eventually generated is as close as possible to the desired force but meets the limitations of the semi-active suspension. The classical implementation of this strategy takes the name of *Karnopp approximation* from the seminal work by Karnopp et al. [29]. There are however more specific strategies for the implementation of state-feedback control to semi-active suspensions, see [30] for a review.

11.4   CAR BODY TILTING SYSTEMS

Active car body tilting is the most mature and widely adopted technology for mechatronic suspensions in railway vehicles. It has been used in service over the past 25 years in many European countries and in Asia. The basic idea is to apply a controlled roll rotation (tilt) of the car body to provide additional compensation of centrifugal forces acting on the passengers, so that a train with a tilting body is able to run through a curve at higher speed than conventional trains. Car body tilt can also be provided using a passive suspension arrangement (the Talgo pendular concept), but the resulting tilt angle is lower than for some active tilting technologies. Furthermore, passive tilting systems exhibit low-frequency fluctuations of lateral acceleration in curve transitions that may cause motion sickness.

Active car body tilt can be realized according to different mechanical arrangements [31, 32]. Early tilting train designs use a tilting bolster and a mechanism with electro-hydraulic or electro-mechanical actuators. This arrangement provides tilt angles up to 6–8°, which means an additional compensation of lateral acceleration between 1 and 1.35 m/s². The drawback is that the tilting bolster and actuation mechanism are heavy and lead to a complicated design of the running gear. The tilting bolster can be fitted either above or below the secondary suspension (see Figure 11.14). In the first case, the secondary suspension needs to react the entire centrifugal force apart from the compensations provided by track cant and a hold-off system is therefore required, while in the *tilt below secondary suspension* arrangement the lateral suspension benefits from the reduction of centrifugal forces provided by tilt. However, the *tilt above secondary suspension* allows a simpler design of the tilting system, especially in terms of the traction link between car body and bogie.

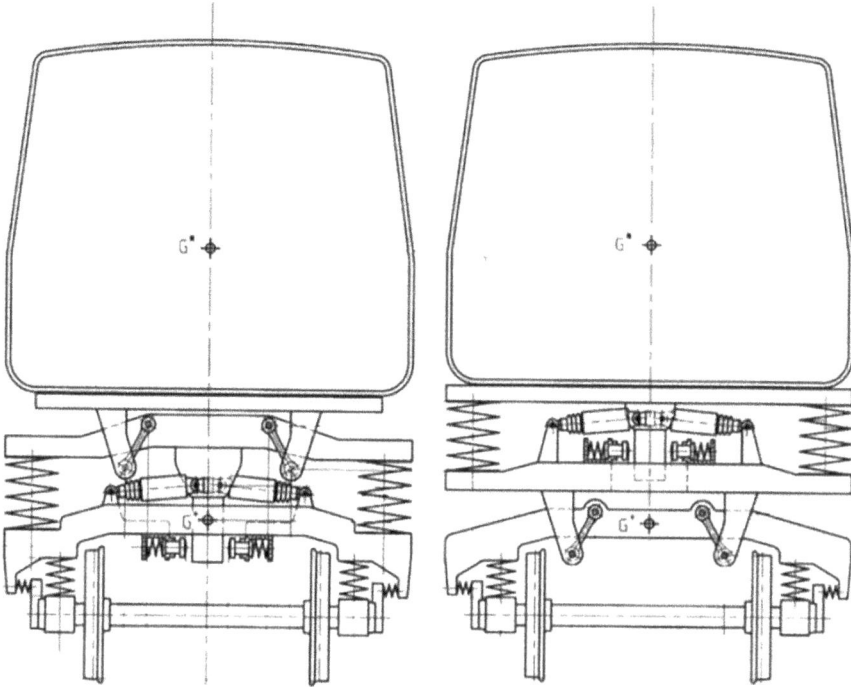

FIGURE 11.14 Tilt above secondary suspension (left) and tilt below secondary suspension (right). (From [32], reproduced with permission.)

Alternatively, active tilt can be applied by directly controlling some components in the secondary suspension, for instance through controlled inflation/deflation of the air springs. The advantage of this concept consists of its ease of implementation: a standard rail vehicle with pneumatic secondary suspension can be provided with car body tilting functionality through a modification of the pneumatic leveling system, with negligible increase of weight and without affecting the design of the bogie. The amount of tilt that can be realized is in the range of 2°, much smaller than that achievable using a tilting bolster and air consumption needs to be carefully considered. Yet, this solution is successfully used in service in Japan and is being considered in other countries to raise commercial speeds on existing high-speed lines. An alternative concept for direct actuation of tilt in the secondary suspension is based on the use of an active anti-roll bar system and is applied in Bombardier's regional Talent trains.

In terms of the control strategy, car body tilt is essentially a reference tracking problem. Ideally, car body tilt should increase gradually while the vehicle negotiates the curve entry, maintain a constant value corresponding to the desired amount of compensation in the full curve and decrease gradually while the vehicle negotiates the exit transition, thereby resulting in a trapezoidal waveform in time of the reference tilt angle. In systems using hydraulic or pneumatic actuation, a feed-forward + feed-back scheme is often adopted to consider the inherent integrative nature of these actuation systems [33]. In terms of the amount of compensation to be applied by the tilting system, there is general recognition that applying full compensation of lateral acceleration is likely to cause motion sickness issues. Therefore, the amount of compensation applied is usually between 40% and 80% of full compensation. Decisions about the speed increase and the optimal amount of compensation are not trivial and need to be taken considering that passenger comfort in curves is not only affected by the non-compensated lateral acceleration, but also by lateral jerk and car body roll speed. This means a higher speed increase for the same amount of car body tilt will lead to higher lateral acceleration, jerk and car body roll, while a higher compensation for the same vehicle speed will lead to lower lateral acceleration and jerk but higher roll velocity. An evaluation of comfort levels quantified in terms of the P_{CT} index defined by EN 12299 for different percentages of speed increase and for different compensation factors is provided in [14].

The main issue with defining the control input for the tilting system is to avoid delays in the detection of the curve and, more generally, to synchronize the tilt command with the position of the vehicle along the curve. Two approaches can be used, the first one based on vehicle mounted sensors and the second one on geo-localization of the train and use of a track database describing the deterministic features of the track (curvature, cant) as functions of the position.

In the first case, the non-compensated lateral acceleration is measured using a bogie-mounted accelerometer (reasons that make unsuitable the use of a car body mounted accelerometer are summarized in [31]). However, the measured acceleration must be low-pass filtered to attenuate high-frequency components related to track irregularities which would lead to disturbance in the actuation of car body tilt. The use of a suitable low-pass filter with cut-off frequency in the order of 1 Hz leads to a significant delay of the signal, resulting in delayed actuation of tilt. The solution to this problem is to use so-called *precedence control*, i.e., the bogie acceleration signal measured on the front vehicle is used to define the tilt reference for all vehicles in the trainset, considering the delay introduced by the low-pass filter and the precedence provided by the distance between the vehicles. Of course, the leading vehicle will not benefit from the precedence and therefore a lower level of comfort can be expected.

The alternative way to define the tilt reference makes use of geo-localization of the train in combination with a track data-base. The use of odometry (i.e., the estimate of the distance run by a wheel by measuring wheel turns) is not suitable for this application because it does not provide an absolute measure of the position and because a drift error accumulates due to deviations from pure rolling of the wheels and variations of wheel radius caused by conicity and by wear. Thus, odometry is combined with the evaluation of the absolute position of the vehicle along the track provided either by a GPS unit or by track balises. Compared to the use of vehicle mounted sensors, this second approach is obviously more complex in terms of implementation but gives more flexibility in the definition of the tilt reference so that improved tilt command for the reduction of motion sickness can be implemented [34].

11.5   ACTIVE SUSPENSIONS FOR NON-CONVENTIONAL VEHICLE ARCHITECTURES

Although most mechatronic suspension concepts have been proposed for vehicles with bogies, there are notable examples of active suspensions for two-axled vehicles. It is widely recognized that the use of active suspensions may allow for substantial simplifications of the vehicle's mechanical structure, potentially leading to large savings in terms of weight and energy consumption. In line with these aims, some ambitious research projects were initiated in recent years to define new architectures for two-axle rail vehicles characterized by a high degree of integration of mechatronic solutions. Worthy of being mentioned are the *Next Generation Train* project run by DLR in Germany [8] and a concept for a two-axle metro vehicle with single stage of suspension offering a substantial reduction of the axle load for the same number of passengers carried [35].

There are two main aims for using mechatronic suspensions in a two-axle passenger vehicle. On one hand, in a vehicle of this type the wheelbase is substantially longer than in one with bogies, which means the trade-off between stability and curving is more challenging for the two-axle vehicle. Therefore, active stability enhancement, active steering, or a combination of the two may be required to achieve a satisfactory overall running behavior of the vehicle. The control strategies that can be adopted in this regard are the same as introduced in Section 11.2.3.1 for a vehicle with bogies.

On the other hand, poor ride quality may be expected for a two-axle vehicle due to the single stage of suspension and can be improved through active control. In this regard, a case study presented in [36] about an existing two-axle vehicle with single stage suspension known for its unsatisfactory ride quality showed prospective benefits from the use of active suspensions implementing either skyhook or LQG control strategies.

REFERENCES

1. R. M. Goodall, Active railway suspensions: Implementation status and technological trends, Vehicle System Dynamics, 28(2–3), 87–117, 1997.
2. S. Bruni, R. Goodall, T. X. Mei, H. Tsunashima, Control and monitoring for railway vehicle dynamics, Vehicle System Dynamics, 45(7–8), 743–779, 2007.
3. B. Fu, R. L. Giossi, R. Persson, S. Stichel, S. Bruni, R. Goodall, Active suspension in railway vehicles: A literature survey, Railway Engineering Science, 28(1), 3–35, 2020.
4. A. H. Wickens. Fundamentals of Rail Vehicle Dynamics: Guidance and Stability, Swets & Zeitlinger, Lisse, the Netherlands, 2003.
5. T. X. Mei, R. M. Goodall, Wheelset control strategies for a two-axle railway vehicle, Vehicle System Dynamics, 33(Suppl), 653–664, 2000.
6. J. T. Pearson et al., Design and experimental implementation of an active stability system for a high-speed bogie, Vehicle System Dynamics, 41(Suppl), 43–52, 2004.
7. Y. Umehara, K. Ishiguri, Y. Yamanaga, S. Kamoshita, Development of electro-hydraulic actuator with fail-safe function for steering system, Quarterly Report of RTRI, 55(3), 131–137, 2014.

8. A. Heckmann, L. Daniel, G. Grether, A. Keck, From Scaled Experiments of Mechatronic Guidance to Multibody Simulations of DLR's Next Generation Train Set, In Dynamics of Vehicles on Roads and Tracks, Proceedings of the 25th IAVSD Symposium, 14–18 August 2017, Rockhampton, Queensland, Australia, 2017.

9. S. Shen, T. X. Mei, R. M. Goodall, J. Pearson, G. Himmelstein, A study of active steering strategies for railway bogie, Vehicle System Dynamics, 41(Suppl), 282–291, 2004.

10. G. Shen, R. Goodall, Active yaw relaxation for improved bogie performance, Vehicle System Dynamics, 28(4–5), 273–289, 1997.

11. T. X. Mei, R. M. Goodall, Stability control of railway bogies using absolute stiffness: Sky-hook spring approach, Vehicle System Dynamics, 44(Suppl 1), 83–92, 2006.

12. A. Stribersky, H. Müller, B. Rath, The development of an integrated suspension control technology for passenger trains, Journal of Rail and Rapid Transit, 212(1), 33–42, 1998.

13. A. Orvnäs, On Active Secondary Suspension in Rail Vehicles to Improve Ride Comfort [Ph.D. dissertation], KTH Royal Institute of Technology, Stockholm, 2011.

14. R. M. Goodall, T. X. Mei, Active Suspensions, in Handbook of Railway Vehicle Dynamics, Second Edition, CRC Press, Boca Raton, FL, 2020.

15. J. Park, Y. Shin, H. Hur, W. You, A practical approach to active lateral suspension for railway vehicles, Measurement and Control, 52(9–10), 1195–1209, 2019.

16. D. H. Wang, W.H. Liao, Semi-active suspension systems for railway vehicles using magnetorheological dampers. Part I: System integration and modelling, Vehicle System Dynamics, 47(11), 1305–1325, 2009.

17. A. Qazizadeh, R. Persson, S. Stichel, On-track tests of active vertical suspension on a passenger train, Vehicle System Dynamics, 53(6), 798–811, 2015.

18. A. Alonso, J. G. Giménez, J. Nieto, J. Vinolas, Air suspension characterisation and effectiveness of a variable area orifice, Vehicle System Dynamics, 48(Suppl 1), 271–286, 2010.

19. Y. Sugahara, A. Kazato, R. Koganei, M. Sampei, S. Nakaura, Suppression of vertical bending and rigid-body-mode vibration in railway vehicle carbody by primary and secondary suspension control: Results of simulations and running tests using Shinkansen vehicle, Journal of Rail and Rapid Transit, 223(6), 517–531, 2009.

20. G. Diana, S. Bruni, F. Cheli, F. Resta, Active control of the running behaviour of a railway vehicle: Stability and curving performances, Vehicle System Dynamics, 37(Suppl), 157–170, 2003.

21. F. Braghin, S. Bruni, F. Resta, Active yaw damper for the improvement of railway vehicle stability and curving performances: Simulations and experimental results, Vehicle System Dynamics, 44(11), 857–869, 2006.

22. X. Wang, E. Di Gialleonardo, B. Liu, S. Bruni, Application of semi-active yaw dampers for the improvement of the stability of high-speed rail vehicles, Vehicle System Dynamics, 2021, 1–28. DOI 10.1080/00423114.2021.1912366.

23. S. A. Simson, C. Cole, Idealized steering for hauling locomotives, Journal of Rail and Rapid Transit, 221(2), 227–236, 2007.

24. S. A. Simson, C. Cole, Simulation of traction curving for active yaw – Force steered bogies in locomotives, Journal of Rail and Rapid Transit, 223(1), 75–84, 2009.

25. S. A. Simson, C. Cole, Simulation of active steering control for curving under traction in hauling locomotives, Vehicle System Dynamics, 49(3), 481–500, 2011.

26. T. Michalek, J. Zelenka, Reduction of lateral forces between the railway vehicle and the track in small-radius curves by means of active elements, Applied and Computational Mechanics, 5, 187–196, 2011

27. T. Hirata, S. Koizumi, R. Takahashi, H_inf control of railroad vehicle active suspension, Automatica, 31(1), 13–24, 1995.

28. S.M. Mousavi Bideleh, T.X. Mei, V. Berbyuk, Robust control and actuator dynamics compensation for railway vehicles, Vehicle System Dynamics, 54(12), 1762–1784, 2016.

29. D. Karnopp, M. Crosby, R. Harwood, Vibration control using semi-active force generators, Journal of Engineering for Industry, 96(2), 619–626, 1974.

30. S. M. Savaresi, C. Poussot-Vassal, C. Spelta, O. Sename, L. Dugard, Semi-Active Suspension Control Design for Vehicles, Butterworth-Heinemann, Paises Bajos, 2010.

31. R. Persson, R.M. Goodall, K. Sasaki, Carbody tilting – Technologies and benefits, Vehicle System Dynamics, 47(8), 949–981, 2009.

32. B. Dalla Chiara, G. Hauser, A. Elia, Tilting trains: Evolution, performances et perspectives, Ingegneria Ferroviaria, 63(7–8), 609–648, 2008. (in Italian)

33. E. Di Gialleonardo, M. Santelia, B. Tian, S. Bruni, A. Zolotas, A simple active carbody roll scheme for hydraulically actuated railway vehicles using internal model control, ISA Transactions, 2021 Mar 08. DOI 10.1016/j.isatra.2021.03.003.

34. R. Persson, B. Kufver, M. Berg, On-track test of tilt control strategies for less motion sickness on tilting trains, Vehicle System Dynamics, 50(7), 1103–1120, 2012.

35. R.L. Giossi, R. Persson, S. Stichel, Gain Scaling for Active Wheelset Steering on Innovative Two-Axle Vehicle, In: M. Klomp et al. (Eds), Advanced Dynamics of Vehicles on Roads and Tracks, Proceedings of the 26th Symposium of the International Association of Vehicle System Dynamics, IAVSD 2019, Springer Lecture Notes in Mechanical Engineering, 2019.

36. A. Pacchioni, R.M. Goodall, S. Bruni, Active suspension for a two-axle railway vehicle, Vehicle System Dynamics, 48(Suppl 1), 105–120, 2010.

12 Real-Time Systems and Simulation

12.1 INTRODUCTION: AIMS OF REAL-TIME STUDIES

The main question that is frequently asked can be formulated as: Why do we need to use real-time systems if the mechatronic system of a railway vehicle can be modeled on a desktop, workstation, or high-performance computer? The answer is that the design of mechatronic systems of rail vehicles requires the performance of verification and validation in an environment that replicates the real rail vehicle operational environment or is as close to the latter as possible. Considering that any physical laboratory or field tests on real rail vehicles are very time consuming and costly processes, it appears reasonable to validate the developed mechatronic system through virtual prototyping and computer simulation in a real-time mode prior to starting any testing program. One useful validation instrument is the application of software-in-the-loop simulation (SILS) or hardware-in-the-loop simulation (HILS) approaches. These approaches provide great potential for rail vehicle development and testing in laboratories and also allow reaching critical or limit conditions with mitigation of the high risks (equipment damage, human life or serious injuries, etc.) involved. In this chapter, we will consider and describe the main components and approaches involved in real-time processes.

12.2 WHAT IS A REAL-TIME SYSTEM?

The term *real-time system* is widely used in the development of a mechatronic system. This term should not be confused with another term which is widely used in software engineering and referred to as a Real-Time Operating System (RTOS). The latter only represents a specific software environment that is designated to serve real-time applications that process data without buffer delays.

The concept of real-time simulation considers the behavior of any mechatronic system based on the specific structural architecture and requires the system to be presented in terms of specific formulations of the dynamics [1].

The general characteristics of a real-time system can be characterized by the following elements [2]:

1. System structure or architecture is responsible for establishing interactions with its environment, the so-called larger system, in a continuous and timely manner. On the one hand, it usually covers all hardware components of the computerized system (microprocessors, memory, peripherals) as well as software components represented by the RTOS and real-time software application. On the other hand, it is represented by analogue-digital or

DOI: 10.1201/9781003028994-12

digital-analogue converters with monitoring signals being obtained from sensors and control signals being sent to actuators that act on targeted elements of the physical system.

2. Real-time responses that are dependent on a RTOS and a real-time application which guarantees that tasks will be performed in a specified time period. The latter is dependent on responses of sensors, actuators, and the computational speed of the real-time application.

3. Highly resource-constrained environments that make the design, in most cases, pre-defined with a limited chance of a change in the system design and a performance optimization. This is commonly done based on strict requirements that formulate a system design that should survive under harsh environmental and operational conditions.

4. Concurrency of the real-time system that should guarantee that several tasks (for example data-processing, computational functions, etc.) are continuously interacting with each other and are not adversely affecting the final outcome and the response time.

5. Predictability of the real-time system which is designated to specify all constraints and to predict the system behavior for specific tasks that should be completed before the pre-defined time deadline in the case of outside situations and unforeseen circumstances that might influence the performance of the real-time system.

6. Reliability that should cover hardware and software components and developed algorithms involved in performing all defined functions in a specified time period.

7. Safety-critical requirements that should prevent, or at least reduce the chance of, system and software algorithm faults and errors. That requires real-time software to be tested in a meticulously intensive manner with the aim of ensuring not only that the software is error-free but also that it completely achieves all required functionalities.

When we talk about a real-time system in terms of rail vehicle mechatronics studies, we usually assume some kind of simplifications that require the abstraction of physical reality and, in most cases, the modeling only includes the following components:

- The physical system
- Electronic control algorithm(s) or unit(s)
- The software environment (RTOS and other specific real-time software applications)
- Communication systems and interfaces

An example of the structure is shown in Figure 12.1. In addition, all these components, when they are in use in a real-time system, should be designed in such a way that any action executed in the RTOS environment for such modeled components should provide a response as fast as it is possible to achieve based on the complexity of those components. In other words, for simulation studies it should be faster than real time or in the expected time period as shown in Figure 12.2.

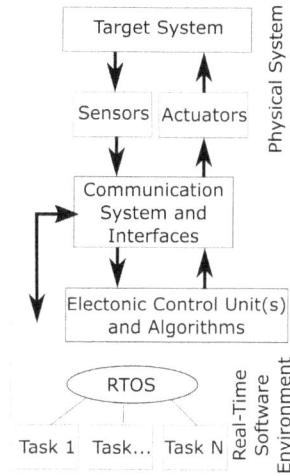

FIGURE 12.1 Structure of the system for real-time applications.

In real-time systems, a physical system can be represented as a simulated or hardware component. The simulated component requires the application of a specialized software or code developed to replicate the rail vehicle dynamics or the dynamics of some components of a rail vehicle. In such software products, the developed mathematical or multi-body model is executed on computer systems running on RTOS and should run faster than real-time. The hardware component under study is set up in a laboratory for physical testing with force or torque actuators (motors can also be considered as torque actuators) attached at each point where the component would connect into the physical system.

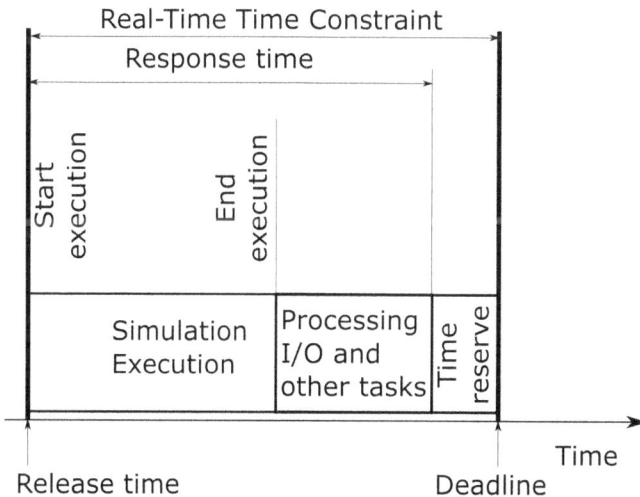

FIGURE 12.2 Real-time execution requirements for simulation studies.

Two common approaches can be considered for the advanced application in real-time systems for the introduction of a rail vehicle as a physical system [3]:

1. Real-time inverse dynamics [4, 5]
2. Real-time multi-body model approach [6, 7]

The application of three-dimensional inverse vehicle-track system models is a more complicated task in comparison with a multi-body model because the load identification problem in most practical cases is ill-posed and, as a result, not all the state variables or initial conditions are known. Therefore, for the common mechatronics developments in rail vehicles, it is quite reasonable to use a multi-body model that allows avoiding parametrical sensitivity.

12.3 REQUIREMENTS FOR THE DEVELOPMENT OF PROGRAMMING CODE FOR A REAL-TIME APPLICATION

The common approaches for programming code development in real-time applications are generally regulated with existing standards.

The first attempt at regulation for the development of real-time applications was made by the Motor Industry Software Reliability Association (MISRA) in 1990, when the project focused on the development of best practice guidelines for safe and secure programs in C and C++ languages running in a real-time mode on embedded control systems and a standalone software system was started. Since then, the MISRA has generated various standards and guidelines that are currently in use [8, 9]. The developed code should comply with the MISRA coding standards in order to assist with the following general development items in the applications:

- Improvement in coding practices by means of prohibiting certain unsafe and insecure language function and constructs
- Minimization and avoidance of unexpected application behavior
- Identification of problematic and infeasible code that might break potential security rules and requirements
- Improvement of portability of the developed code by means of the avoidance of specific compiler- or platform-related functions and constructs
- Reduction of program complexity
- Improvement of program testability
- Compliance with functional safety and security standards

The ability to quickly and easily analyze software with reference to MISRA coding standards delivers tangible benefits to development teams in terms of code quality, consistency, and reduced time-to-deployment. The MISRA project lead to the development of ISO 26262 [10] that is extensively in use in the automotive industry which considers the MISRA guidelines as an appropriate sub-set of the C language. The case of the application of ISO 26262 in the railway systems is discussed in [11].

In the case of development and testing of a real-time code for railway applications, there are two standards: EN 50128 [12] and IEC 62279 [13]. The content of both publications is identical and requires use of coding standards such as MISRA and CERT [14, 15].

The software safety standard EN 50128 "Railway applications. Communications, signaling and processing systems" was originally developed by the European Committee for Electrotechnical Standardization (CENELEC). This standard provides procedures and technical requirements for the development of programmable electronic systems for use in railway control and protection applications and it should be considered as relevant for the application of real-time code in mechatronics tasks for railway applications.

The CERT standards for C and C++ languages are created by the Software Engineering Institute (SEI) for Embedded Developers. The Software Engineering Institute is funded by the U.S. Department of Defense and the Department of Homeland Security and is located at Carnegie Mellon University which is responsible for publishing CERT Secure Coding Standards. These standards provide rules and recommendations that target the avoidance of exploitable vulnerabilities through safe and secure coding practices.

Finally, it can be a good practice to develop real-time codes with an application-specific verification workflow that is applicable to the relevant mechatronic system considering standardized software architecture and its area of application. Examples of the analysis of standardized software architectures for design methods based on the component models for embedded and real-time models used in the RailCab project can be found in [16, 17]. However, this area has not been well developed yet for railway tasks and requirements should be further formulated and developed based on the existing standards mentioned above.

12.4 REQUIREMENTS FOR THE DEVELOPMENT OF REAL-TIME MULTI-BODY MODELS

The requirements for development of real-time multi-body models are generally based on three basic components:

1. Implementation of physical models through numerical modeling and simulation approaches
2. The computing power of a real-time simulation platform
3. Real-time model validation and verification

The implementation requirements are based on two major requirements:

1. Model complexity versus computational speed
2. Real-time numerical integrators versus computational time step versus accuracy

It is well known that the computational speed is completely dependent on the number of simulation input variables, which are also defined as state variables in continuous-time models [1]. Therefore, the definition of their numbers should lead to the formulation of the system with differential-algebraic equations (DAEs) or ordinary differential equations (ODEs). In multi-body models, the equations of motion are generally represented by second-order DAEs that can be reduced to equivalent second-order ODEs under condition of the usage of a minimum set of generalized coordinates [18]. This further also leads to an understanding of the complexity of dynamic systems and to the technical design of models and the proper development of fundamental assumptions that relate to the rejection of unneeded or meaningless states and those with a reduced degree of dynamic model complexity.

Real-time numerical integration means that the equations of motion used in multi-body models should be numerically solved in a fixed time frame that requires performing at least one function evaluation of the right-hand side and one numerical time integration step within one cycle [18]. In other words, real-time numerical integrators are related to the understanding of time requirements by means of time discretization in order to find a proper integration scheme with fixed or variable step size in order to achieve time execution requirements as shown in Figure 12.2.

There are a great number of well-approved real-time application methods for model development and real-time time integration schemes, however there is no unique approach for the generation of real-time capable models automatically from the existing complex multi-body models, preserving the existing model topology and parameterization [18]. Therefore, there is a significant amount of manual work that must be performed in order to transform a standard multi-body model into a model that can satisfy the existing real-time constrains. These constraints can be fulfilled in a number of different ways [3]:

1. Purchasing hardware with ample execution capability for the current unoptimized system.
2. In-house development of the model code from scratch which is optimized for the model to be used. This approach is very important for a safety-critical (also called hard) real-time system that must execute a real-time model in a such a way that the model execution time meets its specified deadline.
3. Research simplifications to the multi-body model to be simulated in the existing multi-body software environment, if possible, or the usage of code generation techniques from the existing multi-body model. The application of such approaches is highly recommended for a soft real-time system that usually allows a real-time model to miss some of its deadlines in order to avoid improper result outputs.

The simplistic concept for the development of a real-time model and its integration into the physical system is shown in Figure 12.3.

Analyzing the real-time approaches what are in use, it is possible to see that the development of real-time multi-body mechanical models requires their verification when these models are integrated into a complex multidisciplinary model. It is necessary to have broad expertise to develop such models as well as appropriate

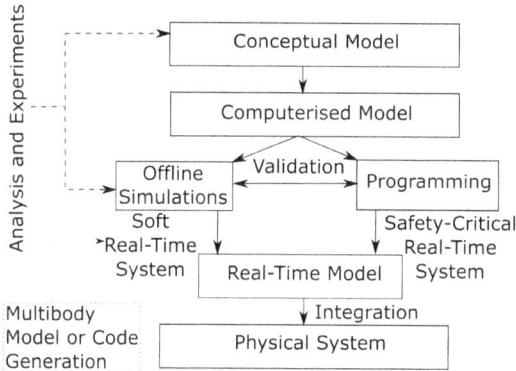

FIGURE 12.3 Simplistic concept of the development of a real-time model for a physical system.

software and hardware products. Some examples of such development have been published by Japanese researchers in [19–21]. However, those examples have been modeled in the Matlab®/Simulink environment, which makes any further changes in the model design too complicated in comparison with the usage of railway multibody software packages.

The mechatronic model of a single wheelset test rig running in a real-time model is described in [22, 23]. This approach shows good results for the development of real-time models in the SimMechanics extension of the Matlab/Simulink package and then transferring this model with Simulink Coder (formerly Real-Time Workshop) into the dSpace environment. However, that approach was focused on the development of its own specific wheel/rail contact model that also makes it complicated for further model extensions (e.g., an extension to a full vehicle model) and a contact model change.

A more advanced approach to produce a real-time bogie test rig model for a heavy haul locomotive was described in [24, 25] and a full freight wagon was described in [7]. The approach used in those references is based on the usage of Gensys multibody software package [26] that runs under the Realtime Linux kernel open-source software.

In general terms, the real-time capable models can be derived automatically from highly complex full vehicle simulation models by means of model reduction methods, which can be formulated as [27]:

- Linearization
- Partial linearization of subsystems
- Decentralized simulation of subsystems
- Reduction of operations by model specific motion equations (symbolic equations)
- Subsystem-based model reduction by neglecting the dynamics of nearly massless bodies in subsystems and concentrating on the quasi-static modes
- Global reduction to relevant dynamic effects

The practical implementation presented in [7] is slightly different to the reduction methods and it is more related to the recommendations for the modifications of the full wagon model in the real-time models which can be summarized as:

- Choice of numerical integrator should be made based on requirements for a relatively accurate result with less computational effort.
- Usage of a simplified contact and track model in the real-time model to eliminate numerous mass, function, and coupling elements of the full wagon model.
- Reduction of the number of elements by means of simplifying user-defined element subroutines. The stiffness and damping elements in all couplings can be replaced by fewer equivalent stiffness and damping elements. The friction elements that require comparatively high simulation times should be investigated and modified or removed in order to reduce calculation time.
- Reduction of storage/sending calculated results by network data. In most cases, the real-time modeling approach creates the output which is directly used in the controller to execute a control mechanism, thus providing the potential for reducing the time needed for storage of data in comparison with a full model. Therefore, in the real-time model, all data storage commands should be deactivated to obtain a reduced calculation time.

After the transformation of the full rail vehicle model into the real-time rail vehicle model, it is necessary to perform a model validation. This can be done in three ways:

- Comparison the full rail vehicle multi-body model results versus those of the real-time rail vehicle multi-body model. This approach is a good and cost-effective one if the full model was previously validated.
- Validation based on an experimental field or laboratory testing program performed on a rail vehicle. This approach is not cost-effective and requires the application of huge resources and is not time friendly.
- Validation based on the model acceptance procedures as described in [28–32]. These approaches are quite reasonable, but they should be used very carefully considering that they might not replicate an actual rail vehicle because they are based on the assumption that, if a rail vehicle model is passing all requirements defined in a standard, then it can be considered as a virtual equivalent of a real rail vehicle. That creates a high probability of the usage of an idealized rail vehicle.

In the case of the code generation techniques, the model is automatically translated to the programming code (e.g., C or C++) as shown in Figure 12.4, and it should then be compiled with a compiler designated for a specific real-time hardware platform (e.g., DSpace [33–37], etc.). However, this approach is more related to the development of

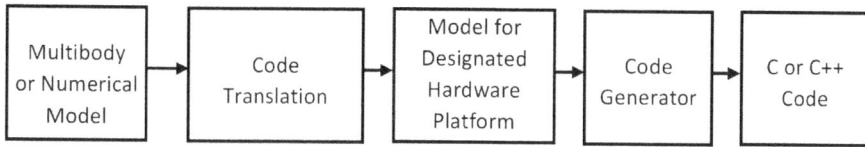

FIGURE 12.4 Example of the code generation concept for the development of a real-time model.

individual models and it is more time consuming in terms of the model validation and acceptance procedures.

In some cases, it is quite reasonable to use a parallel computing technique [38] in order to improve computational performance. One of the approaches that can be of benefit in this direction is parallel computing of wheel-rail contact when the numerical approaches for the calculation of creep forces are needed and look-up tables are not practical to be used (e.g., traction studies). Reference [39] used a Field Programmable Gate Arrays (FPGAs) to parallelize the calculation of creep forces based on the Fastsim algorithm developed by J.J. Kalker. The idea in [39] with real-time simulation is that the wheel-rail contact laws must be used in a hardware-in-the-loop approach for experimental studies of the latest vehicle dynamic and control technologies, however it can also be implemented in a distributed model of a rail vehicle where the communication between model components is performed by means of a high-throughput and low-latency Ethernet link. A quite different approach presented in [40] describes how to use OpenMP [41] to parallelize multiple contact points of all wheel-rail interfaces of a locomotive model in terms of one multi-body model developed in Gensys on one PC computer with the multiple cores. However, the studies for the real-time implementation of this approach are still in progress [42] and additional studies are recommended.

Summarizing the above, it is possible to see that the development of a real-time model is a complicated process and all design aspects should be studied and the risks of applying any simplifications should be estimated at the development stage.

12.5   REAL-TIME PROTOTYPING AND TESTING

Real-time prototyping is a multi-engineering discipline which requires modeling a physical component or a whole system, simulating and, if necessary, visualizing its dynamic behavior under real-world operating scenarios. In the case of real-time testing, there is some extension of simulation boundaries that allows a connection of actual hardware (sensors, actuators, test rigs, etc.) to the simulation system that runs a model or models at required simulation speeds, respecting precise real-time timing requirements. Real-time simulation and testing are commonly connected with two approaches: SILS and HILS.

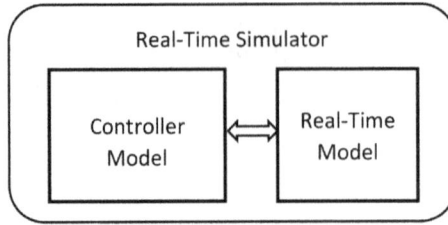

FIGURE 12.5   Example of the software-in-the-loop simulation concept.

12.5.1   SOFTWARE-IN-THE-LOOP APPROACH

The software-in-the-loop approach does not require a physical system for investigation, testing, and evaluation of dynamic behavior and responses of the system. All associated tasks are performed within a computer that is commonly called a real-time simulator. This approach is designated to build and test a physical system prototype through realistic simulation processes without building an actual hardware prototype. This allows engineers or researchers to analyze the full behavior of a complex mechatronic system with various design parameters that can be changed and refined in order to get a completely optimized system prior its final design and production stages. This approach also allows significantly reducing the costs and time of development of a complex mechatronic system. However, initial costs are still involved in the SILS process and they are associated with the purchase of a reliable real-time simulator with an appropriate software package. The commercial real-time simulators are flexible and a development process is logically structured that allows to fine-tune and monitor a system on the fly, but such systems sometimes lack flexibility, especially for research purposes when a modification of software in RTOS is required.

The classical example of SILS implementation in a real-time simulator is shown in Figure 12.5. Such a concept was used in [36, 37].

12.5.2   HARDWARE-IN-THE-LOOP APPROACH

The hardware-in-the-loop approach requires the use of an actual hardware element (or elements) of a physical system. The application of a real-time model can be implemented in two variations. The first variation is when a real-time model is running on a real-time simulator and it is in use instead of a physical plant. This approach is used for testing of actual hardware (e.g., an electronic control unit that can include a unit production or embedded controls implementation) with pre-defined operational scenarios. The example of this concept is shown in Figure 12.6(a) and its practical application can be found in [20, 21].

The second concept is when a real-time model is in use inside of a controller running on a real-time simulator which is connected to a real plant, i.e., a hardware element of a physical system. This approach is commonly in use in component testing and its concept is shown in Figure 12.6(b).

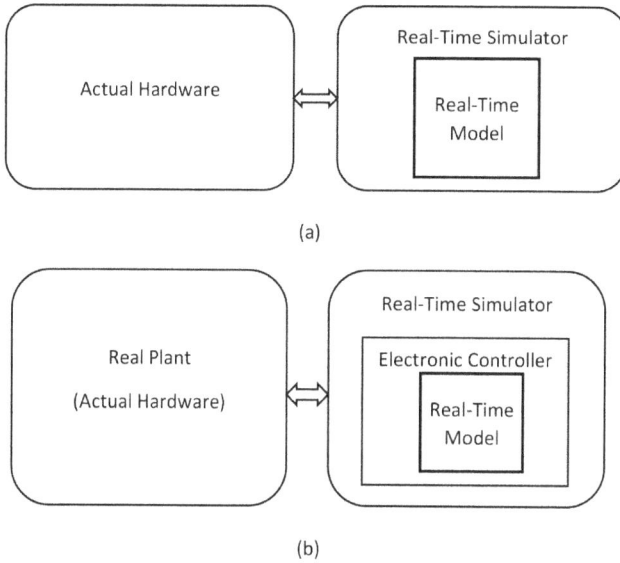

(a)

(b)

FIGURE 12.6 Examples of the HILS concept: (a) actual hardware testing approach, (b) component testing approach.

12.6 CASE STUDY: DEVELOPMENT OF A REAL-TIME MULTI-BODY MODEL

The following task commonly arises when it is necessary to investigate a rail vehicle's dynamic performance under various traction conditions and with different traction control strategies in use. Considering that field testing is a high cost and time-consuming procedure, it is better to perform such testing exercises on existing full-scale roller rigs [43]. However, prior to proceeding with the introduction of any new control strategy into a traction control unit, it seems quite reasonable to use the SILS approach (as shown in Figure 12.5) with a rail vehicle/roller rig real-time model for testing of all desired control strategies. In this case, the real-time model is one of the main resources required for the building of verified real-time computer systems which can reproduce the behavior of a rail vehicle and its component parts. In this case study, we need to develop a multi-body model in the Gensys multi-body software package [26] that can achieve a more advanced simulation level where the calculation time response for the dynamic behavior of the mechanical system (full-scale rail vehicle test rig) under the traction mode will be executed in a real-time mode.

A hypothetical powered rail vehicle is chosen for this case study. Figure 12.7 shows the standard gauge rail vehicle test rig model developed in Gensys. The model consists of a car body, two bogie frames, four wheelsets, and eight rollers. In order to reduce the calculation time and to obtain more stable behavior of the mechanical system, some constraints have been set on these bodies (see Table 12.1). The basic parameters for the test rig model shown in Figure 12.7 are given in Table 12.2. For the modeling of vehicle dynamics behavior, the contact model used in the Gensys

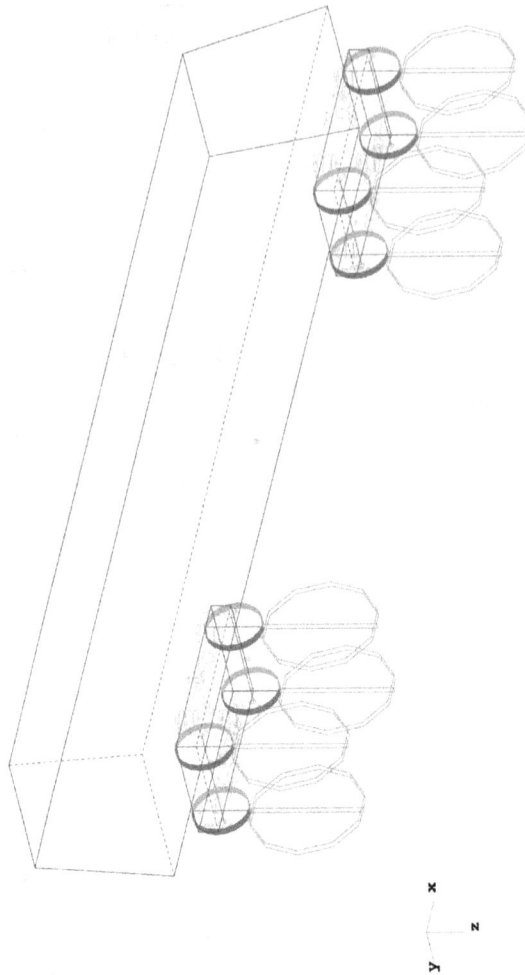

FIGURE 12.7   Test rig model in Gensys multi-body software.

TABLE 12.1
Constraints on Bodies

	x – Longitudinal	y – Lateral	z – Vertical	f – Roll	k – Pitch	p – Yaw
Car body	√, $x = 0$	√	√	√	√	√
Bogie frame	√	√	√	√	√	√
Axle	√	√	√	√	√, $k = 0$	√
Roller	E	e	e	e	√	e

Note: √, degree considered; e, degree eliminated; and $x = 0$ and $k = 0$ refer to longitudinal translation
displacement plus pitch rotation being fixed to be equal to zero.

TABLE 12.2
Parameters for the Bogie Test Rig Model

Parameter	Value
Wheel/roller spacing	1.4 m
Bogie spacing	6.5 m
Rollers	
Center of gravity, vertical	1 m
Mass	5000 kg
Moment of inertia, pitch	2500 kg.m^2
Car Body	
Center of gravity, vertical	2.0 m
Mass	52000 kg
Moment of inertia, roll	80001 kg.m^2
Moment of inertia, pitch	1285447 kg.m^2
Moment of inertia, yaw	1285447 kg.m^2
Bogie Frames	
Center of gravity, vertical	0.7 m
Mass	10000 kg
Moment of inertia, roll	3000 kg.m^2
Moment of inertia, pitch	10000 kg.m^2
Moment of inertia, yaw	15000 kg.m^2
Axles (the traction motor mass shared between bogie and axles)	
Center of gravity, vertical	0.5 m
Mass	2000 kg
Moment of inertia, roll	1789 kg.m^2
Moment of inertia, pitch	1200 kg.m^2
Moment of inertia, yaw	1789 kg.m^2
Secondary Suspension	
Longitudinal stiffness (per spring element)	900 kN/m
Lateral stiffness (per spring element)	600 kN/m
Vertical stiffness (per spring element)	600 kN/m
Vertical damper	40 kN.s/m
Longitudinal traction rod stiffness	25,000 kN/m
Longitudinal traction rod damper	100 kN.s/m
Primary Suspension	
Longitudinal stiffness (per axle box)	20,000 kN/m
Lateral stiffness (per axle box)	20,000 kN/m
Vertical stiffness (per axle box)	1200 kN/m
Longitudinal damper (per axle box)	20 kN.s/m
Lateral damper (per axle box)	20 kN.s/m
Vertical damper (per axle box)	5 kN.s/m
Vertical viscous damper:	
• Stiffness	5650 kN/m
• Damper	60 kN.s/m

TABLE 12.3
Model Parameters for Adjustment of the Polach Model

Parameter	Dry Friction Condition	Dry Friction Condition
μ_S	0.55	0.3
A	0.4	0.4
B (s/m)	0.6	0.2
k_A	1.0	0.3
k_S	0.4	0.1

multi-body software is based on the Polach theory [44]. The main parameters of this theory (μ_s is the maximum coefficient of friction; A is the ratio of limit friction coefficient μ_∞ at infinity slip velocity to maximum friction coefficient μ_s, k_A and k_S are model parameters for different friction conditions) are defined in Table 12.3. The implementation of wheel/roller contact in the model has been done with the standard contact coupling *creep_polach2*. When modeling the contact surface between wheel and roller with this coupling, the radius of the rollers was taken into account. New ENS1002 wheel and UIC60 rail profiles have been used. In the wheel/rail contact modeling, a three-point contact approach is used. The contact between wheel and rail roller is modeled with three spring and damper elements, which are defined as being normal to the three wheel/rail contact surfaces, respectively.

In order to introduce the traction control system into the Gensys model, a special subroutine with a simplified traction system based on the bogie traction control strategy of one inverter per bogie and the application of a PI controller has been implemented [45].

The model has been verified using the procedure published in [6] and no anomalies have been found in the model behavior. The model was checked with standard on-line integrators.

For achieving the aim of this case study, it is necessary to choose an appropriate real-time integrator. A good review for real-time integrators has been described and classified by Klee and Allen [46]. Based on this data, the comparison between potential integrators available in MATLAB with the standard integrators available in Gensys is presented in Table 12.4. Taking into account that it is necessary to reach the best trade-off between calculation speed and accuracy, the two step Runge-Kutta (heun) method numerical integrator has been chosen.

In order to provide simulation in the real-time mode, the generic kernel of the Linux 64-bit operating system has been replaced with the real-time kernel. A computer equipped with an Intel Xeon Gold 6152 processor @ 2.10 GHz with 64 GB of RAM has been used for the real-time simulation with the Gensys 19.5 multi-body software.

The calculation time has been estimated with the special time estimator realized in Gensys:

$$t_{tout} = t_{lsys} + t_{coupl} + t_{func} + t_{mass} + t_{cnstr} + t_{integ} + t_{ds} \tag{12.1}$$

TABLE 12.4

Comparison of Existing Numerical Integrators for Real-Time Models [6]

Real-Time Integrators in MATLAB [15]	Equivalent in Gensys
RK-1 (explicit Euler)	e1
RK-2 (improved Euler)	heun
RK-2 (modified Euler)	mp2
RK-3 (real-time compatible)	-
AB-1	e1
AB-2	-
AB-3	-
AB-4	-
AM-2	-

where t_{lsys} is the computational time spent on the position definition of local coordinate systems regarding the global coordinate system, t_{coupl} is the computational time spent on commands for coupling elements (coupling elements are elements of various types that connect masses to each other; e.g., a coupling-element can be a coil-spring, a rubber bushing, a hydraulic damper, a bumpstop, a friction element, etc.), t_{func} is the computational time required for calculation of defined functions in the model script, t_{mass} is the computational time spent on mass-commands (a mass-command creates an inertia in the model for the mass of the rail vehicle body, bogie, wheelset, etc.), t_{cnstr} is the computational time spent on constraint commands, t_{integ} is the computational time required for calculation inside of the numerical integrator and t_{ds} is the computational time required for output data storage.

The testing procedure for this case study requires performing a two-stage investigation for a rail vehicle running with a constant speed of 20 km/h. The following is the operational scenario for the simulation:

- Motor torque increases to maximum during the first second, after which it remains at the maximum reference torque value.
- Total simulation time is 40 s.
- Friction condition is dry during the first 20 s, after which it is wet for a time period of 10 s and then it is dry until the end of the simulation.
- Calculation time step is set to 1 ms for the solver in order to meet the real-time requirements for mechanical systems as defined in [18, 47].
- Slip threshold is set to 0.05.

At the first stage, it is necessary to estimate rail vehicle dynamic behavior results. The simplified PI adhesion control model results in differences in the axle load distribution and in slight differences in the distribution of motor traction torques between wheelsets as shown in Figure 12.8. This leads to variations in slip values and traction coefficients between wheels as shown in Figure 12.9. These variations are

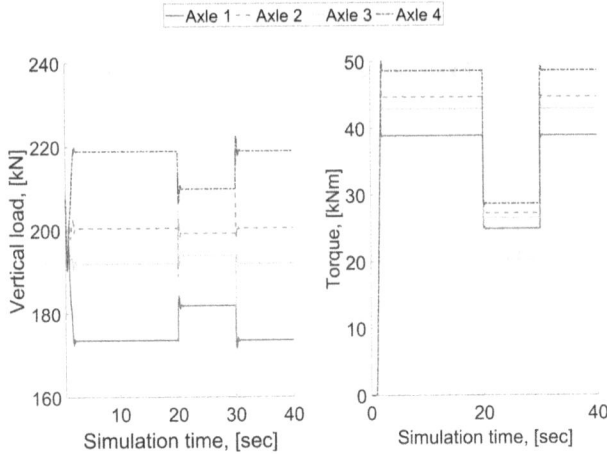

FIGURE 12.8 Comparison of vertical forces and torques acting on the axles in time domain.

caused by the absence in the simplified model of any induction machine dynamics that forces the equalization of wheelset slip. However, the dynamic behavior of the system is adequate, and it confirms the findings published in [48].

The acceptable dynamic behavior of the rail vehicle running on the rail rollers allows us to go to the second stage, where it is necessary to estimate time results as has been done in [6]. The results obtained in this case study, presented in Figure 12.10, show that, for the co-simulation with time steps of 1 ms, real-time implementation was achievable. It also indicates quite low computational loads indicating that the model could be easily used for a real-time simulation. Note that results for t_{ds} (see Equation 12.1) are subject to the number of stored or transmitted (through a communication interface) parameters at each computational time step.

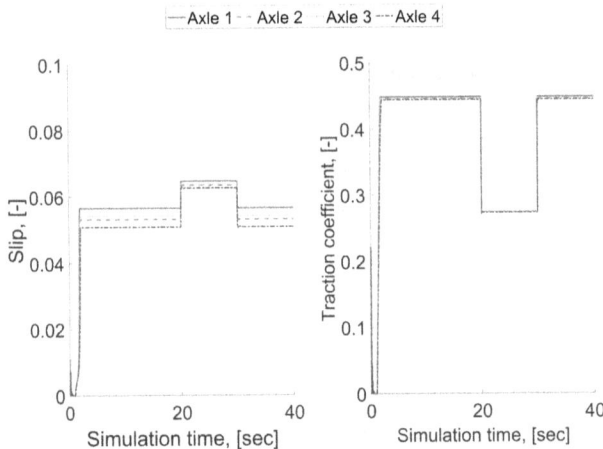

FIGURE 12.9 Comparison of slips and traction coefficients acting on the axles in time domain.

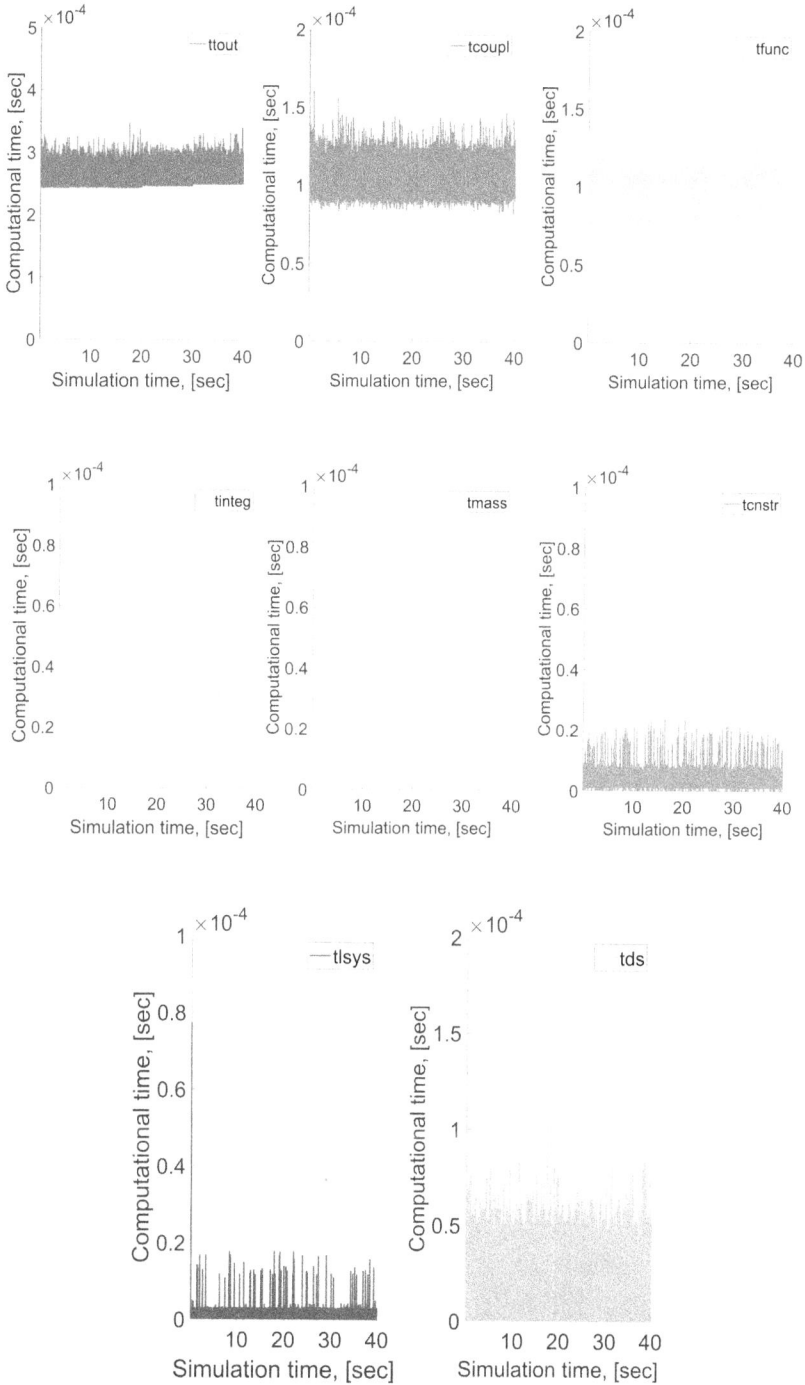

FIGURE 12.10 Time results for the real-time simulation in Gensys with the integration time step of 1 ms.

The results obtained in this case study show that the described approach can be applied for real-time multi-body system software written for Linux platforms. It makes it easier for users to find any *bottleneck* in their models and doesn't require special complex modular computer systems. The traction control strategies can also be easily studied with this approach.

REFERENCES

1. J. Scharpf, R. Hopler, J. Hillyard, Automotive Real-Time Simulation. Modelling and Applications, In: K. Popovichi, P. J. Mosterman (Eds), Real-Time Simulation Technologies. Principles, Methodologies, and Applications, CRC Press, Boca Raton, FL, 501–521, 2013.
2. J. Sun, Real-Time Embedded Systems, John Willey & Sons, Hoboken, NJ, 2017.
3. C. Bosomworth, M. Spiryagin, C. Cole, S. Alahakoon, M. Hayman, Challenges and Solutions for Integrating Simulation into a Transportation Device, In: A. Naweed, M. Wardaszko, E. Leigh, S. Meijer (Eds), Intersections in Simulation and Gaming, Springer, Cham, Switzerland, 317–330, 2018.
4. F. Xia, C. Cole, P. Wolfs, Grey box-based inverse wagon model to predict wheel-rail contact forces from measured wagon body responses, Vehicle System Dynamics, 46(Suppl), 469479, 2008.
5. Y. Sun, C. Cole, M. Spiryagin, Monitoring vertical wheel-rail contact forces based on freight wagon inverse modelling, Advances in Mechanical Engineering, 7, 1–11, 2013.
6. M. Spiryagin, Y. Q. Sun, C. Cole, T. McSweeney, S. Simson, I. Persson, Development of a real-time bogie test rig model based on railway specialised multibody software, Vehicle System Dynamics, 51, 236–250, 2013.
7. M. Spiryagin, S. S. N. Ahmad, C. Cole, Y. Q. Sun, T. McSweeney, Wagon Multibody Model and Its Real-Time Application, In: P. Flores, F. Viadero (Eds), New Trends in Mechanism and Machine Science: From Fundamentals to Industrial, Springer, Cham, Switzerland, 523–532, 2015.
8. MISRA – The Motor Industry Software Reliability Association. https://www.misra.org.uk/.
9. R. Bagnara, A. Bagnara, P. M. Hill, The MISRA C Coding Standard and Its Role in the Development and Analysis of Safety- and Security-Critical Embedded Software, In: Podelski A. (Ed), Static Analysis. SAS 2018. Lecture Notes in Computer Science, 11002, Springer, Cham, 2018.
10. ISO TC22/SC32/WG08. ISO 26262 2nd Ed. (FDIS) Road Vehicles – Functional Safety Parts 112. ISO, Geneva, Switzerland, 2018.
11. M. Rothhämel, Fail-Operational Vehicle Dynamics for Autonomous Operation—A Review Through Other Engineering Domains, In: M. Spiryagin, T. Gordon, C. Cole, T. McSweeney (Eds), Proceedings of 25th IAVSD Symposium, 1, Rockhampton, Australia, 14–18 August 2017, CRC Press/Balkema, 491–496, 2017.
12. CENELEC – EN 50128. Railway applications – Communication, signalling and processing systems – Software for railway control and protection systems, 2011.
13. IEC/TC 9. IEC 62279: Railway applications – Communications, signalling and processing systems –Software for railway control and protection systems, 2015.
14. SEI CERT C Coding Standard. Accessed on 14/04/2020. Available at: https://wiki.sei.cmu.edu/confluence/display/c/SEI+CERT+C+Coding+Standard.
15. SEI CERT C++ Coding Standard. Accessed on 14/04/2020. Available at: https://wiki.sei.cmu.edu/confluence/pages/viewpage.action?pageId=88046682.
16. S. Becker, S. Dziwok, C. Gerking, W. Schäfer, C. Heinzemann, S. Thiele, M. Meyer, C. Priesterjahn, U. Pohlmann, M. Tichy, The MechatronicUML Design Method – Process and Language for Platform-Independent Modeling. Report tr-ri-14-337, Version 0.4, Heinz Nixdorf Institute, University of Paderborn, March 2014.

17. C. Heinzemann, D. Schubert, S. Dziwok, U. Pohlmann, C. Priesterjahn, C. Brenner, W. Schäfer. RailCab Convoys: An Exemplar for Using Self-Adaptation in Cyber-Physical Systems. Report tr-ri-15-344, Software Engineering Group, Heinz Nixdorf Institute, University of Paderborn, January 2015.

18. M. Arnold, B. Burgermeister, C. Führer, G. Hippmann, G. Rill, Numerical methods in vehicle system dynamics: State of the art and current developments, Vehicle System Dynamics, 49(7), 1159–1207, 2011.

19. Y. Umehara, K. Sasaki, N. Watanabe, Y. Maki, K. Tezuka, A Study of Virtual Running Test of Railway Vehicle, Proceedings of the 8th World Congress on Railway Research – WCRR 2008, Seoul, Korea, May 2008.

20. Y. Maki, T. Shimomura, K. Sasaki, Building a railway vehicle model for hardware-in-the-loop simulation, Quarterly Report of RTRI, 50(4), 193–198, 2009.

21. N. Watanabe, Y. Maki, T. Shimomura, K. Sasaki, T. Tohtake, H. Morishita, Hardware-in-the-loop simulation system for duplication of actual running conditions of a multiple-car train consist, Quarterly Report of RTRI, 52(1), 1–6, 2011.

22. N. Bosso, A. Gugliotta, A. Somà, M. Spiryagin, Model of scaled test rig for real time applications, ABCM Symposium Series in Mechatronics, 5, 1288–1298, 2012.

23. N. Bosso, M. Spiryagin, A. Gugliotta, A. Soma, Mechatronic Modelling of Real-Time Wheel-Rail Contact, Springer, Berlin, 2013.

24. M. Spiryagin, C. Cole, Y. Q. Sun, M. McClanachan, V. Spiryagin, T. McSweeney, Design and Simulation of Rail Vehicles, Ground Vehicle Engineering Series, CRC Press, Boca Raton, FL, 2014.

25. M. Spiryagin, C. Cole, Y. Q. Sun, T. McSweeney, Mechatronic Real-Time Multibody Model of Bogie Test Rig, In the Proceedings of 23rd International Symposium on Dynamics of Vehicles on Roads and Tracks (IAVSD 2013), Qingdao, China, 2013.

26. The Gensys Homepage, AB DEsolver, Östersund, SWEDEN. http://www.gensys.se.

27. A. Eichberger, W. Rulka, Process save reduction by macro joint approach: The key to real time and efficient vehicle simulation, Vehicle System Dynamics, 41(5), 401–413, 2004.

28. M. Spiryagin, A. George, S. Ahmad, K. Rathakrishnan, Y. Sun, C. Cole, Wagon Model Acceptance Procedure Using Australian Standards, In: M. Dhanasekar, T. Constable, D. Schonfeld (Eds), The Proceedings of CORE2012: Global Perspectives; Conference on Railway Engineering, 10–12 September 2012, Brisbane, Australia. RTSA, Barton, ACT, 343–350, 2012.

29. M. Spiryagin, A. George, Y. Sun, C. Cole, T. McSweeney, S. Simson, Investigation of locomotive multibody modelling issues and results assessment based on the locomotive model acceptance procedure, Proceedings of the Institution of Mechanical Engineers Part F: Journal of Rail and Rapid Transit, 227(5), 453–468, 2013.

30. G. Götz, O. Polach, Verification and validation of simulations in a rail vehicle certification context, International Journal of Rail Transportation, 6(2), 83–100, 2018.

31. G. Götz, O. Polach, Influence of varying the input parameters on the results of model validation, Proceedings of the Institution of Mechanical Engineers, Part F: Journal of Rail and Rapid Transit, 231(5), 598609, 2017.

32. O. Polach, A. Böttcher, D. Vannucci, J. Sima, H. Schelle, H. Chollet, G. Götz, M. Garcia Prada, D. Nicklisch, L. Mazzola, M. Berg, M. Osman, Validation of simulation models in context of railway vehicle acceptance, Proceedings of the Institution of Mechanical Engineers, Part F: Journal of Rail and Rapid Transit. Rail and Rapid Transit, 229(6), 729–754, 2015.

33. L. Pugi, M. Malvezzi, A. Tarasconi, A. Palazzolo, G. Cocci, M. Violani, HIL simulation of WSP systems on MI-6 test rig, Vehicle System Dynamics, 44(Suppl 1), 843–852, 2006.

34. E. Meli, M. Malvezzi, S. Papini, L. Pugi, M. Rinchi, A. Rindi, A railway vehicle multibody model for real-time applications, Vehicle System Dynamics, 46(12), 1083–1105, 2008.

35. C. G. Kang, H. Y. Kim, M. S. Kim, B. C. Goo, Real-Time Simulations of a Railroad Brake System Using a dSPACE Board, Proceedings of the ICROS-SICE International Joint Conference 2009, Fukuoka, Japan, 4073–4078, 2009.

36. N. Bosso, A. Gugliotta, A. Somà, M. Spiryagin, Model of Scaled Test Rig for Real Time Applications, Proceedings of the 21st Brazilian Congress of Mechanical Engineering COBEM2011, Natal, RN, Brazil, 2011.

37. N. Bosso, M. Spiryagin, A. Gugliotta, A. Somà, Mechatronic Modeling of Real-Time Wheel-Rail Contact, Springer, Berlin, Germany, 2013.

38. D. Negrut, R. Serban, H. Mazhar, T. Heyn, Parallel computing in multibody system dynamics: Why, when and how, ASME Journal of Computational and Nonlinear Dynamics, 9(4), 041007, 2014.

39. T. Mei, Y. Zhou, Systems-on-chip approach for real-time simulation of wheel-rail contact laws, Vehicle System Dynamics, 51(4), 542–553, 2013.

40. Q. Wu, M. Spiryagin, C. Cole, I. Persson, C. Bosomworth, Parallel computing of wheel-rail contact, Journal of Rail and Rapid Transit, 2019. https://doi.org/10.1177/0954409719880737.

41. B. Barney. OpenMP. [Internet]. Lawrence Livermore National Laboratory, Livermore, CA, 2018. Accessed on 12/09/2019. Available at https://computing.llnl.gov/tutorials/openMP/.

42. S. Shrestha, M. Spiryagin, Q. Wu, Real-time multibody modeling and simulation of a scaled test rig for wheel-rail adhesion monitoring. JMTR-D-20-00014 (accepted).

43. P. D. Allen et al., Roller Rigs, Handbook of Railway Vehicle Dynamics, Chapter 19. Second Edition, In: S. Iwnicki, M. Spiryagin, C. Cole, T. McSweeney (Eds), CRC Press, Taylor & Francis Group, Boca Raton, FL, 761–823, 2019.

44. O. Polach, Creep forces in simulations of traction vehicles running on adhesion limit, Wear, 258, 992–1000, 2005.

45. M. Spiryagin, P. Wolfs, C. Cole, V. Spiryagin, Y. Q. Sun, T. McSweeney, Design and Simulation of Heavy Haul Locomotives, Ground Vehicle Engineering Series, CRC Press, Boca Raton, FL, 2017.

46. H. Klee, R. Allen, Simulation of Dynamic Systems with MATLAB and Simulink, Second Edition, CRC Press, 2011.

47. J. Bélanger, P. Venne, J. N. Paquin, The What, Where and Why of Real-Time Simulation. Available online (accessed on 20 May 2020): https://blobtestweb.opal-rt.com/medias/L00161_0436.pdf.

48. M. Spiryagin, P. Wolfs, F. Szanto, C. Cole, Simplified and advanced modelling of traction control systems of heavy-haul locomotives, Vehicle System Dynamics, 53(5), 672–691, 2015.

13 System Integration

When discussing system integration, it assumes that the systems should be considered as not being stand-alone systems. In the real-world environment, the mechatronic system should be studied as a part of the broader complex system. Considering this, it is reasonable to say that the various mechatronic systems that are required to operate seamlessly are forming a system of systems. Taking into account that the integration of the system of systems with a new or re-designed system may lead to unpredictable or unintended behavior of any of the systems included in the system of the systems, it is necessary to follow specific principles of integration and existing standards in order to be sure that the whole system is completely functional as per relevant technical specifications and requirements and that it can remain safe under different rail vehicle operational conditions.

The development of a rail vehicle mechatronic product involves multiple disciplines and it results in multiple stakeholders being involved in the design of the appropriate mechatronic systems. In terms of engineering capabilities, different types of engineers might be involved in the process related to the development of a new design or the modification of an existing design. Those engineers can have connections with asset management, automation, data processing, electrical, electronic, instrumentation, mechanical, maintenance, process, project management, software, and systems engineering. To handle such a variety of knowledge inputs, it is necessary to go from a simplistic mechatronic design to a larger scale mechatronic system considering integration approaches that should not be restricted only to the field of expertise covered by mechatronic engineers. In summary and considering the definition published in [1], it is possible to say that the mechatronic system integration should not only work with a design of one system, but it should deal with various integration aspects, starting with viewpoints from people at all levels, and finishing with tightly coupled models and affiliated tasks for achieving a proper and high-level of reliability, workability, and functionality as well as safe performance and outcomes.

In this chapter, the simplistic introduction of the integration concept is provided in order to illustrate existing approaches as well as to discuss challenges related to the specific views on rail vehicle mechatronic system design and the applicable standards for such tasks.

13.1 INTERPRETATION OF SYSTEM INTEGRATION

There are several interpretations of the definition for system integration. In [2], system integration is defined as a technique that brings together all component systems (elements) into one whole system, ensuring that the all systems (elements) function together properly, and this whole system can be functionally integrated with another existing system of systems. The aim of this broad definition is to cover four major

DOI: 10.1201/9781003028994-13

aspects that consider not only rail vehicles but also industry aspects for a whole railway infrastructure system. Those major aspects associated with integrating complex systems include [3, 4]:

- Project management
- Design and implementation of interfaces between systems
- System integration testing activities
- Risk assessment
- Delivery into an operational system

Within the objectives of this book and the focus on rail vehicle mechatronics, the best way to define the term *systems integration* is that it comprises the technical activities required to simultaneously design and develop the rail vehicle mechatronic system considering the assembly of all implemented subsystems and validation of the system behavior and performance against its design requirements. The proposed definition then allows formulating tasks, previously not covered in the earlier chapters, for newly developed or existing systems to be integrated into one system as:

- Design and implementation of interfaces between systems
- System integration testing activities (verification and validation)

The term *interface* is commonly in use in a system's engineering integration process for the mechatronic system. In the publication [5], it is defined as "the logical or physical relationship integrating the system components of one mechatronic system or the components with their environment" [5, 6]. The interfaces in the design and implementation tasks are generally defined as being of three types:

- Functional interface
- Physical interface
- Information system interface

A functional interface is commonly required at the initial design stage or at the system modification stage when the functions of one subsystem system interact with the functions of another subsystem. For example, if it is necessary to achieve better traction effort from a locomotive by means of the integration of an adhesion traction control subsystem and a variable axle load subsystem [5], the parameters when both systems should be operated at the same time should be defined by design engineers in order to allow systems to work in parallel without any malfunctioning that might cause damage of the railway track.

A physical interface represents existing communication equipment or hardware devices that are designated to link two or more subsystems and establish a communication interface. It can sometimes be defined as a communication network. In some cases, it can be integrated in the control system concept. An example of such a system is the modular control system Siemens Bahn Automatisierungs System (SIBAS) designed to control vehicle control units, traction control units, braking control units,

Design specification		Design solution		Design realization		Evaluation processes		Product delivery
- Stakeholders' viewpoints; - System requirements.		- Alternative design solutions; - Final design solution.		- Product implementation; - Product integration[1].		- Product verification; - Product validation.		- Delivery into operational system.

[1] Development of system interactions and system interfaces are considered at this stage.

FIGURE 13.1 Example of implementation of the system integration approach in the system design process.

auxiliary power supplies, etc. All these units should be considered as subsystems and SIBAS allows those subsystems to communicate through special communication buses that are part of the system's hardware.

For communication purposes, it is necessary to use the information system interface which uses specialized software for the exchange of information. This software should allow avoiding any interferences between subsystems and to be properly protected from cybersecurity threats. The software used for an information system and its subsystems used in rail vehicles is considered a safety critical system software because any malfunctioning or failure can cause serious safety problems and situations.

To avoid any design, safety and reliability issues, the system integration testing activities should be performed based on the existing testing methods and safety analysis techniques, commonly defined in the relevant standards, in order to prove that the whole mechatronic system design is safe, workable, and reliable. In the publication [7], the system integration process is fully integrated in the design realization process and system testing activities are specified as part of evaluation processes where product verification includes functional, environmental, operational testing in an integration test environment, and product validation assumes operational testing in a field operational environment. The whole system integration approach can be presented as shown in Figure 13.1.

13.2 INTER-DISCIPLINARY APPROACH FOR DESIGN AND EVALUATION PROCESSES

The generic product development process used during the design phase of rail vehicles and rail vehicle systems [8], as shown in Figure 13.2, generally does not specify any information on the interdisciplinary approaches used in this field. Similar approaches can be found in the publication [9] with the application of the W-model for the development of mechatronic models. The W-model approach, developed based on design phases stated in [8], is shown in Figure 13.3. The model represents the five process steps that should be performed sequentially in the order from 1 to 5. Again, it also does not cover any interdisciplinary technique for design and evaluation processes.

However, in the generic mechatronic design areas, the design process for interdisciplinary-based systems is commonly constructed with the usage of so-called V-models. These models are related to design stages and associated with product

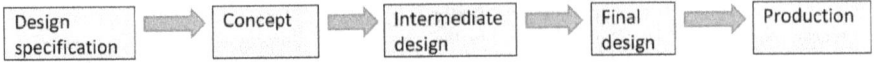

(a)

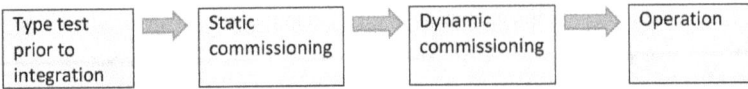

(b)

FIGURE 13.2 Generic product development process of a rail vehicle (a) design stage (b) verification and validation stage.

states and they are corresponded as macro-level models while the problem-solving process by the individual designer is defined as a micro-level model. These both types of models are well established in in the domain of software engineering [10].

In the railway industry, there are some more advanced V-models [11, 12] in use that cover all aspects of railway operations that include not only rollingstock but also other major assets in a whole railway network system lifecycle. Such models are outside the scope of this book because they are a subject for systems engineering studies.

In terms of mechatronic systems, it is still better to consider a V-model. The V-model should include the following design stages [13, 14]: requirements (specification of physical and logical characteristics of the mechatronic system), system design (cross-domain solution concepts with the division into sub-functions or modules), domain specific design (conjoint development of a solution by the involved interdisciplinary domains), modeling and model analysis (mathematical modeling, simulation and computer-aided design activities to justify the system characteristics

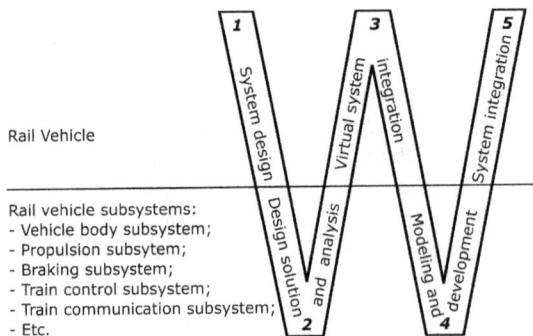

FIGURE 13.3 W-model design development process.

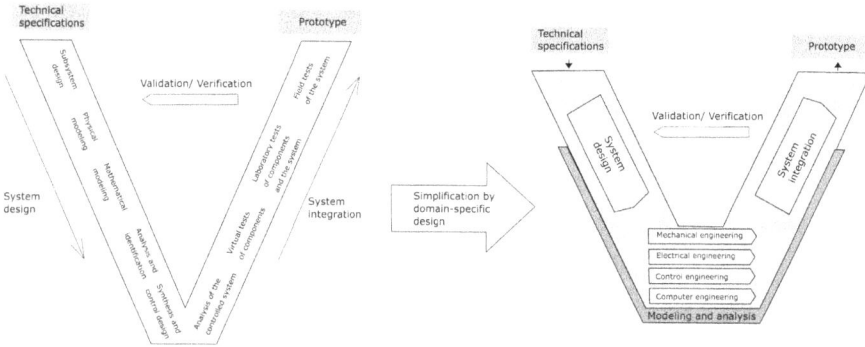

FIGURE 13.4 Examples of V-models for the design of a mechatronic system of a rail vehicle.

and design), system integration (all specific domains integrated in an overall system), and assurance of properties (verification and validation).

The V-model example of the rail vehicle design process can be built and delivered following details in the publication [15]. This model, shown in Figure 13.4, is a mechatronic design cycle used in the development of a mechatronic modular rail vehicle. The development of a concept of a rail vehicle is based on the decoupling of the rail vehicle into three modules such as the drive/brake module, suspension/tilt module, and support/guidance module. The decoupling is made for the improvement in ride comfort and cornering safety. If we consider the decoupling process, it is possible to say that the study was performed with an interdisciplinary design process for the design of a mechatronic system.

13.3 SYSTEMS INTEGRATION ACTIVITIES

The systems engineering model requires performing the verification of all technical specification requirements. Some good reviews on how it is applicable in V-models are published in [10, 13]. However, the current practices in a rail vehicle design process might be different. The common systems integration activities are generally defined by development and production phases and each of these phases consist of its own system integration activities. The latter confirms that the product cannot be finalized in one cycle of the development process. The cycles can be defined as:

- Modeling cycle
- Laboratory cycle
- First prototyping cycle
- Pre-product cycle
- Product cycle

The example of the V-model approach based on several cycles with many stages of system integration activities is shown in Figure 13.5.

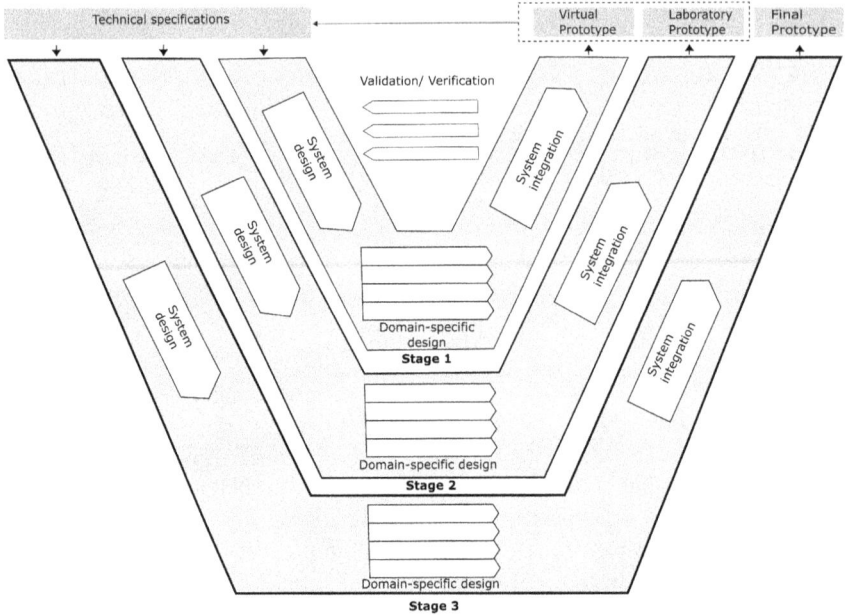

FIGURE 13.5 V-model with several system integration stages for the design of a mechatronic system.

13.4 RAIL VEHICLE SPECIFIC STANDARDS AND GUIDELINES

Standards for the design and testing of each component can significantly differ based on the chosen standards and technical requirements and specifications provided by the railway operators at the purchasing stages. However, the final roadworthiness acceptance of the rail vehicle is done not by the acceptance of each component but is done based on the acceptance of the whole system.

The roadworthiness acceptance is generally considered by the region where the designed or modified vehicle is to be put into rail network operation. For example, the Rail Industry Safety and Standards Board (RISSB) in Australia are producing a national set of standards and guidelines for the design, maintenance, and operation of rail vehicles. However, those standards can be overwritten by local railway operators or government agencies (e.g., by the Asset Standards Authority in Transport for NSW).

In the United States, there are two official organizations, namely the Federal Railroad Authority (FRA) and the Association of American Railroads (AAR) that are involved in this process. The FRA is in charge of the development of regulations, while the AAR provides guidelines to rail industry operators and manufacturers.

Similar approaches are in use in other countries such as China, India, and Russia where those countries have their own localized requirements applicable for their railway national infrastructures.

The approach is quite different in the European Union where there are a great number of common European standards which are issued in addition to the standards and regulations of individual countries. For example, the European Union Agency for Railways (EUAR) issue the Technical Specifications for Interoperability of rail vehicles that allows them to be operated in different countries of the European Union.

It is also necessary to mention that a lot of countries adopt the standards of the International Union of Railways (UIC) and the International Organization for Standardization (ISO) in their rail vehicle design practice.

Considering that the development of mechatronics systems is associated with a great number of simulation studies, it is necessary to consider the standards-based validation and verification of the rail vehicle acceptance through the simulation experimental program as described in [16–19].

However, there are some other concerns about the application of the standards as summarized in [20]:

> Complying with standards can be a substantial cost for suppliers of rolling stock and equipment, and for their customers. Providing evidence of compliance, as required by the Systems Engineering model, can be very time consuming. …. Guide books and reference volumes may well be useful documents, but there is no compulsion to follow their advice. A standard, however, states how things must be done. And generally, the supplier must prove that product is in accordance with the standard.

REFERENCES

1. M. Törngren, A. Qamar, M. Biehl, F. Loiret, J. El-Khoury, Integrating viewpoints in the development of mechatronic products, Mechatronics, 24, 745–762, 2014.
2. Transport for NSW. ASA TMU AM06014 GU – Guide to Systems Integration. Version 1.0. Issued date: 14 March 2018, Sydney, NSW, Australia. Available at https://www.transport.nsw.gov.au/system/files/media/asa_standards/2018/t-mu-am-06014-gu.pdf.
3. EN 50126:2017 – Railway Applications – The Specification and Demonstration of Reliability, Availability, Maintainability and Safety (RAMS).
4. ISO/IEC 15288: 2015 Systems and software engineering – System life cycle processes.
5. C. Zheng, M. Bricogne, J. L. Duigou, P. Hehenberger, B. Eynard, Knowledge-based engineering for multidisciplinary systems: Integrated design based on interface model, Concurrent Engineering: Research and Applications, 26(2), 157–170, 2018.
6. C. Zheng, P. Hehenberger, J. L. Duigou, M. Bricogne, B. Eynard, Multidisciplinary design methodology for mechatronic systems based on interface model, Research in Engineering Design, 28(3), 333356, 2017.
7. National Aeronautics and Space Administration. Chapter 5. Product realization. In NASA Systems Engineering Handbook. NASA/SP-2016-6105 Rev 2, 2017. Available at https://www.nasa.gov/seh/5-realization.
8. Verein Deutscher Ingenieure, VDB-Guideline Quality Engineering During Design Phase of Rail Vehicles and Rail Vehicle Systems, Verband Der Bahnindustrie in Deutschland (VDB) e. V., Berlin-Mitte, Germany, September 2015.
9. G. Barbieri, C. Fantuzzi, R. Borsari, A model-based design methodology for the development of mechatronic systems, Mechatronics, 24(7), 833–843, 2014.
10. I. Graessler, J. Hentze, T. Bruckmann, V-Models for Interdisciplinary Systems Engineering, In: D. Marjanović, M. Štorga, S. Škec, N. Bojčetić, N. Pavković (Eds), Proceedings of the DESIGN 2018 – 15th International Design Conference, University of Zagreb, Dubrovnik, Croatia, 747–756, 21–24 May 2018.

11. U.S. Department of Transportation, Systems Engineering Guidebook for Intelligent Transportation Systems: Version 3.0, U.S. Department of Transportation, Washington D.C., 2009. Available at https://www.fhwa.dot.gov/cadiv/segb/files/segbversion3.pdf.

12. NSW Government. Transport for NSW. Guide – Systems Engineering- T MU AM 06006 GU. Version 2.0, 9 March 2018. Available at https://www.transport.nsw.gov.au/industry/asset-standards-authority/find-a-standard/systems-engineering-guide-2.

13. J. Gausemeier, S. Moehringer, New Guideline VDI 2206 – A Flexible Procedure Model for the Design of Mechatronic Systems, In: A. Folkeson, K. Gralen, M. Norell, U. Sellgren (Eds), Stockholm, 19–21 August Proceedings of 14th International Conference on Engineering Design (ICED 03), The Design Society, UK, 2003.

14. Verein Deutscher Ingenieure, VDI 2206 – Entwicklungsmethodik Für Mechatronische Systeme (design Methodology for Mechatronic Systems), Beuth Verlag, Berlin, 2004.

15. X. Liu-Henke, J. Lückel, K.-P. Jäker, Development of an active suspension/tilt system for a mechatronic railway carriage. IFAC conference on mechatronic systems, Darmstadt, Germany, 18-20 September 2000, IFAC Proceedings Volumes, 33(26), 283–288, 2000.

16. M. Spiryagin, A. George, Y. Sun, C. Cole, T. McSweeney, S. Simson, Investigation of locomotive multibody modelling issues and results assessment based on the locomotive model acceptance procedure, Proceedings of the Institution of Mechanical Engineers Part F: Journal of Rail and Rapid Transit, 227(5), 453–468, 2013.

17. O. Polach, A. Bottcher, D. Vannuci, J. Sima et al., Validation of simulation models in the context of railway vehicle acceptance, Proceedings of the Institution of Mechanical Engineers Part F: Journal of Rail and Rapid Transit, 229(6), 729–754, 2014.

18. G. Götz, O. Polach, Verification and validation of simulations in a rail vehicle certification context, International Journal of Rail Transportation, 6(2), 83–100, 2018.

19. M. Juris, Shift2Rail – PLASA2. Deliverable D 4.1. Virtual Certification: State of the art, gap analysis and barriers identification, benefits for the Rail Industry. Report H2020-S2RJU-CFM-2018. 04 December 2019. Available at: https://projects.shift2rail.org/download.aspx?id=ec959443-fc1b-4c66-9a08-cba84326cc1e.

20. F. Szanto, Do We Need More Standards?, Proceedings of AusRAIL PLUS 2017, Rail's Digital Revolution, Brisbane, Australia, 1–3, 21–23 November 2017.

14 Practical Examples and Case Studies

14.1 CASE A: SIMPLIFIED MODELS OF RAILWAY VEHICLE LATERAL DYNAMICS FOR SUSPENSION CONTROL STUDIES

The aim of this first case study is to develop two simple mathematical models of the lateral dynamics of a railway vehicle. These models are useful to point out the basic mechanisms of the running stability of a railway vehicle, while the use of more complex multi-body models (not addressed here) is recommended for a quantitative analysis.

We start by deriving the classic 2 degrees of freedom (DOFs) model for a single wheelset with primary suspensions, considering linearized wheel/rail contact forces. This model explains the basic mechanism of hunting instability and was used in Chapter 11 to illustrate the basic principles of active stabilization of the running gear. Then, we extend the study to define a 6-DOF linear model of a 2-axle bogie with primary and secondary suspensions. This 6-DOF model and similar ones have frequently been used as the basis for defining active control strategies in railway vehicles; see [1–3] for a review of these studies.

14.1.1 THE 2 DEGREES OF FREEDOM WHEELSET MODEL

The simplest model that can be used to demonstrate the hunting motion of a railway vehicle considers one single wheelset attached to a bogie through springs and dampers representing the primary suspension as shown in Figure 14.1. The bogie frame is assumed to move forward with constant speed V without undergoing any lateral direction movement or yaw rotation. The model considers the motion of the wheelset in the horizontal plane and assumes the wheelset center of mass (COM) has the same forward speed V as the bogie frame. Therefore, the model has 2 DOFs, the independent co-ordinates being the lateral COM displacement y and the yaw angle ψ.

We perform this study under the following simplifying assumptions:

- The wheelset moves at constant speed V along tangent track, with no traction or braking forces applied.
- The rails are considered to be fixed to an inertial reference.
- Track irregularities (i.e., deviations from the ideal track geometry) are not considered.
- Symmetric profiles are assumed for the left and right wheels and for the left and right rails.
- The effect of spin creepage on the creep forces and the effect of the spin moment are neglected.
- The roll angle produced by a lateral displacement of the wheelset is neglected.

DOI: 10.1201/9781003028994-14

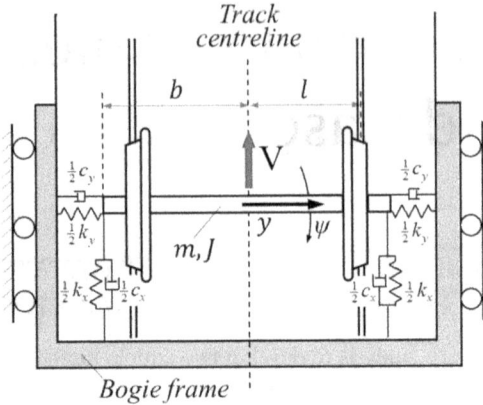

FIGURE 14.1 Model of the lateral dynamics of a railway wheelset with primary suspension having 2 DOFs.

Figure 14.2 shows the wheelset in a generic displaced position. The forces acting on the wheelset are:

- The elastic and viscous forces generated by the primary suspension, applied on the axle at distance b (Figure 14.1) from the wheelset's COM.
- The creep forces acting on the left and right wheels F_{xL}, F_{yL}, F_{xR}, F_{yR}.
- The flange forces, approximated to $\frac{W}{2}\tan(\delta)$ where W is the axle load and δ is the contact angle; we neglect here the small yaw moment created by flange forces for a non-zero yaw angle ψ of the wheelset.
- The equations of motion for the 2 DOFs model are:

$$m\ddot{y}+c_y\dot{y}+k_yy=\left(F_{xL}+F_{xR}\right)\sin\left(\psi\right)+\left(F_{yL}+F_{yR}\right)\cos\left(\psi\right)+\frac{W}{2}\left(\tan\left(\delta_L\right)-\tan\left(\delta_R\right)\right)$$

$$J\ddot{\psi}+c_xb^2\dot{\psi}+k_xb^2\psi=\left(F_{xL}-F_{xR}\right)l \tag{14.1}$$

- We define the primary yaw stiffness and primary yaw damping parameters as:

$$c_\psi=c_xb^2$$
$$k_\psi=k_xb^2 \tag{14.2}$$

- For small values of the yaw angle σ and considering Equation (14.2), Equation (14.1) becomes:

$$m\ddot{y}+c_y\dot{y}+k_yy=\left(F_{xL}+F_{xR}\right)\psi+\left(F_{yL}+F_{yR}\right)+\frac{W}{2}\left(\tan\left(\delta_L\right)-\tan\left(\delta_R\right)\right)$$

$$J\ddot{\psi}+c_\psi\dot{\psi}+k_\psi\psi=\left(F_{xL}-F_{xR}\right)l \tag{14.3}$$

$$\frac{k_x}{2}b\psi + \frac{c_y}{2}b\dot{\psi}$$

$$\frac{k_y}{2}y + \frac{c_y}{2}\dot{y}$$

V

y

ψ

$$\frac{k_y}{2}y + \frac{c_y}{2}\dot{y}$$

$$\frac{k_x}{2}b\psi + \frac{c_y}{2}b\dot{\psi}$$

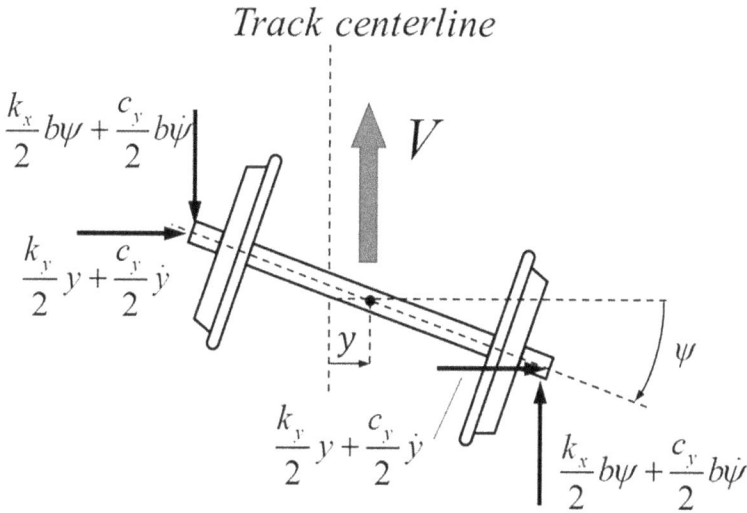

a) Coordinates and forces from primary suspension

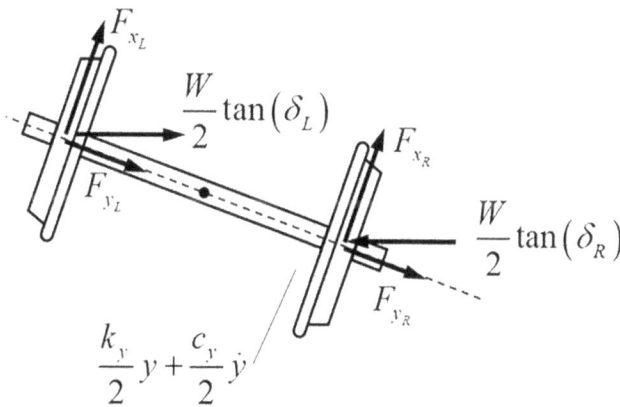

F_{x_L}

$$\frac{W}{2}\tan(\delta_L)$$

F_{x_R}

F_{y_L}

$$\frac{W}{2}\tan(\delta_R)$$

F_{y_R}

$$\frac{k_y}{2}y + \frac{c_y}{2}\dot{y}$$

b) Creep forces and wheel flange forces

FIGURE 14.2 Forces acting on the single wheelset model: (a) forces from the primary suspension, (b) creep forces and wheel flange forces.

Owing to the assumptions made, the creep forces F_{x_L}, F_{y_R}, F_{x_R}, F_{y_R} can be expressed as non-linear functions of the longitudinal and lateral creepages acting on the left and right wheels v_{x_L}, v_{y_L}, v_{x_R}, v_{y_R} according to one of the contact force theories introduced in Chapter 2, giving:

$$F_{x_L} = F_{x_L}\left(v_{x_L}, v_{y_L}\right); \ F_{y_L} = F_{y_L}\left(v_{x_L}, v_{y_L}\right)$$

$$F_{x_R} = F_{x_R}\left(v_{x_R}, v_{y_R}\right); \ F_{y_R} = F_{y_R}\left(v_{x_R}, v_{y_R}\right)$$

$$(14.4)$$

Recalling the definition of the creepages according to Equation (2.13) in Chapter 2 and considering that body 2 (the rail) is not moving, the longitudinal and lateral creepages for the left and right wheels are:

$$v_{x_L} = \frac{V\cos(\psi) + \dot{y}\sin(\psi) + h\dot{\psi} - r_L\omega}{V} \cong \frac{V + h\dot{\psi} - r_L\omega}{V}$$

$$v_{x_R} = \frac{V\cos(\psi) + \dot{y}\sin(\psi) - h\dot{\psi} - r_R\omega}{V} \cong \frac{V - h\dot{\psi} - r_R\omega}{V} \tag{14.5}$$

$$v_{y_L} = v_{y_R} = \frac{-V\sin(\psi) + \dot{y}\cos(\psi)}{V} \cong \frac{-V\psi + \dot{y}}{V}$$

where r_L and r_R are the instantaneous rolling radii of the left and right wheels and ω is the rolling angular speed of the wheelset. As introduced in Chapter 2 (see Figure 2.2 in Chapter 2), the rolling radii of the two wheels and the contact angles δ_L and δ_R appearing in Equation (14.3) are non-linear functions of the lateral displacement of the wheelset relative to the track y_{rel}. Since we neglect here the effect of lateral alignment irregularities in the track and we consider the rails as fixed to the inertial reference, the lateral displacement of the wheelset relative to the track equals the absolute lateral displacement of the wheelset, i.e., $y_{rel} = y$, hence:

$$r_L = r_L(y); r_R = r_R(y)$$
$$\delta_L = \delta_L(y); \delta_R = \delta_R(y) \tag{14.6}$$

For a generic shape of the wheel and rail profiles, the non-linear functions in Equation (14.6) can be represented by wheel-rail geometry functions stored as a contact table.

Equations (14.3)–(14.6) altogether define the non-linear equations of motion for the single wheelset with 2 DOFs. From these equations, a linearized model of the wheelset with 2 DOFs can be derived, which serves as a basis to clarify the mechanism of hunting instability and to introduce the principle of active vehicle stabilization, see Chapter 11. To this aim, we replace the non-linear creepage-creep force relationship Equation (14.4) by a linear approximation according to Kalker's linear model of creep forces already introduced in Chapter 2, neglecting the effect of spin creepage φ and spin moment:

$$F_{x_L} = -f_{11}v_{x_L}; \quad F_{x_R} = -f_{11}v_{x_R}$$
$$F_{y_L} = -f_{22}v_{y_L}; \quad F_{y_R} = -f_{22}v_{y_R} \tag{14.7}$$

Furthermore, we use linear approximations for the expressions in Equation (14.6) relating the rolling radii of the two wheels with the wheelset lateral displacement:

$$r_L = r_0 + \lambda y; r_L = r_0 - \lambda y \tag{14.8}$$

where r_0 is the nominal rolling radius of the two wheels (i.e., the rolling radius of both wheels when the wheelset has no lateral offset with respect to the track) and λ is the equivalent conicity of the wheel/rail profiles arising from a linear approximation of the non-linear relationships in Equation (14.6). For a sufficiently small lateral displacement of the wheelset, a linearized relationship can also be used to express the difference of the tangents of the contact angles as:

$$\tan(\delta_R) - \tan(\delta_L) = \frac{2\varepsilon y}{l} \tag{14.9}$$

For wheels having a conical profile, the equivalent conicity parameter λ and the proportionality coefficient ε in Equation (14.9) take the same value [4]. However, in the further development of the equations, we consider the general case with ε distinct from λ.

Substituting Equation (14.8) into the expression of creepages at Equation (14.5) and recalling that the wheelset forward speed V and rolling angular speed ω are related by:

$$V = r_0 \omega$$

we obtain:

$$v_{xL} = -v_{xR} = \frac{l}{V}\dot{\psi} + \frac{\lambda}{r_0}y$$

$$v_{yL} = v_{yR} = \frac{\dot{y}}{V} - \psi \tag{14.10}$$

Introducing Equations (14.7), (14.9), and (14.10) into Equation (14.3), we obtain:

$$m\ddot{y} + c_y\dot{y} + k_y y = -2f_{22}\left(\frac{\dot{y}}{V} - \psi\right) - \frac{W\varepsilon}{l}y$$

$$J\ddot{\psi} + c_\psi\dot{\psi} + k_\psi\psi = -2f_{11}\left(\frac{l}{V}\dot{\psi} + \frac{\lambda}{r_0}y\right)l$$

which can be re-arranged as:

$$m\ddot{y} + \left(c_y + \frac{2f_{22}}{V}\right)\dot{y} + \left(k_y + \frac{W\varepsilon}{l}\right)y - 2f_{22}\psi = 0$$

$$J\ddot{\psi} + \left(c_\psi + \frac{2f_{11}l^2}{V}\right)\dot{\psi} + \frac{2f_{11}\lambda l}{r_0}y + k_\psi\psi = 0 \tag{14.11}$$

Equation (14.11) can be rewritten in matrix form as:

$$m\ddot{y} + (c_s + c_c)\dot{y} + (k_s + k_c + k_g)y = 0 \qquad (14.12)$$

with:

$$y = \left\{ \begin{array}{c} y \\ \psi \end{array} \right\}$$

$$m = \left[\begin{array}{cc} m & 0 \\ 0 & J \end{array} \right]$$

$$c_s = \left[\begin{array}{cc} c_y & 0 \\ 0 & c_\psi \end{array} \right]; \; c_c = \frac{1}{V} \left[\begin{array}{cc} 2f_{22} & 0 \\ 0 & 2f_{11}l^2 \end{array} \right]$$

$$k_s = \left[\begin{array}{cc} k_y & 0 \\ 0 & k_\psi \end{array} \right]; \; k_c = \left[\begin{array}{cc} 0 & -2f_{22} \\ \dfrac{2f_{11}l\lambda}{r_0} & 0 \end{array} \right]; \; k_g = \left[\begin{array}{cc} \dfrac{W\varepsilon}{l} & 0 \\ 0 & 0 \end{array} \right]$$

where m is the mass matrix, $c_s + c_c$ is the total damping matrix, with c_s being the term from the primary suspensions and c_c being the term due to the creep forces, $k_s + k_c + k_g$ is the total stiffness matrix, with k_s being the term from the primary suspensions, k_c being the term due to the creep forces and k_g being the gravitational stiffness matrix. We see that the total stiffness matrix is non-symmetric due to the k_c term; this shows that the positional component of the state-dependent forces associated with wheel/rail contact is non-conservative. In [5] it is shown that an undamped 2 DOFs system for which the off-diagonal terms of the stiffness matrix are of opposite sign may suffer an unstable motion in the form of an expansive oscillation due to energy introduced in the motion of the system by the non-conservative state-dependent forces. In the case of the 2 DOFs wheelset considered here, damping is provided both by the suspensions through matrix c_s and by the creep forces through matrix c_c. These dissipative effects may overcome the positive flow of energy caused by non-conservative forces, resulting in a stable motion for the wheelset. However, we note the terms in matrix c_c are inversely proportional to the speed V of the wheelset, while the damping terms related to the primary suspensions are usually so low that they are often neglected. We conclude that, at low speed, energy dissipation is large enough to stabilize the motion of the wheelset but, with increasing speed, damping effects become weaker until the energy introduced by position-dependent creep forces cannot be completely dissipated. In this condition, an unstable motion of the wheelset takes place in the form of an expansive oscillation involving a combination of lateral displacement and yaw rotation. The threshold speed above which instability takes placed is called the *critical speed* of the wheelset. We note further that the magnitude of one of the off-diagonal terms in the k_c matrix is proportional to the value of the conicity parameter λ. Owing to the analysis in [5], we conclude that an increase of wheel conicity leads to a decrease of the wheelset's critical speed.

In order to find the critical speed value, the state vector x of the wheelset is introduced as a 4-rows column vector formed by vectors $\dot{y}$ and y:

$$x = \left\{ \begin{array}{c} \dot{y} \\ y \end{array} \right\} \tag{14.13}$$

and Equation (14.12) is rewritten in state-space form:

$$\dot{x} = Ax \tag{14.14}$$

where A is the system's state matrix. The solution of this equation is sought in the well-known form:

$$x = X_0 e^{st}$$

where s is the complex-valued characteristic exponent with real part μ and imaginary part ω:

$$s = \mu + j\omega$$

The values of the characteristic exponents of the system are obtained as the eigenvalues of the state matrix A:

$$\det(sI - A) = 0$$

Since the state matrix A depends on the wheelset speed V, the values of the characteristic exponents are affected by the speed and can be represented in a root locus as shown in Figure 14.3. It is observed that two conjugate branches of the diagram cross the imaginary axis at some speed value and move in the half-space corresponding to a positive real part of the characteristic exponent, which corresponds to the onset of the unstable motion.

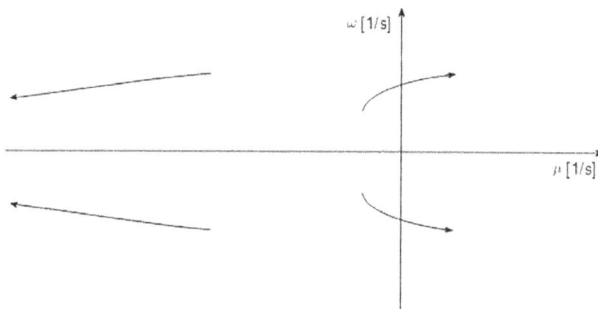

FIGURE 14.3 Root locus of the characteristic exponents of the 2 DOFs wheelset for varying speed values.

TABLE 14.1

Parameters of the Wheelset Model with 2 DOFs

Parameter	Symbol	Value
Wheelset mass	m_w	1500 kg
Wheelset yaw moment of inertia	J_w	800 kg m^2
Primary longitudinal damping coefficient (per wheelset)	c_x	0
Primary lateral damping coefficient (per wheelset)	c_y	0
Primary longitudinal stiffness (per wheelset)	k_x	8.0E6 N/m
Primary lateral stiffness (per wheelset)	k_y	8.0E6 N/m
Nominal wheel radius	r_0	0.45 m
Semi-lateral distance of primary suspensions	b	1.0 m
Semi-lateral distance of wheels	l	0.75 m
Kalker's linear theory coefficient f_{11}	f_{11}	1.0E7 N
Kalker's linear theory coefficient f_{22}	f_{22}	1.0E7 N
Conicity (and gravitational stiffness coefficient)	λ	0.3 -
Axle load	W	1.1E5 N

We consider a numerical example using the data of the wheelset model in Table 14.1.

The trend with speed of the real part μ and imaginary part ω of the characteristic exponent with the largest real part are shown in Figure 14.4. This plot shows that the wheelset becomes unstable at speeds above 335 km/h approximately, and the circular frequency of the self-excited oscillation is 87 rad/s approximately, corresponding to a frequency of 13.8 Hz.

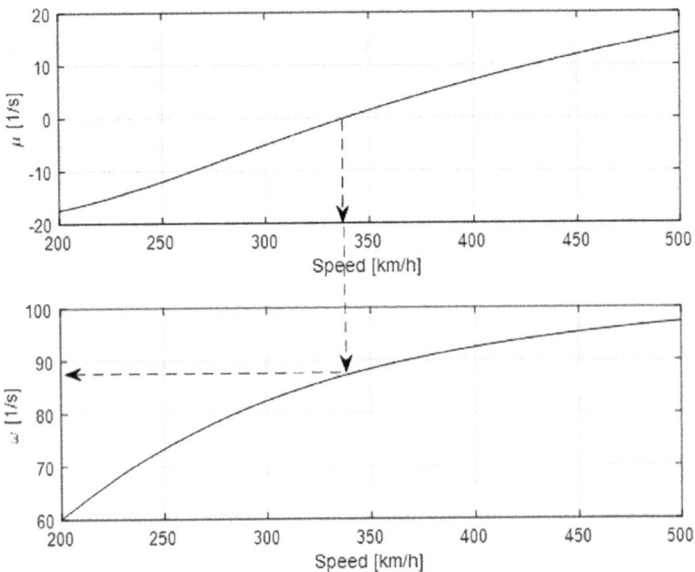

FIGURE 14.4 Trend with speed of the real (top) and imaginary (bottom) parts of the characteristic exponent with the largest real part for the 2 DOFs wheelset.

It is worth mentioning that a non-linear critical speed of the vehicle can be defined considering the non-linear equations of motion of the wheelset at Equation (14.1) together with a non-linear creepage-creep force relationship in the form of Equation (14.4) and a non-linear dependence of the rolling radius and contact angle on the lateral movement of the wheelset according to Equation (14.4). In this case, the solution is sought in the form of an orbitally stable periodic motion of the wheelset and the non-linear critical speed is defined as the lowest speed at which a periodic solution with non-zero amplitude occurs, see [6] for more details. The concepts of linear and non-linear critical speed are clarified by Figure 14.5, in which the bifurcation diagram for the wheelset model with 2 DOFs is shown qualitatively. In this diagram, the amplitude of possible periodic motions of the system are plotted as a function of the speed V. The thick solid lines represent the stable branches of the diagram and the thick dashed lines indicate the unstable branches of the diagram. It is observed that a speed range exists in which two stable periodic motions of the wheelset are possible, one with zero amplitude and one with finite amplitude; in this speed range the motion of the wheelset will be *attracted* by one of the two possible stable motions, depending on the initial conditions applied. At lower speeds, only the solution with zero amplitude exists, whereas, at very high speed, only the periodic solution with non-zero amplitude exists. The linear critical speed evaluated according to the linear stability analysis reported above is the largest speed value for which the zero-amplitude solution is found and this is called the *linear critical speed* of the wheelset. The lowest speed at which a periodic solution with non-zero amplitude is found is called the non-*linear critical speed* of the wheelset. Due to the particular shape of the bifurcation diagram, the non-linear critical speed is usually lower than the linear critical speed, which means a non-linear stability analysis should be performed for a railway vehicle rather than a linear stability analysis, see [7] for more details. In this case study, however, we focus rather on a linear stability analysis as this can be performed using simpler mathematical tools and because linearized equations are more meaningful than non-linear equations in the perspective of designing a control application for the vehicle.

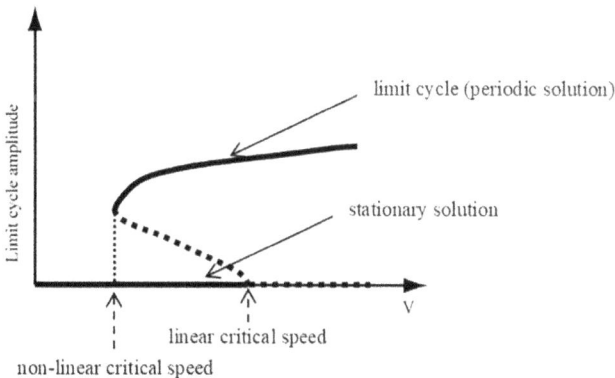

FIGURE 14.5 Qualitative bifurcation diagram for the wheelset model with 2 DOFs.

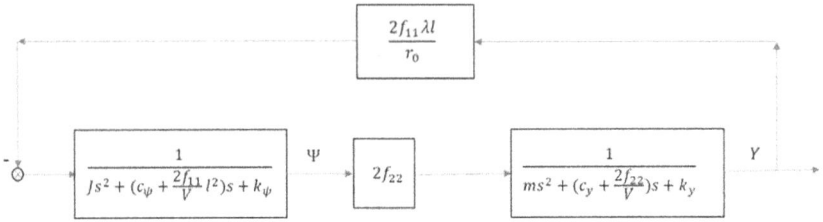

FIGURE 14.6 Representation of the equations of motion of the wheelset with 2 DOFs in block diagram form, showing the natural feedback created by the creep forces.

Finally, we consider again Equation (14.11) and we use the Laplace transform to rewrite the equations of motion of the wheelset in the form of algebraic equations. After a trivial treatment, the equations take the form:

$$\left[ms^2 + \left(c_y + \frac{2f_{22}}{V} \right)s + \left(k_y + \frac{w\varepsilon}{l} \right) \right] Y = 2f_{22}\psi$$

$$\left[Js^2 + \left(c_\psi + \frac{2f_{11}l^2}{V} \right)s + k_\psi \right] \psi = -\frac{2f_{11}\lambda l}{r_0} Y$$

(14.15)

where Y and Ψ are the Laplace transforms of the wheelset's lateral displacement y and yaw rotation ψ, respectively and s is the Laplace variable. In this form, the equations of the 2-DOF wheelset can be represented by a block diagram, as shown in Figure 14.6 (note the gravitational stiffness term is omitted in the block diagram). This representation shows the natural feedback effect which is produced by the effect of creep forces in combination with the use of conical wheels and a solid axle providing a rigid connection of the wheels. This representation of the wheelset's equations of motion is used in Chapter 11 of this book to introduce the concept of active stabilization by means of active primary suspensions.

14.1.2 THE 6 DEGREES OF FREEDOM BOGIE MODEL

A more detailed model of a railway vehicle is obtained considering the dynamics of a complete bogie consisting of a rigid frame connected by primary suspensions to two wheelsets, as shown in Figure 14.7. The secondary suspension of the vehicle is also considered in the model, but the car body is assumed to move forward with constant speed V without undergoing any movement in lateral direction. Therefore, the model has 6 DOFs, the independent co-ordinates being the lateral COM displacement and the yaw rotation of the bogie frame y_b and ψ_b, together with the lateral displacement and yaw rotation of the front and rear wheelsets y_f, ψ_f, y_r, and ψ_r. With respect to the 2 DOFs model introduced previously, this model reproduces more accurately the lateral dynamics of the vehicle, which typically involves significant lateral and yaw motion of the bogie frame, not considered by the 2 DOFs model. Furthermore,

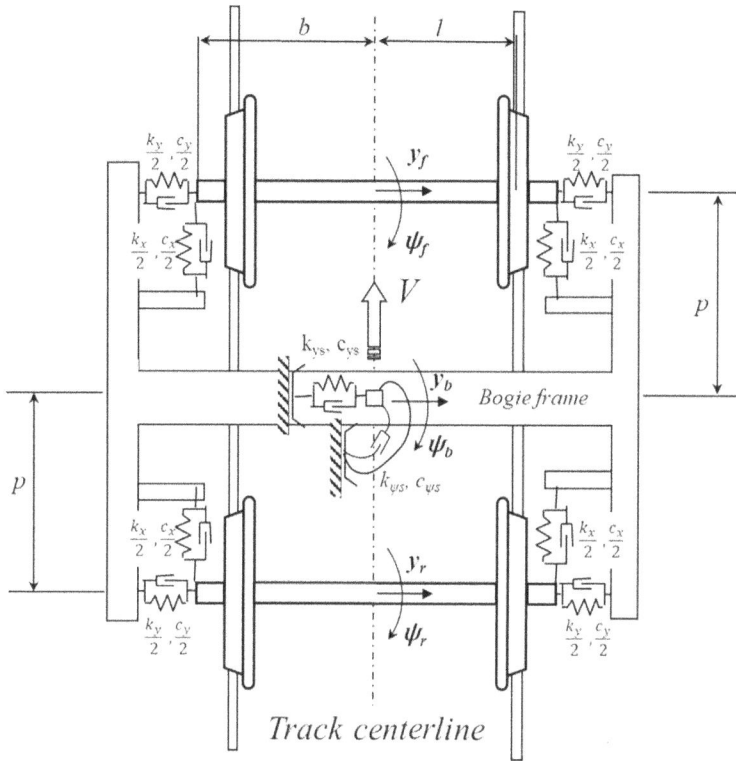

FIGURE 14.7 Model of the lateral dynamics of a bogie with primary and secondary suspensions having 6 DOFs.

the 6 DOFs model describes more accurately the effect of fundamental design parameters of the bogie such as longitudinal and lateral stiffnesses of the primary suspensions, the bogie wheelbase, as well as the effect of including yaw dampers in the secondary suspension.

The secondary yaw stiffness $k_{\psi s}$ comes from the shear stiffness of the secondary suspension, considering the lateral and longitudinal distances between the single springs of the secondary suspension. We assume here a symmetric layout of the springs comprising the secondary suspension, so that there is no elastic coupling between the lateral and yaw movements of the bogie frame. In the same way, the secondary yaw damping coefficient $c_{\psi s}$ arises from a longitudinal distance between secondary lateral dampers, and from the lateral distance between yaw dampers in the case where these devices are installed in the bogie.

The equations of motion for the bogie frame can be easily derived from Newton's second law considering the elastic and viscous forces generated by the primary and secondary suspensions. The equations of motion for the leading and trailing wheelset are the same as already derived for the 2-DOFs model but contain additional stiffness and damping terms due to the coupling between the motion of the bogie frame and of the wheelsets caused by the primary suspensions. Therefore, the equations of the 6-DOFs model also take the matrix form (see Equation 14.12), with vector y being the 6-row vector:

$$y = \left\{ \begin{array}{cccccc} y_b & \psi_b & y_f & \psi_f & y_r & \psi_r \end{array} \right\}^T$$

while the matrices involved are of the 6th order and take the following expressions:

$$m = \text{diag}\left[\begin{array}{cccccc} m_b & J_b & m_f & J_f & m_r & J_r \end{array} \right]$$

$$c_s = \begin{bmatrix} c_{ys} + 2c_y & 0 & -c_y & 0 & -c_y & 0 \\ 0 & c_{\psi s} + 2c_x b^2 + 2c_y p^2 & -c_y p & -c_x b^2 & c_y p & -c_x b^2 \\ -c_y & -c_y p & c_y & 0 & 0 & 0 \\ 0 & -c_x b^2 & 0 & c_x b^2 & 0 & 0 \\ -c_y & c_y p & 0 & 0 & c_y & 0 \\ 0 & -c_x b^2 & 0 & 0 & 0 & c_x b^2 \end{bmatrix} \tag{14.16}$$

$$c_c = \frac{1}{V}\text{diag}\left[\begin{array}{cccccc} 0 & 0 & 2f_{22} & 2f_{11}l^2 & 2f_{22} & 2f_{11}l^2 \end{array} \right] \tag{14.17}$$

$$k_s = \begin{bmatrix} k_{ys} + 2k_y & 0 & -k_y & 0 & -k_y & 0 \\ 0 & k_{\psi s} + 2k_x b^2 + 2k_y p^2 & -k_y p & -k_x b^2 & k_y p & -k_x b^2 \\ -k_y & -k_y p & k_y & 0 & 0 & 0 \\ 0 & -k_x b^2 & 0 & k_x b^2 & 0 & 0 \\ -k_y & k_y p & 0 & 0 & k_y & 0 \\ 0 & -k_x b^2 & 0 & 0 & 0 & k_x b^2 \end{bmatrix} \tag{14.18}$$

$$k_c = \begin{bmatrix} 0 & 0 & 0 & 0 & 0 & 0 \\ 0 & 0 & 0 & 0 & 0 & 0 \\ 0 & 0 & 0 & -2f_{22} & 0 & 0 \\ 0 & 0 & \dfrac{2f_{11}l\lambda}{r_0} & 0 & 0 & 0 \\ 0 & 0 & 0 & 0 & 0 & -2f_{22} \\ 0 & 0 & 0 & 0 & \dfrac{2f_{11}l\lambda}{r_0} & 0 \end{bmatrix} \tag{14.19}$$

$$k_g = \text{diag}\left[\begin{array}{cccccc} 0 & 0 & \dfrac{W\varepsilon}{l} & 0 & \dfrac{W\varepsilon}{l} & 0 \end{array} \right] \tag{14.20}$$

TABLE 14.2
Parameters of the Bogie Model with 6 DOFs

Parameter	Symbol	Value
Bogie frame mass	m_b	2600 kg
Bogie frame yaw moment of inertia	J_b	3300 kg m^2
Wheelset mass	m_w	1500 kg
Wheelset yaw moment of inertia	J_w	800 kg m^2
Primary longitudinal damping coefficient (per wheelset)	c_x	0.
Primary lateral damping coefficient (per wheelset)	c_y	0.
Secondary lateral damping coefficient (per bogie)	c_{ys}	2.0E4 Ns/m
Secondary yaw damping coefficient (per bogie)	$c_{\psi s}$	0.
Primary longitudinal stiffness (per wheelset)	k_x	8.0E6 N/m
Primary lateral stiffness (per wheelset)	k_y	8.0E6 N/m
Secondary lateral stiffness (per bogie)	k_{ys}	1.0E6 N/m
Secondary yaw stiffness (per bogie)	$k_{\psi s}$	2.0E6 N/m
Nominal wheel radius	r_0	0.45 m
Bogie semi-wheelbase	p	1.1 m
Semi-lateral distance of primary suspensions	b	1.0 m
Semi-lateral distance of wheels	l	0.75 m
Kalker's linear theory coefficient f$_{11}$	f_{11}	1.0E7 N
Kalker's linear theory coefficient f$_{22}$	f_{22}	1.0E7 N
Conicity (and gravitational stiffness coefficient)	λ	0.3
Axle load	W	1.1E5 N

We analyze the stability behavior of the 6-DOFs bogie model considering the set of model parameters in Table 14.2

Figure 14.8 shows the root locus diagram for the bogie model with 6 DOFs. Compared to the wheelset model with 2 DOFs, more branches of the diagram are obtained, consistently with the increased number of DOFs of the model. However, it is still observed that the two conjugate branches with the smallest absolute value of

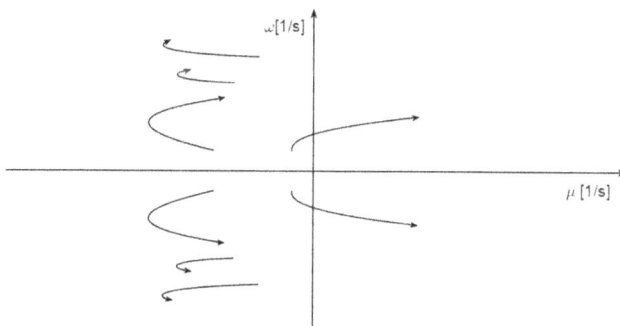

FIGURE 14.8   Root locus of the characteristic exponents of the 6 DOFs model of the bogie for varying speed values.

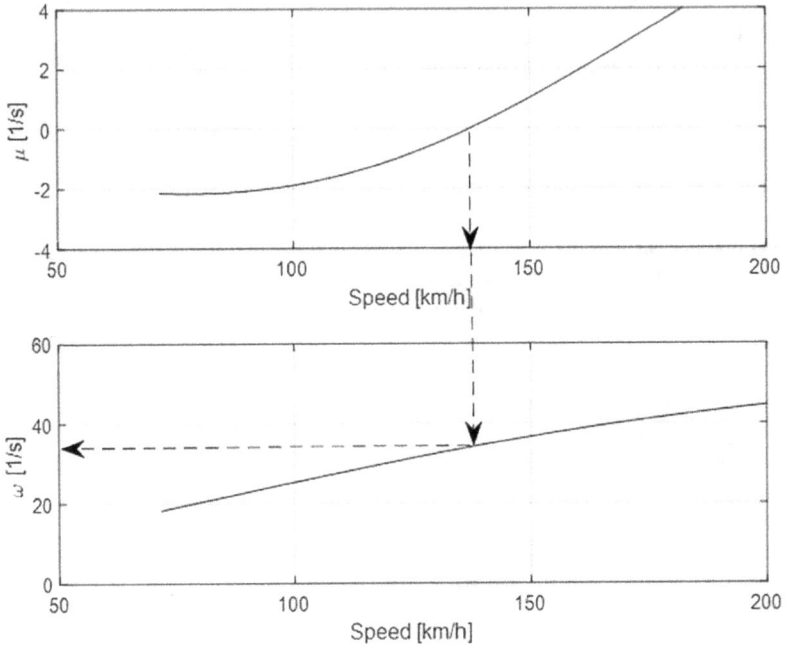

FIGURE 14.9   Trend with speed of the real (top) and imaginary (bottom) part the characteristic exponent with largest real part for the 6 DOFs model of the bogie.

the real part move toward the imaginary axis for increasing vehicle speed, until they fall in the semi-space of positive real parts, determining the unstable behavior of the bogie. Note that two pairs of conjugate branches are not shown in Figure 14.8 as they fall too far into the negative real parts region of the diagram.

Figure 14.9 shows the trend with speed of the real and imaginary parts of the characteristic exponent with the largest real part for the bogie model with 6 DOFs. The critical speed for this model is 137 km/h approximately, much lower than that obtained for the 2 DOFs wheelset model despite the same set of vehicle parameters being considered in the two examples. This is because the 6 DOFs model provides a more accurate representation of the actual hunting motion of the vehicle, while the simple 2 DOFs model, although useful to clarify the nature of hunting instability, significantly over-estimates the vehicle's critical speed. The frequency of the unstable hunting motion is approximately 5.4 Hz, also much lower than that obtained from the 2 DOFs model.

Before concluding this case study, we use again the 6 DOFs model to investigate the effect of primary suspension stiffness in longitudinal and lateral directions. Starting again from the parameters of the 6 DOFs model in Table 14.2, we explore different combinations of the longitudinal and lateral primary stiffnesses in the range $k_x = 2 \div 100$ MN/m and $k_y = 2 \div 60$ MN/m. Figure 14.10 shows the values of the linear critical speed V_{crit} for different combinations of primary stiffness parameters. It is observed that the largest critical speed is not obtained for the maximum values

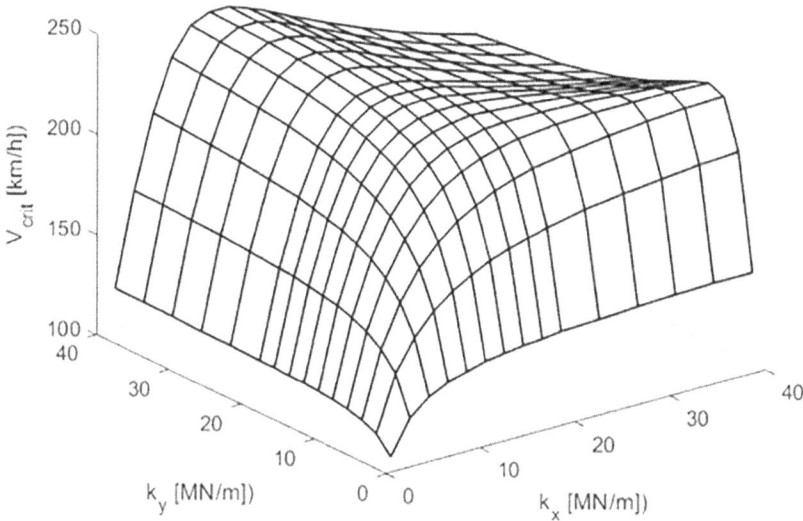

FIGURE 14.10 Value of the linear critical speed for different combinations of primary stiffness parameters

of primary suspension stiffness, but is rather obtained for a combination of sufficiently large lateral stiffness k_y and intermediate longitudinal stiffness k_x or, vice versa, for large longitudinal stiffness k_x and intermediate lateral stiffness k_y. Note that the values of parameters k_x and k_y maximizing the critical speed are specific to the example considered here and could be different for a bogie with a different design or for a different conicity value.

14.2   CASE B: MODELING OF A BOGIE WITH ACTIVE STEERING SYSTEM

As introduced in previous chapters of this book, the design of a railway bogie is faced with a fundamental conflict between stability and curving, and mechatronic suspensions can be used to achieve an overall increase of vehicle performance [1–3]. One implementation of active primary suspensions is active bogie steering which is presented here as a case study for a trailed (i.e., unpowered) passenger vehicle with two bogies and four solid-axle wheelsets.

14.2.1   Basic Principle of Active Steering System for Solid-Axle Wheelset

A basic mechanical layout of the active steering system is shown in Figure 14.11. The traction rods connecting the axle-boxes to each side beam of the bogie frame are replaced by controlled actuators. Rubber end mounts are placed in series to each actuator at both sides, providing a high longitudinal stiffness when the actuator is

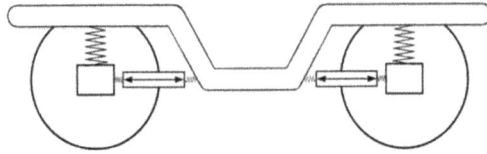

FIGURE 14.11   Mechanical layout of the active steering system.

kept at a fixed length; in this way, the desired level of stability can be ensured. The shear stiffness of the coil springs in the primary suspension provides an additional stiffness in the longitudinal direction, in parallel to the actuators, but this is kept low as the actuators need to counteract the elastic force generated by the spring and, therefore, an excessively high stiffness would imply a larger force demand for the actuators.

The principle of active steering in curves is to generate a yaw angle between the two wheelsets in the same bogie by controlling the elongation of the actuators. In this way, track shift forces and wheel wear can be reduced effectively. The movement of the actuation system is defined based on the local track curvature. Considering typical lengths of curve transitions in the order of tens or hundreds of meters and vehicle speeds in the order of tens of m/s, it follows that wheelset steering movements take place in a low frequency range, below 1 Hz. Therefore, a low-pass filter (LPF) can be introduced in the control loop to remove high-frequency disturbances caused by track irregularities.

Suitable control strategies for active steering have been introduced in Chapter 11. Perfect steering [3, 8] provides superior performance in terms of reducing wear effects and track shift forces, but is difficult to implement in a real application as it involves the measurement or estimation of contact force parameters which is highly challenging [9]. Radial control of the wheelsets is an alternative solution, enabling a much easier implementation and nearly the same performances compared to perfect steering. Therefore, this radial control strategy is adopted in the present case study.

The schematic diagram of radial control is shown in Figure 14.12, a change of length is defined for the controlled actuators so that a relative yaw rotation is produced between the two wheelsets in the bogie to compensate the difference in the radial direction at the two wheelsets. Thanks to the steering action, the wheelsets take a nearly radial attitude, apart from minor deviations caused by the small yaw rotation of the bogie and by small deflections in the rubber mounts. The reference displacement of the actuators ΔL is defined as:

$$\Delta L = \pm \frac{b}{R} \cdot a \qquad (14.21)$$

where a and b are respectively the semi-gauge of the actuators and the semi-wheelbase of the bogie and R is the radius of the centerline of the curve.

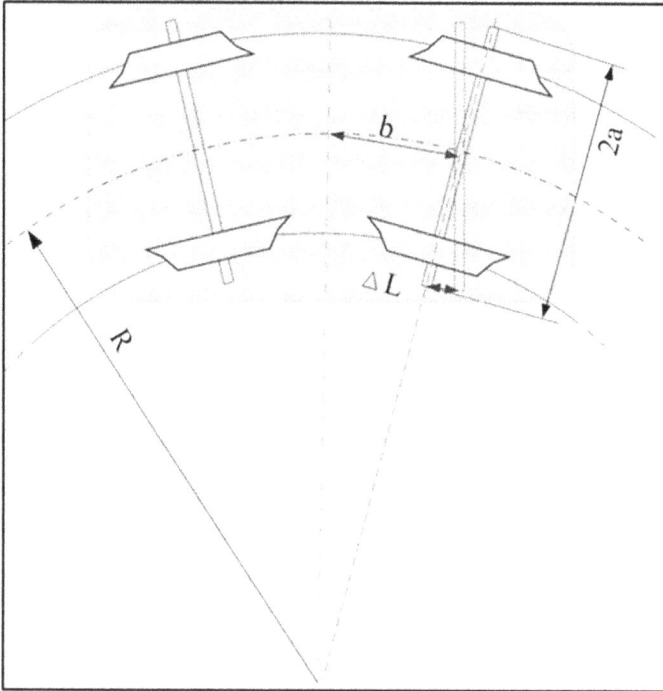

FIGURE 14.12 Schematic diagram of radial control.

The implementation of radial control requires the estimate of local track curvature. This can be obtained using two different schemes. Scheme 1 estimates the track curvature in real-time using the measured yaw angular velocity of the bogie $\dot{\sigma}$ and the forward velocity of the vehicle V which, in turn, can be either directly measured/estimated or obtained from the train communication network. According to this scheme, track curvature is estimated from:

$$1/R = \frac{\dot{\sigma}}{V} \qquad (14.22)$$

The measured yaw angular speed is affected by disturbances from track irregularities, so a LPF is applied to improve the estimation of the curvature. However, the LPF introduces a delay in the estimation and hence in the control action, leading to a reduced performance of the active control scheme. To alleviate this problem, precedence control is adopted in which the curvature is estimated according to Equation (14.22) using a sensor (typically a gyroscope) mounted on the leading bogie and the curvature signal is used to define the reference elongation for the actuators in other bogies of the train set, considering the distance between the bogies and the speed of the vehicle. As long as the precedence of the estimated curvature caused by the

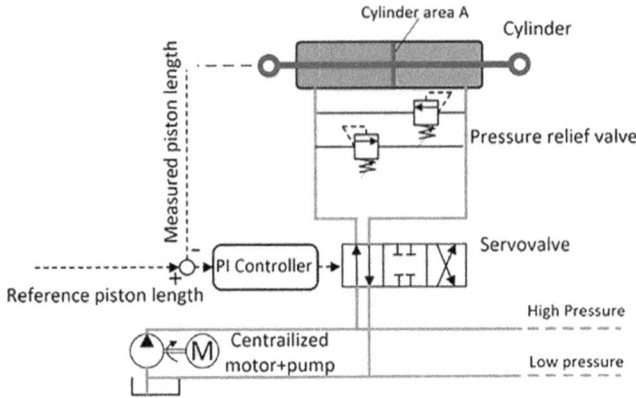

FIGURE 14.13   Schematic diagram of the HSA hydraulic circuit.

distance between the bogies is larger than the delay introduced by the LPF, this technique can successfully synchronize the control action with the actual local value of curvature for all bogies except the leading one.

In Scheme 2, track curvature comes from the geo-localization of the vehicle, obtained using GPS or balises, in conjunction with a track data-base containing information on track deterministic features such as curvature and cant as function of the kilometric point. In this scheme, the geo-localization system provides the current position of the vehicle and track curvature is obtained from the look-up of the track database. In this case study, both schemes are adopted to estimate the curvature of the track and the two approaches are compared in Section 14.2.4.

Regarding the actuator technology, a hydraulic servo actuator (HSA) [10] is considered in this case study. The circuit of the HSA model is illustrated in Figure 14.13. It mainly consists of a double-acting hydraulic cylinder, a *3-position 4-way* servo-valve, two pressure relief valves, the centralized motor and pump, and pipelines. The motor and pump establish a suitable pressure difference between the high-pressure and low-pressure branches of the hydraulic circuit, shown respectively in red and blue colors in the figure. In the servo-valve, the movement of the spool controls the opening area and the flow of pressurized oil in the two chambers of the cylinder, extending or retracting the actuator. The motion of the spool is in turn controlled by a PI controller, the controlled variable being the length of the piston. Pressure relief valves are implemented to limit the maximum pressure in the chambers of the actuator.

The maximum force and moving speed of the cylinder are two basic specifications that should be guaranteed in the design of the HSA. The requirement for the maximum force of the cylinder F_{max} can be satisfied by the proper setting of the piston cross-sectional area A and the pressure difference between the two branches of the hydraulic circuit $\Delta P = P_h - P_l$:

$$F_{max} = A \cdot (P_h - P_l) = A \cdot \Delta P \qquad (14.23)$$

Once the piston area A is defined, the maximum moving speed of the cylinder v_{max} is determined by the maximum flow rate, i.e., $v_{max} = q_{max}/A$. Equation (14.24) establishes a relation between the maximum flow rate q_{max} and the maximum opening area of the orifice a_{max} of the servo-valve. In this equation, the characteristics of fluid oil and servo-valve are involved, including the flow discharge coefficient of the valve C_d, fluid density ρ, and the minimum pressure for turbulent flow p_{cr}.

$$q_{max} = C_d \cdot a_{max} \sqrt{\frac{2}{\rho}} \cdot \frac{\Delta P}{\left(\Delta P^2 + p_{cr}^2\right)^{1/4}} \tag{14.24}$$

The pressure p_{cr} in turn is defined according to Equation (14.25),

$$p_{cr} = \frac{\rho}{2} \left(\frac{Re_{cr} \cdot \nu}{C_d} \sqrt{\frac{\pi}{4a_{max}}} \right)^2 \tag{14.25}$$

where symbols Re_{cr} and ν denote respectively the critical Reynolds number and fluid kinematic viscosity.

In this case study, the modeling of the vehicle with the active steering system is realized using Simpack-Simulink in co-simulation. A single trailed vehicle with two bogies and four axles is modeled in Simpack while the model of the controller and of the HSA is defined in Simulink as shown in Figure 14.14 for control scheme 1.

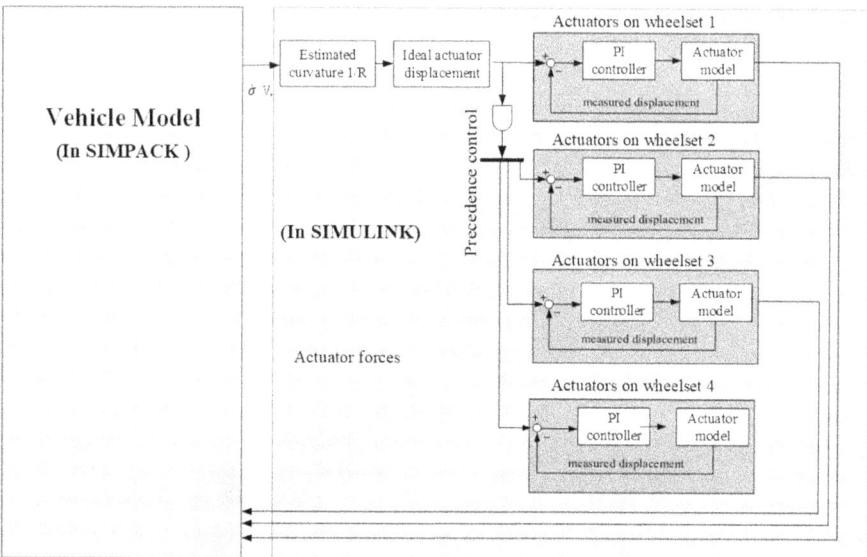

FIGURE 14.14 Schematic diagram of the co-simulation model (Scheme 1).

The Simpack model exports to Simulink marker variables representing the vehicle speed and the yaw angular velocity of the front bogie, while the Simulink model elaborates these quantities to define the forces in the eight hydraulic actuators and returns these variables to the Simpack model. In the following sub-sections, the Simpack model and the Simulink model are described in more detail.

14.2.2 Vehicle Model Built in Simpack

The vehicle model built in Simpack contains one car-body, two trailer bogies and four wheelsets (see Figure 14.15). The vehicle model is used to simulate both the behavior of a conventional vehicle equipped with passive primary suspensions and the behavior of the vehicle equipped with the active steering system, using the Simpack-Simulink co-simulation. The simulation of the conventional vehicle provides a baseline to assess the performance improvements provided by the active steering system.

A realistic set of parameters is defined for the model, considering the case of a trailed passenger vehicle designed for a maximum speed 160 km/h. The parameters of the vehicle model are summarized in Table 14.3. In the primary suspension, the coil springs above the axle-box bear the vertical load and provide the vertical stiffness of the suspension, together with a small amount of longitudinal and lateral stiffness produced by the shear stiffness of the springs. The traction rods between the axle-boxes and bogie frame transfer the longitudinal forces and provide the primary yaw stiffness. The secondary suspension is realized by air-springs, modeled here by linear elastic springs in vertical, lateral, and longitudinal directions. Lateral and vertical dampers are also fitted in the secondary suspension. Two yaw dampers are installed to guarantee vehicle stability. It should be noted that, in this case study, no active stabilization control scheme is implemented for the vehicle equipped with the active steering system, so the stability of the vehicle entirely relies on the passive

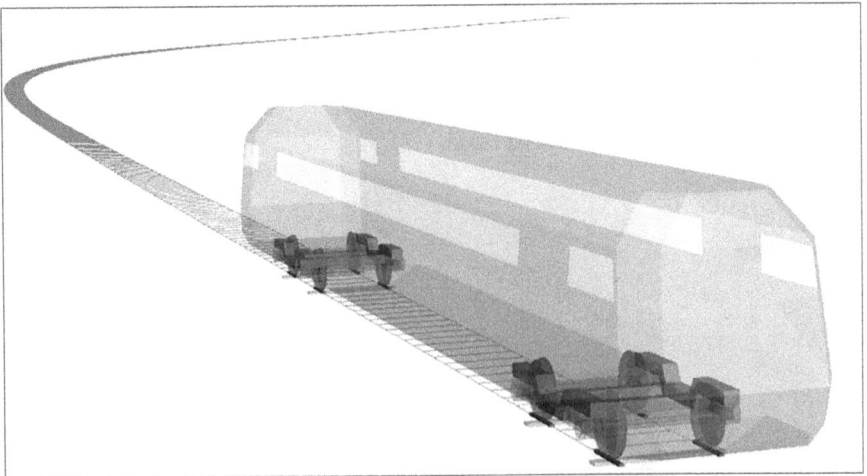

FIGURE 14.15 Vehicle model defined in Simpack.

TABLE 14.3
Parameters of the Vehicle Model

Parameters of the Vehicle Dynamics Model	Value (unit)
Axle load (tare condition)	110 kN
Distance between bogie wheelsets	2.5 m
Distance between bogie centers	16 m
Diameter of wheel (new)	860 mm
Wheel and rail profile	S1002/UIC60
Rail cant	1:40
Mass of car body	30,000 kg
Mass of bogie frame	3,000 kg
Mass of wheel-set	1,800 kg
Stiffness of primary coil spring in x/y direction	1.2 MN/m
Stiffness of primary coil spring in z direction	1.2 MN/m
Stiffness of air-spring in x/y directions	0.15 MN/m
Stiffness of air-spring in z direction	0.25 MN/m
Longitudinal stiffness of the traction rod (passive scheme)	10 MN/m
Longitudinal stiffness of actuator bushing (active scheme)	50 MN/m
Secondary vertical damper	30 kN/m/s
Secondary lateral damper	60 kN/m/s
Secondary yaw damper	200 kN/m/s
Equivalent stiffness of anti-roll bar	1.5 MN/rad

design of the suspensions. When the active steering system is applied, the forces simulating the traction rods are disabled and replaced by the actuator forces generated by the Simulink model.

14.2.3 CONTROLLER AND ACTUATOR MODEL IN SIMULINK

The co-simulation model in the Simulink environment for the vehicle with active steering using Scheme 1 (sensor-based) to define the steering action is shown in Figure 14.16. The block in the top left corner of the figure represents the Simpack model. It exports the vehicle speed and yaw angular velocity of the leading bogie to the controller and the velocities of the two ends of each actuator, on the axle-box and on the bogie frame, which are the inputs of the actuator models and are also used to perform position control of the actuators. The reference displacements of the actuators are generated by the module in the top right corner of the figure. The vehicle speed and yaw angular velocity signals are low-pass filtered to mitigate the effect of high frequency disturbances and then used to define track curvature and the reference length for the actuators of the leading wheelset of the vehicle according to Equations (14.21) and (14.22), implemented by a MATLAB® function. The references for the leading wheelset are sent to the next block on the right in the reference generator to produce the references for the other axles according to the precedence rule.

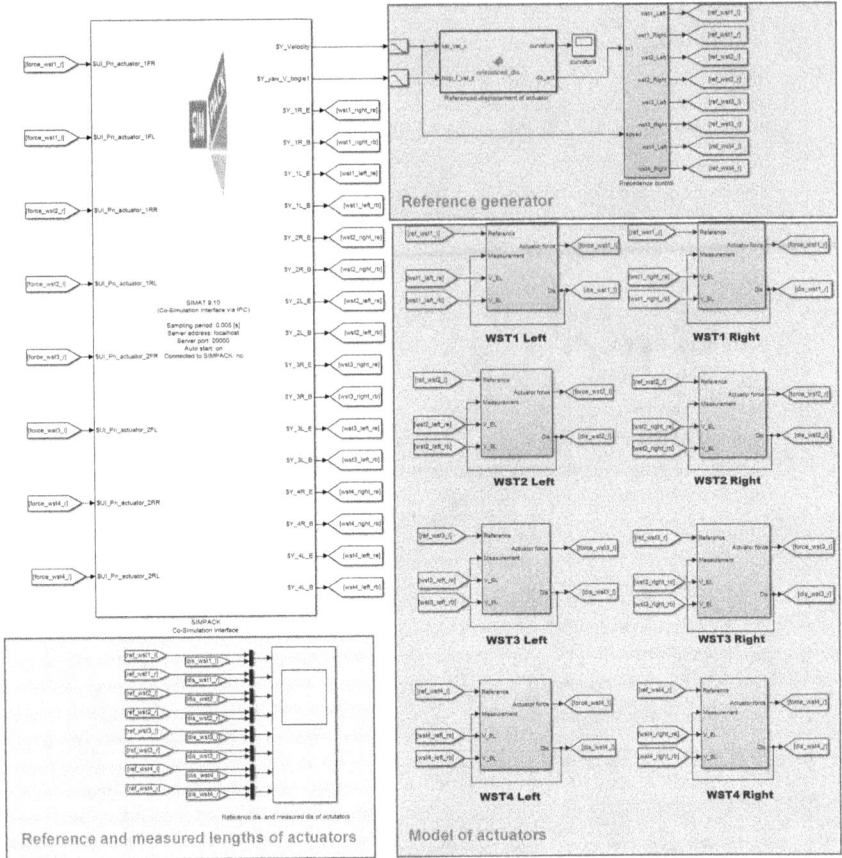

FIGURE 14.16 Co-simulation model in Simulink.

The eight blocks in the bottom right corner of Figure 14.16 are the HSA models. The model of each actuator receives as inputs the longitudinal speed of the end mounts and the reference length elaborated by the controller and produces as outputs the force and displacement of the actuator. The force signals are fed back to the Simpack model in the top left corner of the figure. Finally, the block in the bottom left corner is a display to monitor the reference displacement and actual lengths of the actuators. From now on, the actuator lengths obtained from the model are called *measured* lengths, as the implementation of the steering system in a real vehicle would require the measurement of these quantities.

The co-simulation model for the vehicle with active steering using Scheme 2 (geo-localization of the vehicle and look-up of a track geometry table) is similar, but the block in the upper right corner of Figure 14.16 is replaced by the block shown in Figure 14.17. In this latter block, the position in the track of the leading bogie is assumed to be provided by a geo-localization system (not modeled) and the curvature of the track is elaborated by a table look-up block using pre-stored data of track geometry. A MATLAB function then elaborates the reference length of the actuators.

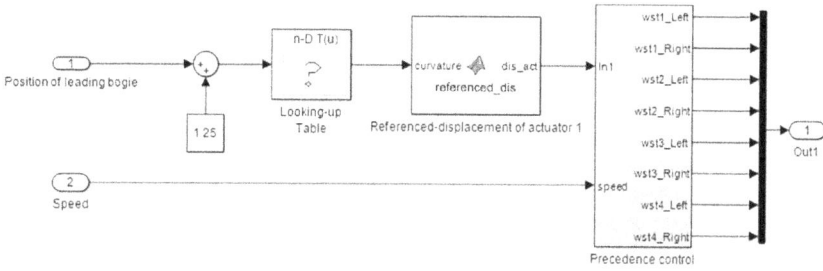

FIGURE 14.17 Reference generator block for Scheme 2.

The expanded view of the HSA model is shown in Figure 14.18. It is defined using Simscape and considers:

- The hydraulic circuit of the HSA (see Figure 14.14), visible in the central portion of the scheme
- The mechanical model of the piston, with springs and dampers representing the end mounts, located in the scheme just above the hydraulic circuit
- The 4-way valve with proportional servo-valve command, located below the hydraulic circuit
- The PI controller of the servo-valve, in the bottom right corner of the scheme

The other blocks in the scheme are used to compute and output other quantities of interest. The basic parameters of this model are listed in Table 14.4.

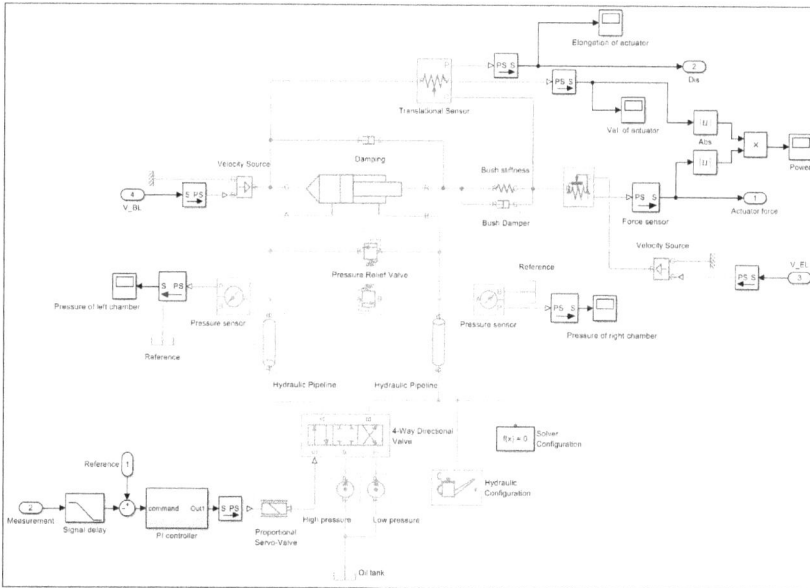

FIGURE 14.18 Simulink model of the HSA using Simscape.

TABLE 14.4

Key Parameters of the HSA

Parameter	Value
Maximum actuation force F_{max} (Design target)	20 kN
Piston area A	5×10^{-3} m^2
Piston stroke (Single side) L_p	10 mm
High pressure P_h	50 bar
Low pressure P_l	10 bar
Maximum moving speed v_{max} (Design target)	20 mm/s
Discharge coefficient C_d	0.7
Fluid density ρ	880 kg/m^3
Critical Reynolds number Re_{cr}	12
Fluid kinematic viscosity v	32×10^{-6} m^2/s
Maximum opening area of the orifice a_{max}	1.5×10^{-6} m^2
Fluid bulk modulus at atm. pressure	1.8×10^9 Pa

14.2.4 SIMULATION SCENARIOS AND RESULTS

The dynamics of the HSA model is tested first, considering the left end of the actuator attached to a fixed point and the other end connected to a serial spring as shown in Figure 14.19. The stiffness of the serial spring is set to 3 MN/m.

A piecewise linear reference displacement is used for the test, and the measured and reference displacement (change of length) of the actuator are compared in Figure 14.20(a). As long as the reference displacement is kept in the ±6 mm range, the actuator provides a fast response to the reference displacement applied and the difference of the two signals is very low, showing a fully satisfactory behavior of the controlled actuator. However, when the reference displacement exceeds 6 mm (last 4 s in the plot), the measured displacement cannot follow the reference value because the force required to deform the spring exceeds the maximum force of 20 kN that can be generated by the actuator (see Table 14.4). The time history of the actuator force is presented in Figure 14.20(b) and this plot clearly shows the force saturation effect. The piston velocity and actuator power signals are shown in Figures 14.20(c) and (d), respectively. Oscillations of the piston velocity are observed at points where

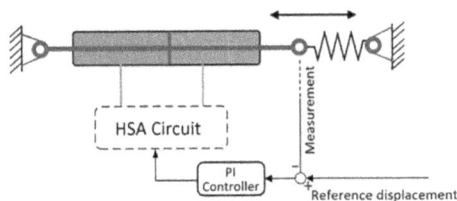

FIGURE 14.19 Simulation scenario to test the HSA model.

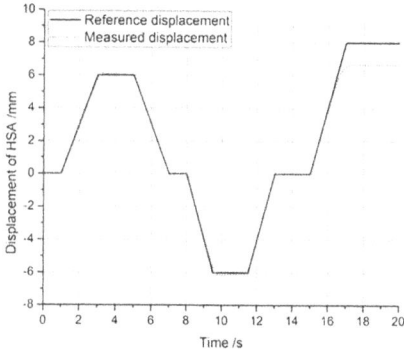

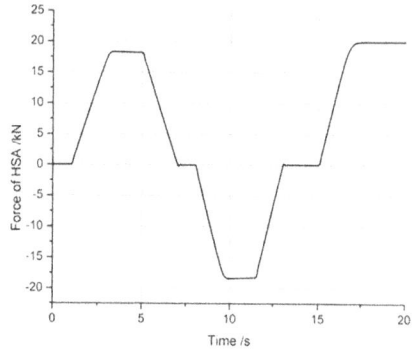

(a) Reference displacement vs measured displacement

(b) Actuator force

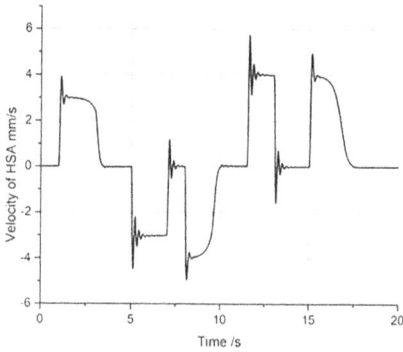

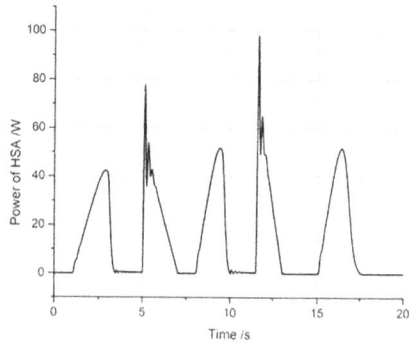

(c) Actuator velocity

(d) Actuator power

FIGURE 14.20 Results of the actuator test scenario: (a) reference and measured displacement, (b) actuator force, (c) actuator velocity, (d) actuator power.

the reference displacement implies a discontinuity in the velocity profile. These oscillations are caused by the flexibility of the end mounts, combined with the internal actuator flexibility caused by oil compressibility which is accounted for in the actuator model (the oil bulk modulus is provided in Table 14.4).

In Figures 14.21–14.24 the curving performance of the vehicle equipped with the active steering system is analyzed and compared to the baseline passive vehicle considering the two schemes proposed to estimate track curvature. The simulations are performed in a track layout starting with a tangent track (length 50 m) which is followed by the entry transition of the curve (length 100 m), a full curve section with radius $R = 400$ m and cant $h = 100$ mm (length 150 m), the exit transition (length 100 m), and finally a tangent track section. A constant speed of 70 km/h is assumed for the vehicle, corresponding to a non-compensated lateral acceleration of 0.3 m/s² approximately. To ease the analysis of results, track irregularities are not considered in the simulations.

The wear number of wheelsets under different schemes is compared in Figure 14.21. Active steering drastically reduces the wear number with respect to the baseline passive vehicle. For Scheme 1 the maximum wear number is approximately 75 N and for Scheme 2 approximately 25 N, to be compared with a maximum wear number of

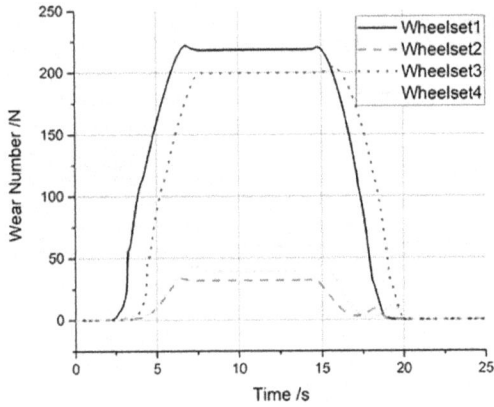

(a) Passive scheme

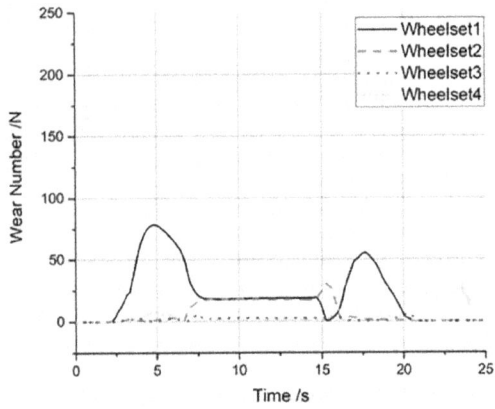

(b) Active steering scheme 1

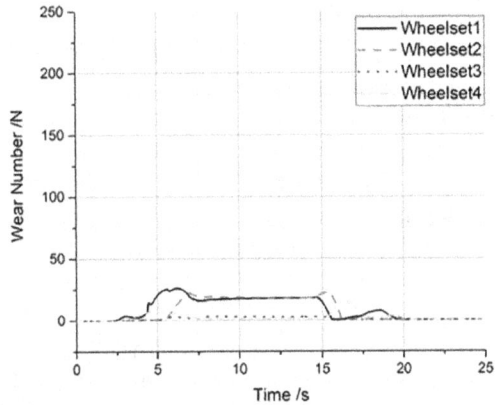

(c) Active steering scheme 2

FIGURE 14.21 Wear number of wheelsets with (a) passive scheme, (b) active scheme 1, and (c) active scheme 2.

approximately 225 N for the passive vehicle. However, using Scheme 1 (sensor-based) to define the track curvature, the wear number of the leading wheelset in the initial part of the entry transition, and final part of the exit transition is much larger than for the other wheelsets and comparable to the value obtained for the same wheelset of the passive vehicle. This is due to the delay in the steering action caused by the LPF, which is not compensated by the precedence in the leading wheelset. Active steering Scheme 2 can effectively eliminate this issue, as shown in Figure 14.21(c) where all the wheelsets have satisfactory performance.

Figure 14.22 presents the angle of attack of wheelsets which helps understanding the principle of active steering. In the passive vehicle, relatively large negative attack angles are generated on the leading wheelsets of the two bogies (wheelsets 1 and 3) due to the difference in the local radial direction at the two wheelsets in the same bogie and to the yaw rotation of the bogie. As a consequence, large creep forces are generated in these two wheelsets, pushing the wheelsets toward the outside of the curve and causing significant wear effects. Using the active steering system, the angles of attack of the wheelsets are decreased to much smaller values, which is a good indicator for an improved wheel-rail contact situation in curves. Note the angle of attack cannot be completely reduced to zero because the yaw angle of the bogies is not controlled by the steering system. The effect of the delay in the steering action applied to wheelset 1 when Scheme 1 is used is also visible in the time history of the angle of attack and justifies the larger wear number values observed in the transitions.

The reduction of the angle of attack produced by the steering system is not only beneficial to reduce wear effects, but also results in a more favorable situation in terms of lateral wheel-rail contact forces. The time history of the track shift forces is shown in Figure 14.23. For the vehicles with active steering, the track shift forces are nearly equal on the four wheelsets, at least in the full curve section, while a largely uneven distribution of the same forces is observed in the passive vehicle. As a result, the maximum absolute value of the track shift forces for the vehicles with active steering is approximately 7 kN for Scheme 1 and less than 4 kN for Scheme 2, against a maximum absolute value exceeding 10 kN for the passive vehicle. Once more, a better performance is observed for Scheme 2 compared to Scheme 1, as delay in the actuation of steering can be fully eliminated.

Finally, Figure 14.24 shows the time histories of the derailment coefficients. These are the ratios of the lateral force over vertical force in the outer wheels of the vehicle and are taken as an indicator of the potential of the wheelset to perform a flange climb derailment [11]. Once more, it is found that the use of active steering, according to both Schemes 1 and 2, leads to a maximum absolute value of the derailment coefficient well below 0.1, while the same value is approximately 0.3 for the passive vehicle. It should be noted that the limit value for the derailment coefficient is usually assumed to be 0.8, so in the curving condition considered here the steering system is not strictly needed to ensure the safe running of the vehicle with respect to derailment. However, there are other more challenging running conditions, particularly curves with greatly reduced radii and weather conditions leading to a higher value of the friction coefficient, which might lead to a dangerous running condition for the passive vehicle. In these conditions, the use of the steering system is expected to significantly enhance running safety, in addition to reducing wear effects.

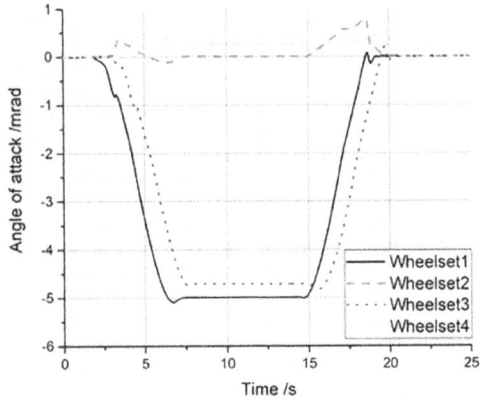

(a) Passive scheme

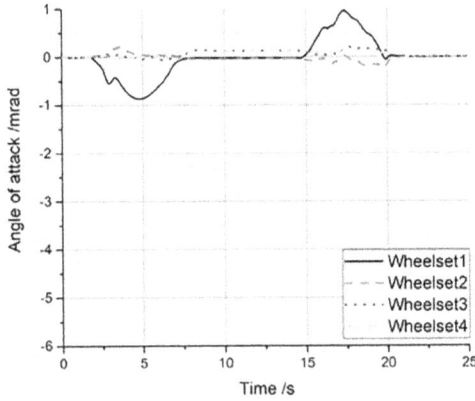

(b) Active steering scheme 1

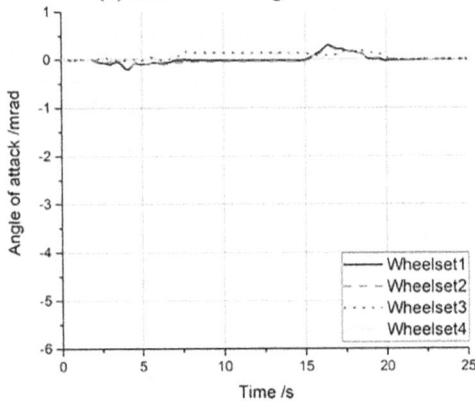

(c) Active steering scheme 2

FIGURE 14.22 Angle of attack of wheelsets with (a) passive scheme, (b) active Scheme 1, and (c) active Scheme 2.

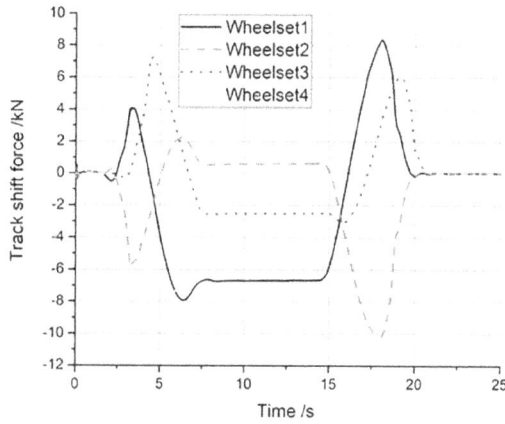

(a) Passive scheme

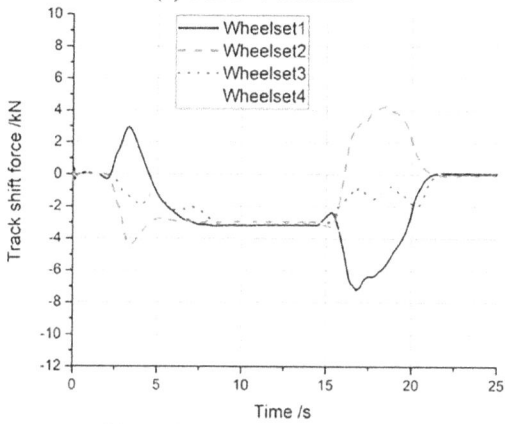

(b) Active steering scheme 1

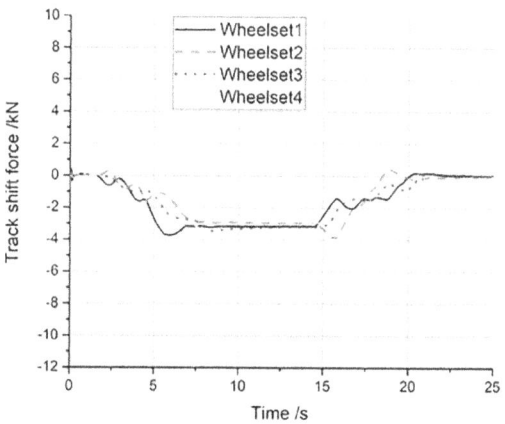

(c) Active steering scheme 2

FIGURE 14.23 Track shift force of wheelsets with (a) passive scheme, (b) active Scheme 1, and (c) active Scheme 2.

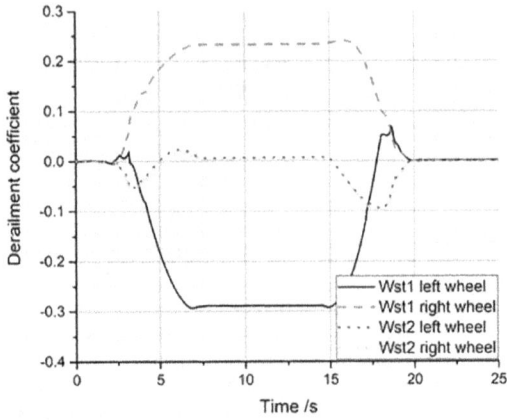

(a) Passive scheme

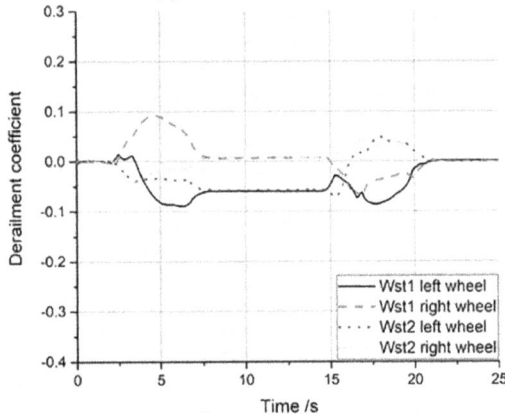

(b) Active steering scheme 1

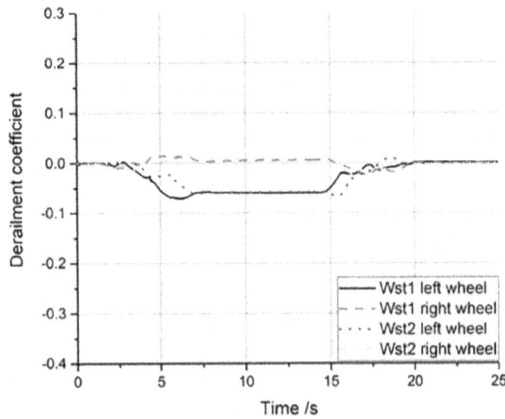

(c) Active steering scheme 2

FIGURE 14.24 Derailment coefficient of wheelsets with (a) passive scheme, (b) active Scheme 1, and (c) active Scheme 2.

14.3 CASE C: MODELING OF A HEAVY HAUL DIESEL-ELECTRIC LOCOMOTIVE TRACTION POWER SYSTEM

This case study requires the development of a heavy haul diesel-electric locomotive model with a bogie traction control strategy based on the co-simulation modeling approach described in Chapter 7. The model outputs should deliver traction effort and dynamic braking effort characteristics for each throttle (notch) and dynamic brake handle positions, respectively.

14.3.1 MODELING CONCEPT

The modeling concept for this study is shown in Figure 14.25 and is based on the co-simulation approach developed in [12–15]. The co-simulation simulation is built based on the shared library co-simulation approach, i.e., the shared library for this Simulink model was built based on the procedure described in Section 6.5.1 of Chapter 7. The mechanical model of a heavy haul locomotive is described in detail in Section 7.6.1.1. The inputs and outputs of the co-simulation interface are the same as used in traction studies and are shown in Figure 14.25, i.e., motor torques, angular speeds of motors, and locomotive linear speed.

14.3.1.1 Modeling of the Power System

An overview of a typical bogie power system design is presented in Figure 3.1 of Chapter 3. The locomotive has two bogies with three axles in a Co-Co configuration. It has one traction inverter for each bogie. Each inverter supplies three traction motors which are connected in parallel. The inverter controls will utilize either direct torque control (DTC) [16–18] or field-oriented control (FOC) [19]. Both methods independently control the machine current component responsible for establishing the air gap flux and the current component responsible for torque production. Both can accurately control the machine torque and are capable of response times well below a hundred milliseconds.

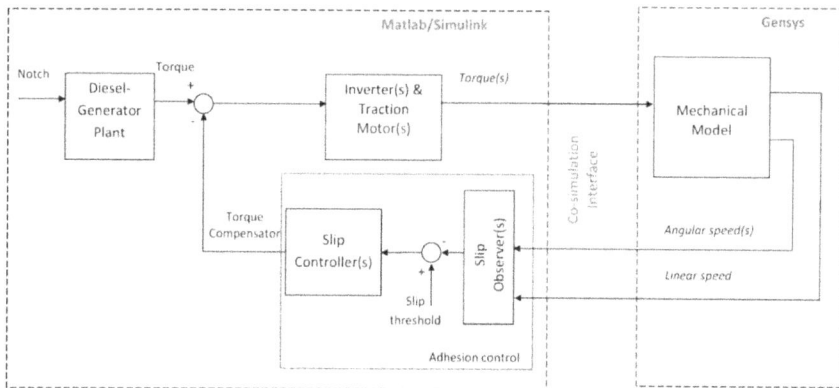

FIGURE 14.25 Co-simulation modeling of a heavy haul diesel-electric locomotive. (From [14], with permission.)

TABLE 14.5
Diesel Engine rpm and Power Capability

Notch Position	Rotational Speed	Power
Idle	200 rpm	0 kW
1	269 rpm	149 kW
2	343 rpm	329 kW
3	490 rpm	746 kW
4	568 rpm	1060 kW
5	651 rpm	1405 kW
6	729 rpm	2042 kW
7	820 rpm	2693 kW
8	904 rpm	3095 kW

The generator subsystem provides the motive power. At any point in time, the generator system will have a power capability that is dependent upon the rotational speed of the diesel prime mover. This is set by the notch position as shown in Table 14.5. These powers are steady state values. As the notch position changes, the diesel prime mover will require many seconds, or even tens of seconds, to respond. It is normal to impose a ramp rate limitation to model the time delay of the locomotive. In this case, a ramp rate limit of 28 rpm/s is imposed which means the diesel engine will require 25 s to accelerate from idle to full speed. While the diesel is transitioning between notch levels, the power is estimated by linear interpolation.

The generator is voltage controlled with the terminal voltage reference depending on the throttle notch setting and the traction motor rotational velocity depending directly upon the locomotive's linear velocity. This relationship is shown in Table 14.6. Linear interpolation is used for intermediate speed values.

TABLE 14.6
Generator Voltage Lookup Table Data

Notch	Locomotive Velocity m/s Traction Machine rpm							
	0 m/s 0 rpm	0.97 m/s 93 rpm	7.66 m/s 735 rpm	13.2 m/s 1265 rpm	14.7 m/s 1415 rpm	26.1 m/s 2500 rpm	27.6 m/s 2655 rpm	36.5 m/s 3500 rpm
Idle	620 V	620 V	620 V	620 V	620 V	620 V	620 V	620 V
1	620 V	620 V	620 V	620 V	620 V	620 V	620 V	620 V
2	880 V	880 V	880 V	880 V	880 V	880 V	880 V	880 V
3	1200 V	1200 V	1295 V	1295 V	1295 V	1295 V	1295 V	1295 V
4	1410 V	1410 V	1540 V	1540 V	1540 V	1540 V	1540 V	1540 V
5	1500 V	1500 V	1760 V	1760 V	1760 V	1760 V	1760 V	1760 V
6	1500 V	1500 V	2130 V	2130 V	2130 V	2130 V	2130 V	2130 V
7	1500 V	1500 V	2400 V	2400 V	2430 V	2430 V	2430 V	2430 V
8	1500 V	1500 V	2400 V	2400 V	2500 V	2500 V	2600 V	2600 V

TABLE 14.7
Traction motor Parameters

Parameter	Value	Per Unit Value
Real power rating	500 kW	0.85
Apparent power rating	585 kVA	1.0
Line to line voltage	2027 Vrms	1.0
Number of phases	3	
Base frequency	29 Hz	1.0
Number of poles	4	
Stator resistance	132 mΩ	1.9%
Stator reactance	3.14 mH	8.1%
Rotor resistance	132 mΩ	1.9%
Rotor reactance	3.14 mH	8.1%
Magnetizing resistance	1240 Ω	176
Magnetizing reactance	117 mH	3.0

Source: From [15], with permission.

The traction motor parameters used in this chapter are shown in Table 14.7. In this study, parts of the MATLAB Simulink AC3 drive module was utilized to build the drive system model [20]. The AC3 model is included in the Simpower system libraries and is an indirect rotor based FOC drive that uses a state space averaged approach.

For a generalized induction machine, the fundamental motor equation is:

$$
\begin{bmatrix} \overline{v}_s(t) \\ \overline{v_r}'(t) \end{bmatrix} = \begin{bmatrix} r_s & 0 \\ 0 & r_r \end{bmatrix} \begin{bmatrix} \overline{i}_s(t) \\ \overline{i_r}'(t) \end{bmatrix} + \frac{d}{dt} \begin{bmatrix} L_s & L_m \\ L_m & L_r \end{bmatrix} \begin{bmatrix} \overline{i}_s(t) \\ \overline{i_r}'(t) \end{bmatrix}
$$

$$
- j\omega_r \begin{bmatrix} 0 & 0 \\ L_m & L_r \end{bmatrix} \begin{bmatrix} \overline{i}_s(t) \\ \overline{i_r}'(t) \end{bmatrix}
$$

(14.26)

where $\overline{v}_s(t)$ and $\overline{i}_s(t)$ are the stator voltage and current space phasors;

$\overline{v_r}'(t)$ and $\overline{i_r}'(t)$ are the rotor voltage and current space phasors referred to the stator side;

r_s and r_r are the stator and rotor resistances referred to the stator side;

L_s and L_r are the stator and rotor leakage inductances referred to the stator side;

L_m is the stator magnetizing inductance.

Two equivalent expressions for the torque production are:

$$t_e(t) = -\frac{3}{2}\left(L_r\, \overline{i_r}'(t) + L_m\, \overline{i_s}(t) \right) \overline{i_r}'(t) \tag{14.27}$$

$$t_e(t) = -\frac{3}{2}\, \overline{\lambda_r}'(t)\, \overline{i_r}'(t) \tag{14.28}$$

where $\overline{\lambda_r}'(t)$ is the rotor flux referred to the stator side.

The indirect form of field-oriented torque control is shown in Figure 14.26. In this case, the drive accepts two control input references. These are the flux reference λ_{ref} and the torque reference T_{ref}. The motor is current controlled in a Direct and Quadrature (DQ) rotating reference frame system that decouples the magnetization of the machine from the torque production. The direct axis current establishes the magnetizing flux while the quadrature axis current is responsible for torque production.

It is normal practice to operate the machine at its rated flux in the low speed or torque limited region [19]. Equation (14.28) gives the highest torque production for a given rotor current. The required machine terminal voltage increases linearly with speed. Once there is insufficient voltage to maintain the rated flux, the machine enters the field weakening or constant horsepower region.

For the machines used in this study, field weakening is introduced above 29 Hz to maintain the stator voltage within the 2915 V_{peak} fundamental voltage that can be achieved with a 2650 V DC bus [15] and a flux of 8.5 Wb peak was set to maintain the motor voltages within the inverter limits.

The torque control system provides torque set points to the inverters in response to the throttle notch settings. Additional limits on torque, based on the locomotive speed, are imposed to maintain the DC power demand below the instantaneous capability of the diesel generator subsystem. For this locomotive, the notch torque settings are shown in Table 14.8. The bogie inverters each receive a torque setting equivalent to 50% of the locomotive notch setting. As previously mentioned, a torque rate ramp limit is set so that the transition from zero to the full rating of the locomotive matches the ramp rate of the diesel prime mover and requires 25 s.

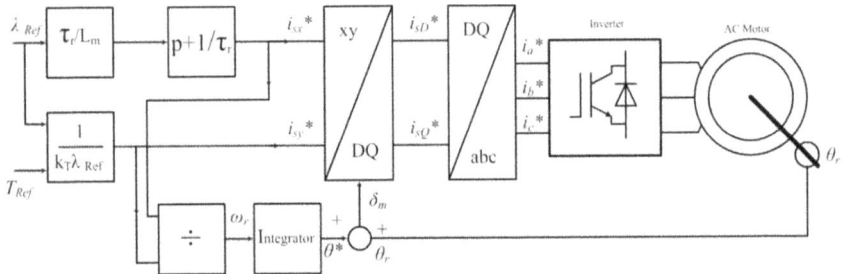

FIGURE 14.26 Indirect field-oriented control. (From [15], with permission.)

TABLE 14.8
Locomotive Torque Limits by Notch

Notch Position	Torque Limit
Idle	0 kNm
1	8.7 kNm
2	17.3 kNm
3	26.0 kNm
4	34.7 kNm
5	43.3 kNm
6	52.0 kNm
7	60.7 kNm
8	69.4 kNm

AC traction inverters are naturally regenerative. The application of a negative torque set point produces a retarding torque and power naturally flows back to the DC bus. This will charge the DC bus capacitors and steadily increase the DC bus voltage. To limit the voltage rise at the DC bus, this power must be dissipated in a brake resistor. The braking resistors are controlled through a DC-DC chopper and are continuously variable up to their power limit.

Dynamic braking in an AC drive is available to almost zero speed with no reduction in torque. To produce torque the machine magnetizing flux has to be maintained. An adequate voltage is required on the DC bus for full torque operation at low speed. It is normal to operate the traction alternator at approximately 50% of rated speed to maintain a lower floor limit for the DC bus voltage during braking. The power requirement is low, typically less than one percent of the machine ratings, as only the magnetization core losses for the AC machines need to be provided. For this locomotive, the dynamic brake handle torque settings are shown in Table 14.9.

TABLE 14.9
Locomotive Torque Limits by Brake Handle Position

Brake Handle Position	Torque Limit
Idle	0.0 kNm
1	−4.7 kNm
2	−9.4 kNm
3	−14.1 kNm
4	−18.8 kNm
5	−23.5 kNm
6	−28.2 kNm
7	−32.9 kNm
8	−37.6 kNm

14.3.1.2 Modeling of the Adhesion Control

Considering that a bogie traction control is modeled in this case, it is reasonable for this study to describe a modeling of this control per a single bogie as shown in Figure 14.27, where T_{ref} is the reference torque (based on the notch position); T_{ref*} is the reference torque generated by the control system; Tin is the input motor torque; T_{motors} is the traction torque applied to the rotor(s) of the traction motor(s) and then through a gear box to the wheelsets; ΔT is the torque reduction; ω_1, ω_2, ω_3 are the angular velocities of the front, middle, and rear axles of the bogie respectively; ω is a maximal angular velocity detected from the angular velocities of the front, middle, and rear axles of the bogie; s_{est} is the estimated longitudinal slip; $s_{threshold}$ is the reference value of the longitudinal slip.

A value of the estimated longitudinal slip is calculated based on the following relation:

$$s_{est} = \frac{\omega \cdot r - V}{V} \tag{14.29}$$

where ω is the maximum axle angular velocity; V is the locomotive speed; r is the nominal rolling radius of the wheels.

In the case of the traction mode, the slip estimator works based on Equation (14.29). A slip error can be defined as:

$$e = s_{threshold} - s_{est} \tag{14.30}$$

As one can see in Figure 14.5, the reference slip has a positive value. The slip control is active for the traction mode when the estimated slip value, s_{est}, is higher than the

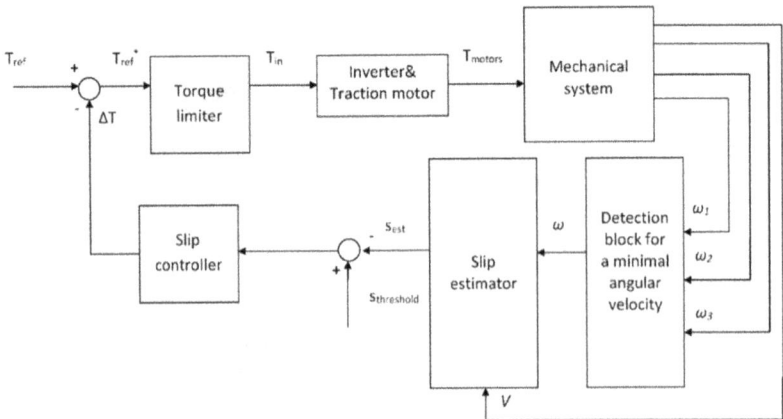

FIGURE 14.27 Adhesion control strategy for a single bogie that uses a bogie traction control. (From [14], with permission.)

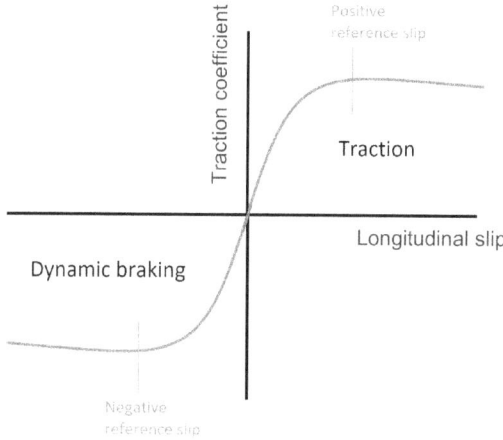

FIGURE 14.28 Slip references in the modeling approach for adhesion control strategy.

reference slip, $s_{\text{threshold}}$, which has a positive value. In this case, the slip error correction, e_c, can be found as:

$$\text{for} \quad e > 0, e_c = 0 \tag{14.31}$$

$$\text{else} \quad e_c = e \tag{14.32}$$

In case of the dynamic braking mode, the slip error can be also calculated by Equation (14.30). As one can see in Figure 14.28, the reference slip has a negative value. The slip control is active for the dynamic braking mode when the estimated slip value, s_{est}, is lower than the optimal slip, $s_{\text{threshold}}$. In this case, the slip error correction, e_c, can be found as:

$$\text{for} \quad e < 0, e_c = 0 \tag{14.33}$$

$$\text{else} \quad e_c = e \tag{14.34}$$

The slip controller is a simple controller with proportional action (P controller), which uses the slip error correction, e_c, as the input to the controller. The control law can be represented by the following equation:

$$\Delta T = K_P \cdot e_c \tag{14.35}$$

where K_P is the proportional gain.

14.3.2 IMPLEMENTATION IN SIMULINK

The full model in Simulink used for this study is shown in Figure 14.29. The model has been developed with the application of elements included in Simscape Electrical libraries. For this model, the discrete solver with a fixed time step of 2e–5 s was used

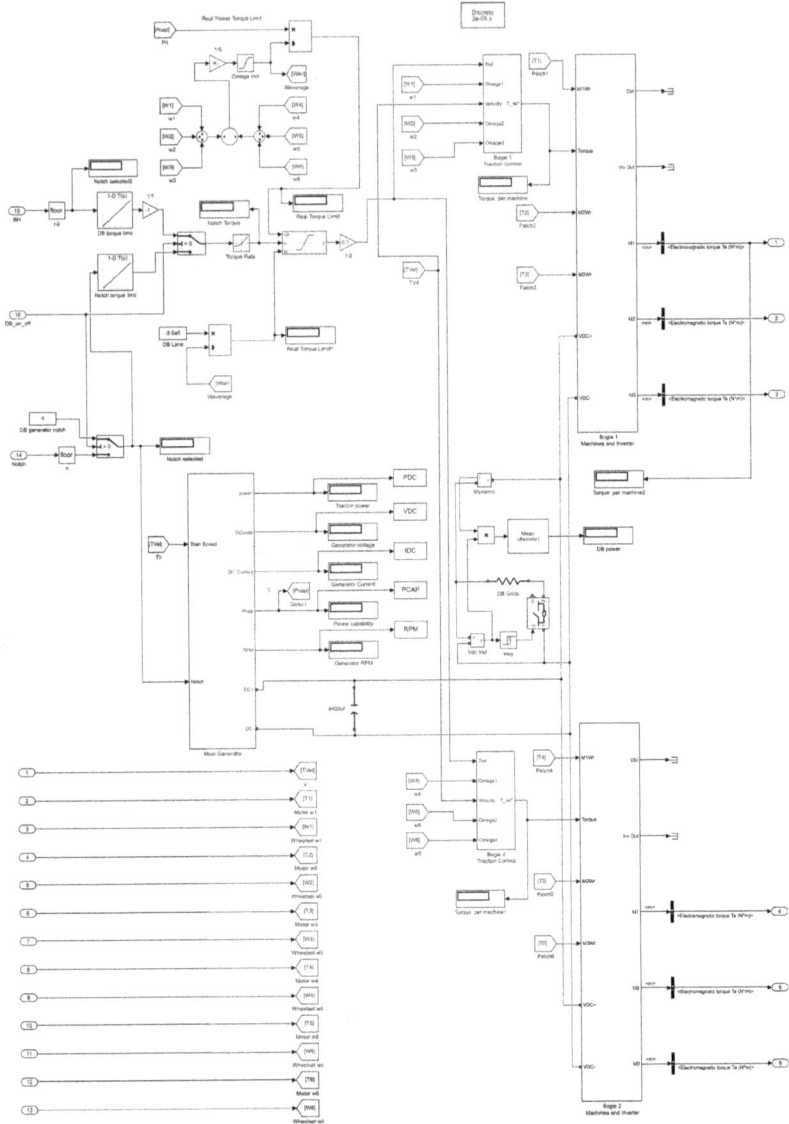

FIGURE 14.29   Full locomotive traction system model in Simulink.

in Simulink. An advanced simulation model for the power traction system of the diesel-electric locomotive includes the following subblocks:

- Traction power is represented by the block called *Main Generator*. The subsystem for this block is shown in Figure 14.30. The main model, shown in Figure 14.29, includes the DC bus capacitor which is responsible for balancing the instantaneous demand of the inverter drives and the output available from the main generator.

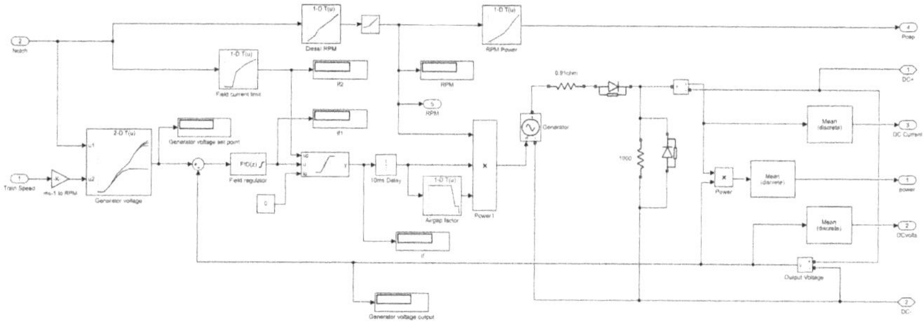

FIGURE 14.30 Traction power subblock in Simulink.

• Two physics-based drive system blocks (see *Machines and Inverter* blocks in Figure 14.29) for the inventers and traction machines (motors). Each block is designated for one bogie and it covers one inverter and three traction motors. The drive is a field-oriented controller (FOC) with torque control that is adapted from the Simulink library as shown in Figure 14.1. The torque controller operates with a 100 µs sampling time. The FOC block uses a 20 µs sampling time. The drive subsystem is shown in Figure 14.31.

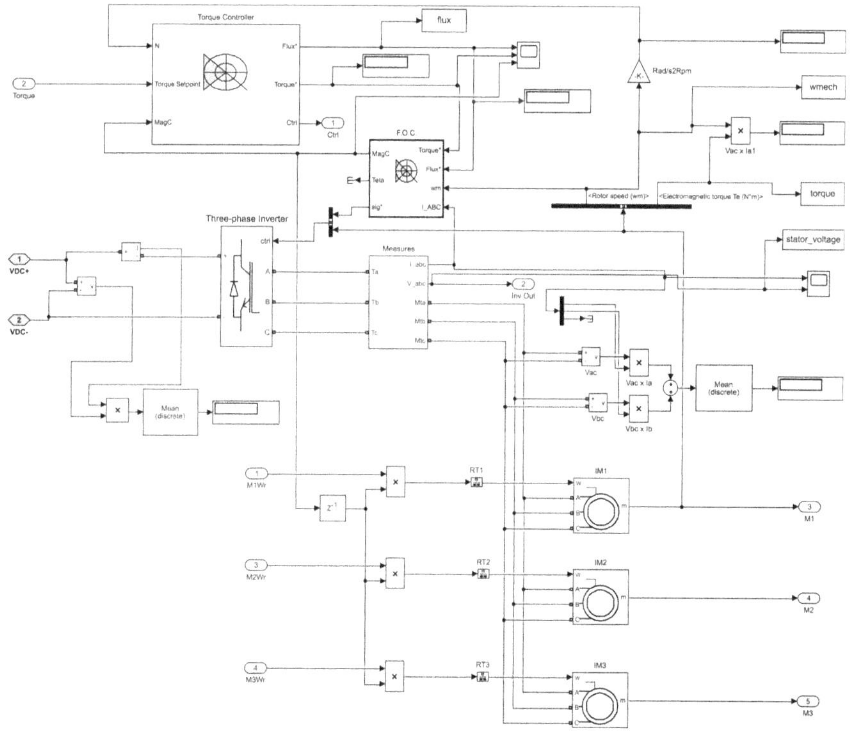

FIGURE 14.31 Drive system model subblock in Simulink.

- The torque management subsystem is represented by two slip controllers for a bogie traction control architecture used in the locomotive. Each controller is shown as a *Traction Control* block and it provides the input motor torque values to the designated *Machines and Inverter* block as shown in Figure 14.29.
- The dynamic brake resistor grid has been implemented inside of the main model as shown in Figure 14.29. In a simulation model, the alternator can be operated at a mid-range notch, typically notch 4, during braking to provide a voltage floor.

For the co-simulation approach between the multi-body software and Simulink, it is necessary to define the physical variables used for the input and output data. The Simulink model of the locomotive has 6 outputs and 16 inputs.

The model outputs are:

1. Torque of traction motor 1;
2. Torque of traction motor 2;
3. Torque of traction motor 3;
4. Torque of traction motor 4;
5. Torque of traction motor 5; and
6. Torque of traction motor 6.

The model inputs are:

1. Locomotive speed;
2. Angular velocity of traction motor 1;
3. Angular velocity of wheelset 1;
4. Angular velocity of traction motor 2;
5. Angular velocity of wheelset 2;
6. Angular velocity of traction motor 3;
7. Angular velocity of wheelset 3;
8. Angular velocity of traction motor 4;
9. Angular velocity of wheelset 4;
10. Angular velocity of traction motor 5;
11. Angular velocity of wheelset 5;
12. Angular velocity of traction motor 6;
13. Angular velocity of wheelset 6;
14. Notch position;
15. Brake Handle position; and
16. DB usage flag (on/off).

For the co-simulation purposes, the model of the traction power system has been represented by a shared library, generated in Simulink as described in Chapter 7, for use in the locomotive multi-body models created in Gensys for each notch or dynamic brake position, in order to perform tasks defined for this case study.

14.3.3 SIMULATION SCENARIOS AND RESULTS

In both traction and dynamic braking effort tests, the total simulation time in each simulation test was set to 235 s. The co-simulation time step was set to 1 ms. The locomotive was running on tangent track with no track irregularities.

In order to deliver the TE characteristics of a locomotive, the locomotive speed in each test (i.e., in each locomotive model with a specified notch position) has been set as a variable parameter that changes from 0 km/h to the maximum (design) speed of the locomotive involved in the test. The speed profile for this test is shown in Figure 14.32. All tests are performed under dry rail friction conditions. In order to accurately reproduce the traction behavior of the locomotive, the equivalent draw bar force has been applied in each multi-body model through a sky-hook element attached to the locomotive car body. In total, 8 simulation cases have been performed in order to deliver the traction effort characteristics presented in Figure 14.33.

In order to deliver the DB characteristics of the full locomotive model, the locomotive speed in each test has been set as a variable parameter that changes from the maximum (design) speed of the locomotive involved in the test to 0 km/h. The speed profile for this test is shown in Figure 14.34. In order to accurately reproduce the braking behavior of the locomotive, the equivalent draw bar force has been applied in each multi-body model through a sky-hook element attached to the locomotive car body. In total, 8 simulation cases have been performed in order to deliver the dynamic braking characteristics presented in Figures 14.35.

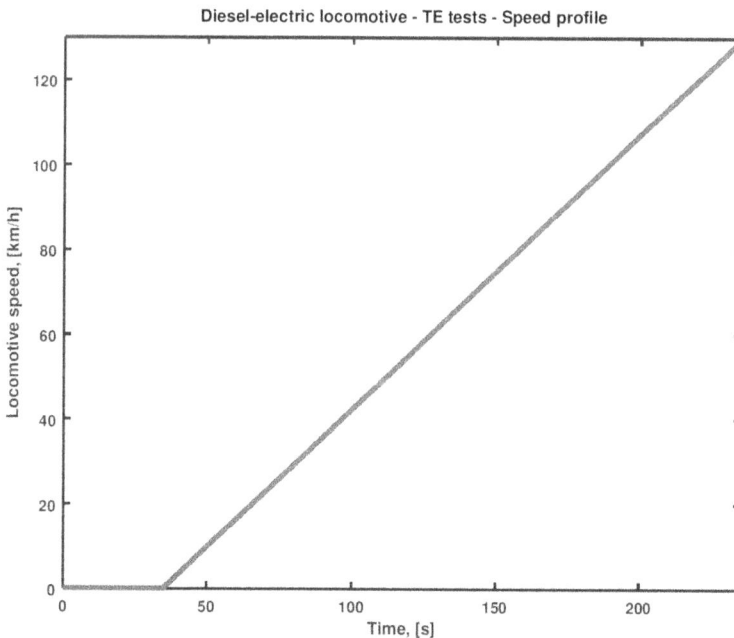

FIGURE 14.32 Speed profile in time domain for TE test cases.

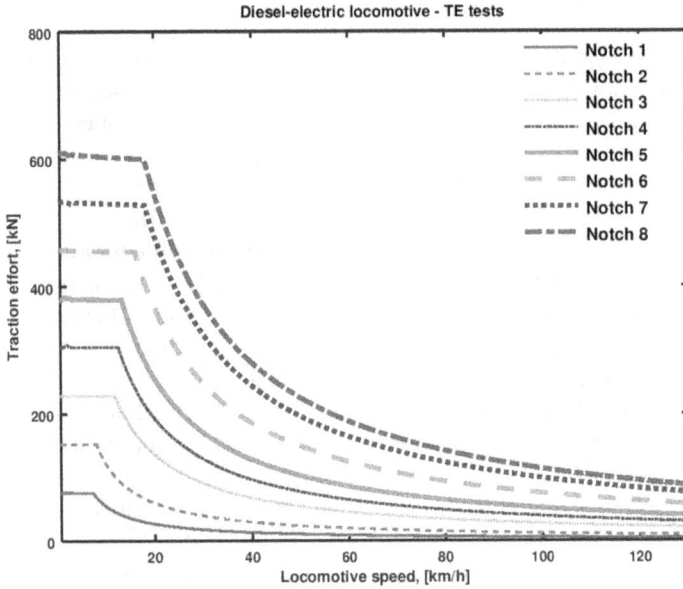

FIGURE 14.33 TE versus locomotive speed characteristics for each notch position for the modeled locomotive.

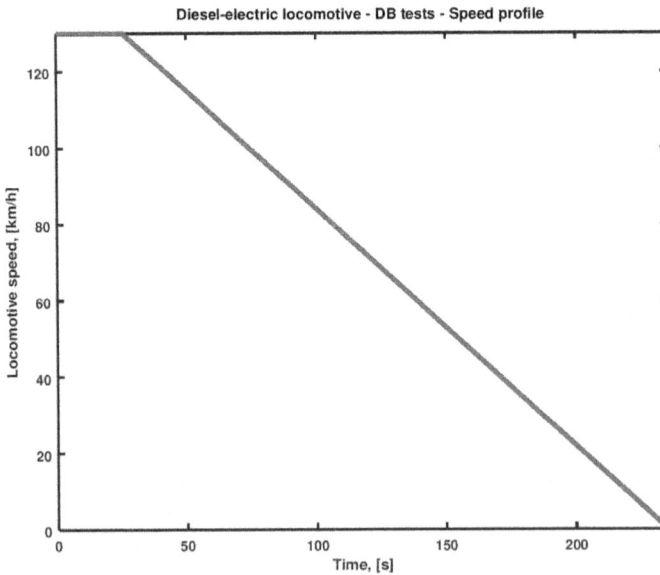

FIGURE 14.34 Speed profile in time domain for DB test cases.

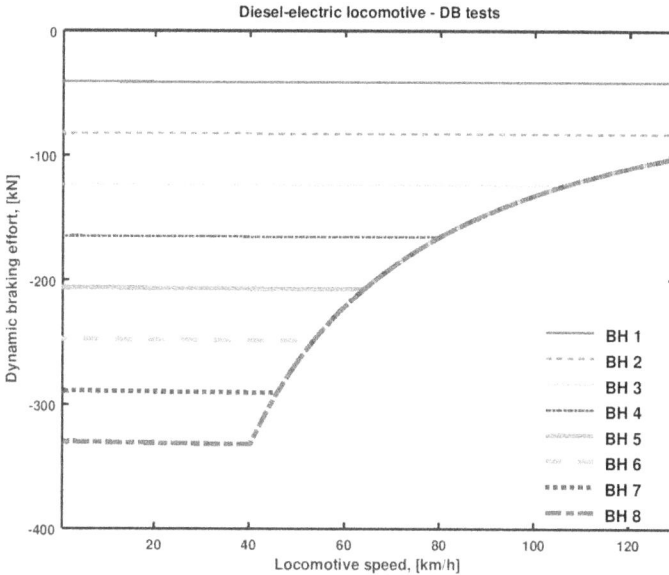

FIGURE 14.35 DB effort versus locomotive speed characteristics for each brake handle position for the modeled locomotive.

14.4 CASE D: MODELING OF A HEAVY HAUL HYBRID LOCOMOTIVE

In this case study, it is necessary to model an energy storage system (ESS) for a hybrid locomotive based on the locomotive design specification used in Section 14.3 – Case C. The methodology used in this case study is the same as in the Case C study. In addition to the Case C deliverables of TE and DB characteristics, this Case D study should also provide the State of Charge (SOC) for the ESS for all simulated cases.

This case study is based on the works published in [21, 22].

14.4.1 LOCOMOTIVE DESIGN MODIFICATION

In the case of the hybrid locomotive, we assume that the battery system mass should be designed to be equal to 50–60 tons inclusive of inverters, and it should fit in the existing locomotive design envelope considering that the diesel engine, its cooling system, the main generator, and fuel tanks are removed to produce the modified hybrid version. After such a re-design, the hybrid locomotive still retains the same weight as used in the Case C study.

The modeling concept for a power traction system of the hybrid locomotive, shown in Figure 14.36, has the energy storage system that replaces three major components of the diesel-electric locomotive, namely the diesel-engine, an alternator, and a rectifier. The comparison of the main characteristics of the locomotives used in Cases C and D is shown in Table 14.10.

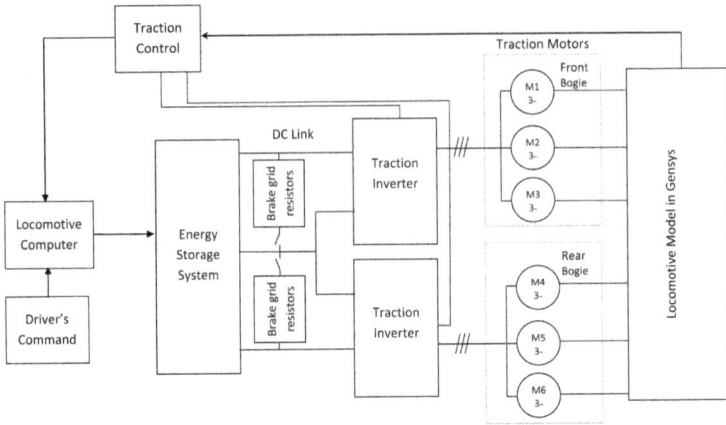

FIGURE 14.36 Modeling concept of hybrid locomotive power traction system.

TABLE 14.10
Main Characteristics of the Diesel Electric and Hybrid Locomotives

General Information	Case C	Case D
Power type	Diesel-electric	Electric (battery)
Wheel arrangement (UIC/AAR)	C_o-C_o / C-C	C_o-C_o / C-C
Dimensions (approximately):		
Length, mm	21,200	21,200
• Width, mm	2950	2950
• Height, mm	4245	4245
• Locomotive weight, tons	136.2	136.2
Axle load, tons	22.7	22.7
Topology of electric power transmission system	AC-DC-AC	DC-AC
Wheel diameter, mm	1066	1066
Power Plant and Battery System Data		
Power output (gross), kW	3356	-
Battery system maximum discharge/ charge capacity, kW		3100/5000
Battery system specification, MWh		5
Performance Figures		
Traction power, kW	3100	3100
Tractive effort:		
Maximum starting traction effort, kN	600	600
Maximum continuous traction effort, kN	520	520
Maximum dynamic braking power, kW	-	5000
Maximum dynamic braking effort, kN	325	325
	(from 50 to near zero km/h)	(from 50 to near zero km/h)

14.4.2 MODELING OF ESS TRACTION SYSTEM FOR THE HYBRID LOCOMOTIVE

The developed ESS model is very close to a direct replacement – as long as it is charged. It has an internal voltage table that is the same as the alternator (see Table 14.6). The model has been developed in a similar manner as a model published in [21]. In this model, the battery has a 15% internal reserve, and it would not foldback until the battery power is 15% higher than the power capability signal. The battery is interfaced to the DC link via a DC-DC converter. The maximum voltage and current of the DC link were set to 2650 V and 1886 A, respectively. The battery unit is predicted to have a mass of 40–50 tons.

The designed ESS can discharge at 3100 kW and accepts current charging at 5000 kW. There are 8 braking and powering notches. Torque and power limits were independently set for each notch via look up tables. Brake notches having the same torque limits and a power limit of equal magnitude to the corresponding powering notches were added in the model – i.e., 8 powering is 3100 kW and −8 braking is −3100 kW. In normal dynamic braking, the DC link voltage, DB design and battery voltage settings are such that the battery would absorb everything, and the DB is only active once the battery is nearly full. If the SoC for the battery pack falls below 5%, the allowable discharge rates will be proportionally reduced to zero at 0.5% SoC. If SoC achieves a 100% signal, it means that the battery is charged up to at least 85%.

14.4.3 IMPLEMENTATION IN SIMULINK

The full model in Simulink used for this study is shown in Figure 14.37. The developed model represents a modification of the model shown in Figure 14.29 and it was also developed with the application of elements included in Simscape Electrical libraries. For this model, the discrete solver with a fixed time step of 2e−5 s was used in Simulink. In comparison with the diesel-electric locomotive model, three modifications were implemented:

- Traction power block (see *Main Generator* in Figure 14.29) was replaced with the battery block called *5 MWh Battery system* as shown in Figure 14.37. The subsystem for this block is shown in Figure 14.38.
- The main model, shown in Figure 14.37, includes some modifications on the DC bus near the capacitor in order to limit voltages which are available to the ESS for the instantaneous demand of the inverter drives.
- The dynamic brake resistor grid scheme has been modified inside of the main model as shown in Figure 14.37. The proposed modification allows to use dynamic braking only in cases when the ESS is fully charged.

For the co-simulation approach between the multi-body software and Simulink, it is necessary to define the physical variables used for the input and output data. The Simulink model of the locomotive has 7 outputs and 16 inputs.

The model outputs are:

1. Torque of traction motor 1;
2. Torque of traction motor 2;

3. Torque of traction motor 3;
4. Torque of traction motor 4;
5. Torque of traction motor 5;
6. Torque of traction motor 6; and
7. State of charge of the ESS.

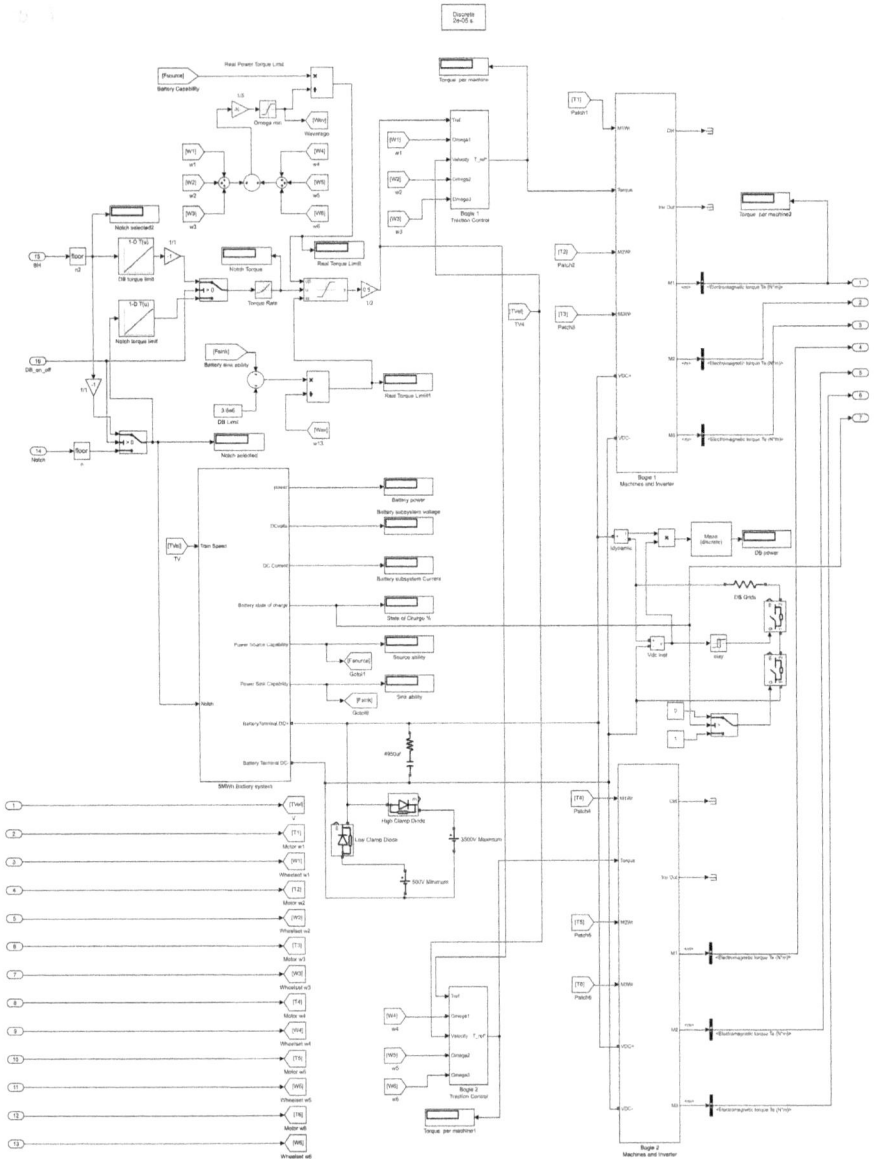

FIGURE 14.37 Full hybrid locomotive traction system model in Simulink.

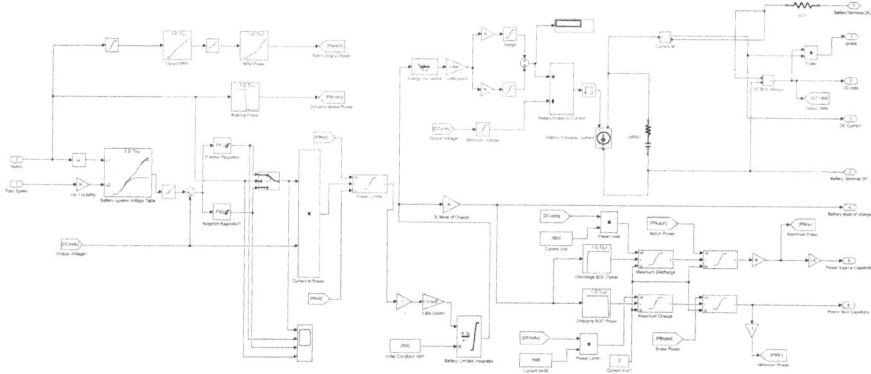

FIGURE 14.38 Energy storage system subblock in Simulink.

The model inputs are:

1. Locomotive speed;
2. Angular velocity of traction motor 1;
3. Angular velocity of wheelset 1;
4. Angular velocity of traction motor 2;
5. Angular velocity of wheelset 2;
6. Angular velocity of traction motor 3;
7. Angular velocity of wheelset 3;
8. Angular velocity of traction motor 4;
9. Angular velocity of wheelset 4;
10. Angular velocity of traction motor 5;
11. Angular velocity of wheelset 5;
12. Angular velocity of traction motor 6;
13. Angular velocity of wheelset 6;
14. Notch position;
15. Brake Handle position; and
16. DB usage flag (on/off).

For the co-simulation purposes, the model of the traction power system has been represented by a shared library, generated in Simulink as described in Chapter 7, for use in the locomotive multi-body models created in Gensys for each notch or dynamic brake position, in order to perform tasks defined for this case study.

14.4.4 SIMULATION SCENARIOS AND RESULTS

In both traction and dynamic braking effort tests, the battery has the initial charge of 50%. The total simulation time in each simulation test was set to 235 s. The co-simulation time step was set to 1 ms. The locomotive was running on tangent track with no track irregularities.

In order to deliver the TE characteristics of a locomotive, the locomotive speed in each test (i.e., in each locomotive model with a specified notch position) has been set as a variable parameter that changes from 0 km/h to the maximum (design) speed of the locomotive involved in the test. The same speed profile was used for these tests as shown in Figure 14.32. All tests are performed under dry rail friction conditions. In order to accurately reproduce the traction behavior of the locomotive, the equivalent draw bar force has been applied in each multi-body model through a sky-hook element attached to the locomotive car body. In total, eight simulation cases have been performed in order to deliver the traction effort characteristics presented in Figure 14.39. The state of charge results for the developed ESS system for these simulation cases are shown in Figure 14.40 and they confirm battery discharging at each notch position.

In order to deliver the DB characteristics of full locomotive models, the locomotive speed in each test has been set as a variable parameter that changes from the maximum (design) speed of the locomotive involved in the test to 0 km/h. The same speed profile was used for these tests as shown in Figure 14.34. In order to accurately reproduce the braking behavior of the locomotive, the equivalent draw bar force has been applied in each multi-body model through a sky-hook element attached to the locomotive car body. In total, eight simulation cases have been performed in order to deliver the dynamic braking characteristics presented in Figures 14.41. The state of charge results for the developed ESS system for these simulation cases are shown in Figure 14.42 and they confirm battery charging at each brake handle position. At the time of ending the simulations, the drop of SoC values is observed because a locomotive approaches zero speed in the braking mode as shown in Figure 14.34.

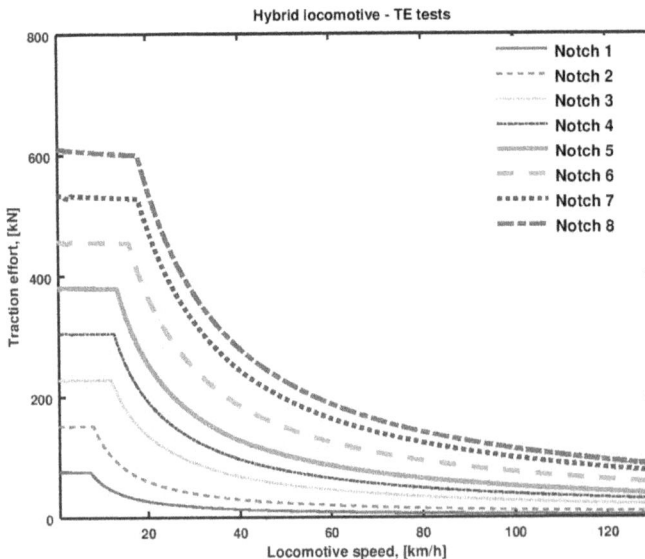

FIGURE 14.39 TE versus locomotive speed characteristics for each notch position of the modeled hybrid locomotive.

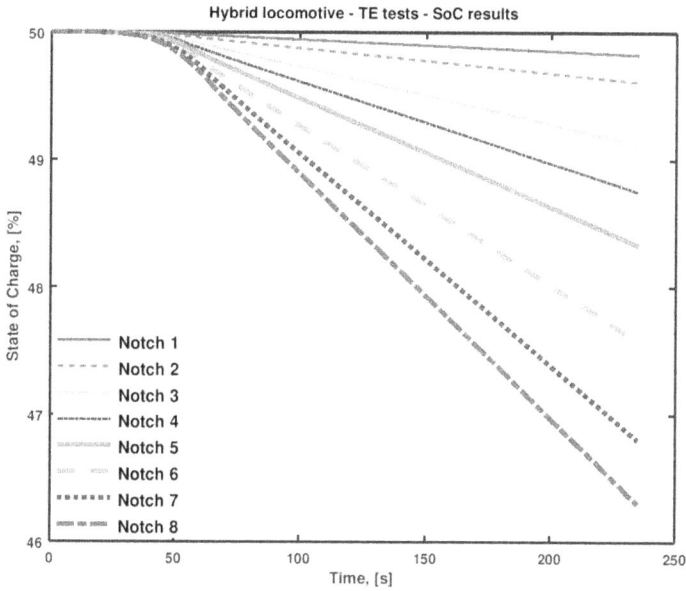

FIGURE 14.40   SoC results versus in time-domain for each notch position of the modeled hybrid locomotive in a traction mode.

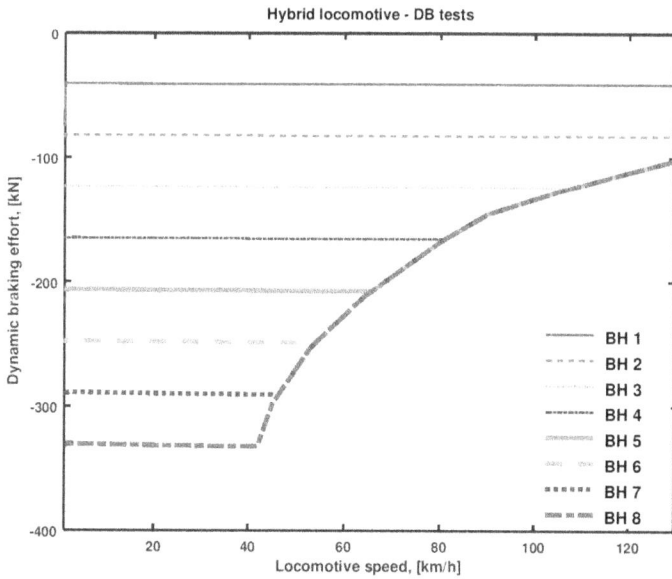

FIGURE 14.41   DB effort versus locomotive speed characteristics for each brake handle position of the modeled hybrid locomotive.

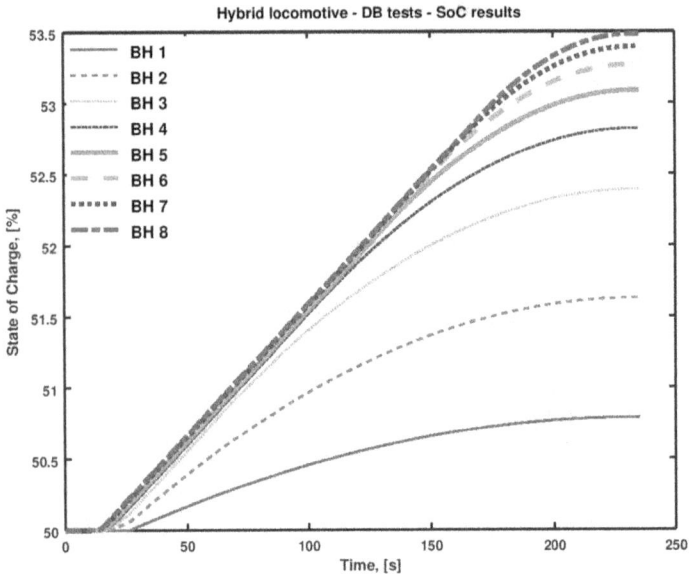

FIGURE 14.42 SoC results versus in time domain for each notch position of the modeled hybrid locomotive in a braking mode.

From the simulation results, it is possible to see that both diesel-powered (i.e., modeled in Case C study) and ESS powered (modeled in this case study) locomotives have the same TE and DB performance. This is mainly due to the battery ESS being powerful enough to mimic the performance of a diesel-powered locomotive.

There is a benefit to be gained in matching the performance of the two types of locomotives as it offers easy train driving control and general train management if the hybrid locomotives are introduced while the standard diesel electric locomotives are still in operation on a railway network. Practical implementation of this concept is described in [22, 23].

REFERENCES

1. R. M. Goodall, T. X. Mei, Active Suspensions, in Handbook of Railway Vehicle Dynamics Second Edition, CRC Press, Boca Raton, FL, 2020.
2. S. Bruni, R. M. Goodall, T. X. Mei, H. Tsunashima, Control and monitoring for railway vehicle dynamics, Vehicle System Dynamics, 45(7–8), 743–779, 2007.
3. B. Fu, R. L. Giossi, R. Persson, S. Stichel, S. Bruni, R. Goodall, Active suspension in railway vehicles: A literature survey, Railway Engineering Science, 28, 3–35, 2020.
4. K. Knothe, S. Stichel, Rail Vehicle Dynamics, Springer, Cham, Switzerland, 2017.
5. F. Cheli, G. Diana, Advanced Dynamics of Mechanical Systems, Springer, Cham, Switzerland, 2015.
6. H. True, On the theory of nonlinear dynamics and its applications in vehicle systems dynamics, Vehicle System Dynamics, 31(5–6), 393–421, 1999.
7. O. Polach, M. Berg, S. Iwnicki, Simulation of Railway Vehicle Dynamics, in Handbook of Railway Vehicle Dynamics Second Edition, CRC Press, Boca Raton, FL, 2020.

8. J. Pérez, J. M. Busturia, R. M. Goodall, Control strategies for active steering of bogie-based railway vehicles, Control Engineering Practice, 10(9), 1005–1012, 2002.

9. B. Fu, S. Hossein-Nia, S. Stichel, S. Bruni, Study on active wheelset steering from the perspective of wheel wear evolution, Vehicle System Dynamics, 2020, 1–24. DOI: 10.1080/00423114.2020.1838569.

10. B. Fu, S. Bruni, Fault-tolerant design and evaluation for a railway bogie active steering system, Vehicle System Dynamics, 2020, 1–25. DOI:10.1080/00423114.2020.1838563.

11. N. Wilson, R. Fries, M. Witte, A. Haigermoser, M. Wrang, J. Evans, A. Orlova, Assessment of safety against derailment using simulations and vehicle acceptance tests: A worldwide comparison of state-of-the-art assessment methods, Vehicle System Dynamics, 49(7), 1113–1157, 2011.

12. M. Spiryagin, S. Simson, C. Cole, I. Persson, Co-simulation of a mechatronic system using Gensys and Simulink, Vehicle System Dynamics, 50(3), 495–507, 2012.

13. M. Spiryagin, P. Wolfs, F. Szanto, C. Cole, Simplified and advanced modelling of traction control systems of heavy-haul locomotives, Vehicle System Dynamics, 53(5), 672–691, 2015.

14. M. Spiryagin, P. Wolfs, C. Cole, S. Stichel, M. Berg, M. Plöchl, Influence of AC system design on the realisation of tractive efforts by high adhesion locomotives, Vehicle System Dynamics, 55(8), 1241–1264, 2017.

15. M. Spiryagin, P. Wolfs, C. Cole, V. Spiryagin, Y. Q. Sun, T. McSweeney, Design and Simulation of Heavy Haul Locomotives, Ground Vehicle Engineering Series, CRC Press, Boca Raton, FL, 2017.

16. M. Depenbrock, Direct self control (DSC) of inverter fed induction machine, IEEE Transactions on Power Electronics, 3(4), 420–429, 1988.

17. G. S. Buja, M.P. Kazmierkowski, Direct torque control of PWM inverter-fed AC motors – A survey, IEEE Transactions on Industrial Electronics, 51(4), 744–757, 2004.

18. D. Casadei, F. Profumo, G. Serra, A. Tani, FOC and DTC: Two viable schemes for induction motors torque control, IEEE Transactions on Power Electronics, 17(5), 779–787, 2002.

19. B. K. Bose, Power Electronics and Variable Frequency Drives – Technology and Applications, IEEE Press, Piscataway, NJ, 1996.

20. R. Mathew, F. Flinders, W. Oghanna, Locomotive "total systems" simulation using SIMULINK, Proceedings of International Conference on Electric Railways in a United Europe, 27–30 March 1995, IET, 202–206.

21. M. Spiryagin, Q. Wu, P. Wolfs, Y. Sun, C. Cole, Comparison of locomotive energy storage systems for heavy-haul operation, International Journal of Rail Transportation, 6(1), 1–15, 2008.

22. M. Spiryagin, Q. Wu, C. Bosomworth, C. Cole, P. Wolfs, M. Hayman, Understanding the impact of high traction hybrid locomotive designs on heavy haul train performance, AUSRAIL PLUS 2019, Sydney, Australia, 3–5 December 2019.

23. M. Spiryagin, P. Wolfs, Q. Wu, C. Bosomworth, C. Cole, T. McSweeney, Rapid charging energy storage system for a hybrid freight locomotive, 2020 Joint Rail Conference, JRC 2020, Paper No: JRC2020-8017, St. Louis, Missouri, 20–22 April 2020.

Index

Note: Page numbers in italic and bold refer to figures and tables, respectively.

For Product Safety Concerns and Information please contact our EU
representative GPSR@taylorandfrancis.com
Taylor & Francis Verlag GmbH, Kaufingerstraße 24, 80331 München, Germany

www.ingramcontent.com/pod-product-compliance
Lightning Source LLC
Chambersburg PA
CBHW060748220326
41598CB00022B/2363